ROCKS & MINERALS

The Collector's Encyclopedia of

ROCKS

& MINERALS

edited by

A F L Deeson MA PhD DSc

researched by

James R Tindall BSc FGS

Annette Rogers M/Sc

Eric Deeson BA MSc

David & Charles : Newton Abbot

A Carter Nash Cameron book
Photographs by Ian Cameron
Design by Tom Carter

*Published by David & Charles (Holdings) Limited, South Devon
House, Newton Abbot, Devon*
*Produced by Carter Nash Cameron Limited, 25 Lloyd Baker
Street, London W.C.1*
*Text set by Page Bros (Norwich) Limited, Mile Cross Lane,
Norwich*
*Colour reproduction by Colour Workshop Limited, Mimram
Road, Hertford*
*Printed by George Pulman & Sons Limited, Watling Street,
Bletchley, Bucks*
*Bound by Robert Hartnoll Limited, Victoria Square, Bodmin,
Cornwall*

INTRODUCTION

This book has been designed primarily to assist the amateur collector, whether he is a beginner or already somewhere along the road. It is hoped that it will also be useful to geological students who require clear and explicit descriptions of all the principal rocks and minerals. It does not claim to be exhaustive but sets out to describe succinctly the major types, together with some others which are less well known but are to be found in Britain, the United States and the principal Commonwealth and English-speaking countries. Nobody knows exactly how many different kinds of rocks and minerals there are in the world – it is certain that many have not yet been named or classified. On the other hand it is also certain that some of the very 'comprehensive' listings that are available contain many duplications or purely local names for varieties already named in another part of the world.

The present work is intended for reference rather than for continuous reading. All entries – including definitions – are listed in alphabetical order. Thus all information is contained in the main structure of the book.

Under Occurrence in each entry, the localities given do not amount to a comprehensive list but are intended only as examples. Additional localities may be found in the list of illustrations.

Within each entry, all rocks and minerals are shown in italics, except for the subject-matter of the particular entry. Words or phrases defined elsewhere are *not* shown in italics, since the number of terms defined would result in visual confusion if this common practice were adopted here.

Definitions are largely confined to geological terms or to terms where the geological meaning differs from that in common usage or in another discipline.

ROCKS
Each entry contains, as applicable, the following information:
Clan Igneous only – diorite, gabbro, granite, ultrabasic, syenite.
Type Igneous – extrusive, hypabyssal, plutonic.
Metamorphic – contact, dynamic, regional.
Sedimentary – clastic, carbonate, evaporites, ferruginous, organic, siliceous.
Grade Metamorphic only.
Colour Principal colour or colours.
Texture
Structures If any.
Essential Minerals/Constituents
Special Features If any.
Occurrence Some important localities where the rock has been found.
Environment Type of conditions in which the rock is usually found.
Varieties and Synonyms If any.

MINERALS
Each entry contains, as applicable, the following information:
Chemical Formula The most commonly accepted chemical formula has been used; some differences occur in geological works because of variations in the samples analysed.
Composition The 'translation' of the chemical formula.
Crystal System The system to which the mineral belongs.
Habit Normal form and shape.

Colour Principal colour or colours (local variations are often found).
Lustre
Streak If any.
Fracture If any.
Cleavage If any.
Hardness Test According to the Mohs test.
Specific Gravity
Special Features If any.
Methods of Identification When fairly simple tests are available, they are indicated. Unless otherwise stated acids to be used are dilute.
Occurrence Some important localities where the mineral has been found.
Environment Type of conditions in which the mineral has been found.
Varieties and Synonyms If any.

CHARTS
Two easy reference charts appear at the beginning of the book. The major minerals are grouped under cations – oxides, chlorides, sulphates, etc. – and the information reproduced in the text is then summarised under 12 headings which detail: mineral name, formula, crystal system, common habit, colour, lustre, streak, fracture, cleavage, hardness, specific gravity, and reaction to acids or water.

A dash appears wherever the information is not available or applicable. All minerals can occur in massive form, but those which are specifically described as 'massive' under habit do not occur in a crystallised form.

Tests are restricted to a mineral's reaction to one of four acids – acetic, hydrochloric, nitric and sulphuric – or water. Chemical formulae are used for the three commonest acids, i.e. HCl for hydrochloric, HNO_3 for nitric, and H_2SO_4 for sulphuric. Unless otherwise stated, acid is dilute.

The major rocks are subdivided by clan and/or type.

ILLUSTRATIONS
The colour photographs, all of which have been specially taken for this book, are integrated with the entries. The list of illustrations gives any necessary descriptions, together with the source and locality of the specimen and the linear magnification of the picture. More than one illustration has been provided for some entries to indicate a range of habit, colour or appearance. On the whole, exceptionally spectacular specimens have been avoided, as they are likely to be found only in the display collections of major museums. The majority of the specimens illustrated are taken from the stocks of mineral dealers and from a teaching collection. For letting us photograph specimens and for their generosity in helping us, we are deeply indebted to the following people and the organisations with which they are connected:
Dr Douglas Bassett and his staff at the National Museum of Wales, Cardiff.
H. L. Douch and R. D. Penhallurick at the County Museum, Truro, Cornwall.
John Turner of Glenjoy Lapidary Supplies, Sun Lane, Wakefield, Yorkshire.
C. Gregory of Rocks & Minerals, 4 Moorcourt Drive, Cheltenham, Gloucestershire.
Mrs J. Beetham of Cheltenham.

CHARTS

IGNEOUS ROCKS

GRANITE CLAN

Plutonic
Adamellite
Ailsyte
Alaskite
Birkremite
Gneissoid Granite
Granite
Granite porphyry
Granodiorite
Graphic granite
Greisen
Kalialaskite
Luxullianite
Northfieldite
Protogine
Unakite

Hypabyssal
Alsbachite
Aplite
Beresite
Brandbergite
Elvan
Felsite
Granophyre
Kalitordrillite
Leopardite
Lindoite
Markfieldite
Microadamellite
Paisanite
Pegmatite
Tollite
Topazite
Tsingtauite

Extrusive
Aporhyolite
Cantalite
Comendite
Cumbraite
Dacite
Khagiarite
Nevadite
Pantellerite
Perlite
Pumice
Rhyolite
Todrillite
Toscanite

SYENITE CLAN

Plutonic
Akerite
Beloeilite
Bigwoodite
Canadite
Congressite
Covite
Deldoradite
Foyaite
Highwoodite
Hilairite
Husebyite
Juvite
Kalisyenite
Lakarpite
Larvikite
Laurdalite
Ledmorite
Litchfieldite
Lugarite
Lusitanite
Malignite
Mangerite
Monmouthite
Nordmarkite
Orthoclasite
Pulaskite
Shonkinite
Sommaite
Syenite
Tonsbergite
Trowlesworthite
Umptekite
Uncompahgrite

Solvsbergite
Sussexite
Tinguite
Ulrichite
Vogesite
Wennebergite

Hypabyssal
Albitite
Albitophyre
Allochetite
Borolanite
Bostonite
Bowralite
Cascadite
Durbachite
Elkhornite
Hedrumite
Heronite
Heumite
Kamperite
Minette
Muniongite
Rhomb porphyry

Extrusive
Apachite
Arsoite
Blairmorite
Ciminite
Drakonite
Gibelite
Jumillite
Keratophyre
Madupite
Mugearite
Orendite
Selagite
Shackanite
Tavolatite
Trachyte
Vulsinite
Woodendite

DIORITE CLAN

Plutonic
Anchorite
Cavalorite
Diorite
Dungannonite
Glenmuirite
Gooderite
Hirnantite
Kentallenite
Laugenite
Oroite
Puglianite
Sebastianite
Tonalite

Hypabyssal
Akenobeite
Andesinite
Beringite
Camptonite
Cuselite
Helsinkite
Holyokeite
Kersantite
Krageroite
Kulaite
Kullaite
Leideite
Maenaite
Mondhaldeite
Plumasite
Timazite
Vaugnerite
Vintlite
Yatalite

Extrusive
Absarokite
Andesite
Hungarite
Kaiwekite
Marloesite
Ordanchite
Ottajanite
Pollenite
Propylite
Santorinite
Skomerite
Volcanite
Weiselbergite

GABBRO CLAN

Plutonic
Allalinite
Anorthosite
Bojite
Corsite
Essexite
Gabbro
Labradite
Norite
Oligoclasite
Ossipite
Rougemontite
Routivarite
Theralite
Troctolite
Yamaskite

Heptorite
Holmite
Kylite
Navite
Odinite
Soggendalite
Teschenite
Tholeiite

Extrusive
Alboranite
Anamesite
Arapahite
Atlantite
Basalt
Buchonite
Caltonite
Domite

Hypabyssal
Aasby Diabase
Alnoite
Ankaramite
Anorthitissite
Baldite
Crinanite
Devonite
Dolerite
Eucrite
Harrisite

Dorgalite
Macedonite
Mimosite
Nonesite
Schonfelsite
Shoshonite
Spilite
Sudburite
Tephrite

SILICA UNDERSATURATED FELDSPATHOID ROCKS

Plutonic	*Extrusive*
Fasinite	Basanite
Fergusite	Kajanite
Ijolite	Leucitite
Italite	Leucitophyre
Melteigite	Nephelinite
Missourite	Wyomingite

Hypabyssal
Arkite
Katzenbuckelite

PYROCLASTIC IGNEOUS ROCKS

Agglomerate
Breccia (volcanic)
Conglomerate (volcanic)
Ignimbrite
Palagonite tuff
Sillar
Tuff

ULTRABASIC

Plutonic
Alexoite
Algarvite
Allivalite
Anabohitsite
Bahiate
Bebedourite
Biotitite
Bronzitfels
Davidite
Dunite
Harzburgite
Hornblendite
Hypersthenite
Jacupirangite
Kiirunavaarite
Madeirite
Montrealite
Peridotite
Perknite
Pyroxenite
Saxonite
Scyelite
Serpentinite
Websterite
Wehrlite

Hypabyssal
Avezakite
Barshawite
Chromitite
Cortlandtite
Cromaltite
Diopsidite
Ilmenitite
Kimberlite
Kvellite
Lherzolite
Monchiquite
Newlandite
Ouachitite
Schriesheimite
Valbellite

Extrusive
Augitite
Batukite
Bielenite
Coppaelite
Picrite
Verite

SEDIMENTARY ROCKS

CLASTIC		CARBONATE ROCKS	FERRUGINOUS ROCKS	SILICEOUS ROCKS	ORGANIC ROCKS
Rudaceous	*Argillaceous*	Algal limestone	Black-band ironstone	Diatomite	Anthracite
Basal conglomerate	Adobe	Anthraconite	Chert ironstone	Lydite	Asphalt
Breccia	Alphitite	Cementstone	Clay ironstone	Malmstone	Bitumen
Calcrete	Alum Shale	Chalk	Ironstone		Bituminous coal
Conglomerate	Argillite	Cipolino	Oolitic ironstone		Boghead coal
Coombe Rock	Bentonite	Coquina	Pisolitic ironstone		Coal
Ferricrete	Blae	Cornstone			Lignite
Molasse	Blue mud	Landscape marble			Oil shale
Placer deposits	Boulder clay	Limestone			
Pseudoconglomerate	Caliche	Magnesite rock			
Scree deposits	Clay	Oolitic limestone			
	Fireclay	Pisolitic limestone			EVAPORITES
Arenaceous	Fuller's earth	Pseudobreccia			Evaporite
Bone bed	Loess	Sideritic limestone			
China stone	Marl	Tufa			
Ganister	Mudstone				
Greensand	Ooze				
Greywacke	Siderite mudstone				
Grit	Terra rossa				
Laterite	Tillite				
Lateritite	Tilloid				
Placer deposits	Turbidite				
Sandstone	Varve clay				
Turbidite	Varvite				

METAMORPHIC ROCKS

REGIONAL	CONTACT	DYNAMIC
High grade	*High grade*	Cataclasite
Amphibolite	Astite	Flaser gabbro
Eclogite	Buchite	Flaser Gneiss
Gneiss	Calc-flinta	Flinty crush rock
Gondite	Catawberite	Kakirite
Migmatite	Charnockite	Mylonite
Ollenite	Cornubianite	Protomylonite
Prasinite	Enderbite	
Quartzite	Epidiorite	
	Epidosite	
Low grade	Hornfels	
Augen schist	Limurite	
Green schist	Porcellanite	
Itabirite	Quartzite	
Kinzigite		
Lavialite	*Low grade*	
Leptynite	Adinole	
Marble	Phyllite	
Mica schist	Skarn	
Murasakite	Slate	
Phyllite	Spotted Slate	
Pinolite		
Schist		

MINERAL	FORMULA	SYSTEM	HABIT	COLOUR	LUSTRE
ELEMENTS					
Antimony	Sb	Hexagonal	Tabular crystals	Tin-white	Metallic
Arsenic	As	Hexagonal	Massive	Tin-white	Metallic
Bismuth	Bi	Hexagonal	Massive	Silver-white	Metallic
Copper	Cu	Cubic	Massive	Red	Metallic
Diamond	C	Cubic	Octahedral crystals	Variable	Adamantine
Gold	Au	Cubic	Massive	Yellow	Metallic
Graphite	C	Hexagonal	Tabular crystals	Greyish-black	Metallic to dull
Iron	Fe	Cubic	Massive	Grey or black	Metallic
Platinum	Pt	Cubic	Grains	Whitish-grey to dark grey	Metallic
Silver	Ag	Cubic	Cubic crystals	Silver-white	Metallic
Sulphur	S_8	Orthorhombic	Tabular crystals	Yellow	Resinous to greasy
Tellurium	Te	Hexagonal	Prismatic or acicular crystals	Tin-white	Metallic
OXIDES					
Anatase	TiO_2	Tetragonal	Pyramidal crystals	Shades of brown	Metallic adamantine
Arsenolite	As_2O_3	Cubic	Octahedral crystals	White	—
Baddeleyite	ZrO_2	Monoclinic	Prismatic crystals	Variable	Greasy to vitreous
Bauxite	$Al_2O_3.2H_2O$	Amorphous	Massive	White or grey	Dull
Bismite	Bi_2O_3	Monoclinic	Massive	Yellow or green	subadamantine to dull
Brannerite	UTi_2O_6	Triclinic	Grains	Black	—
Brookite	TiO_2	Orthorhombic	Tabular crystals	Yellowish-brown to black	Metallic adamantine
Cassiterite	SnO_2	Tetragonal	Prismatic crystals	Yellowish-brown to brownish-black	Metallic adamantine
Chromite	$FeCr_2O_4$	Cubic	Massive	Black	Metallic
Columbite	$(Fe, Mn)(Nb, Ta)_2O_6$	Orthorhombic	Prismatic crystals	Black to brownish-black	Submetallic to dull greasy
Corundum	Al_2O_3	Hexagonal	Pyramidal crystals	Variable	Adamantine to vitreous
Cuprite	Cu_2O	Cubic	Octahedral crystals	Red or black	Adamantine to earthy
Diaspore	$AlO.OH$	Orthorhombic	Acicular crystals	Variable	Pearly to vitreous
Franklinite	$(Zn, Mn, Fe^{2+})(Fe^{3+}, Mn^{3+})_2O_4$	Cubic	Octahedral crystals	Black	Metallic
Goethite	$FeO.OH$	Orthorhombic	Platy crystals	Black or brown	Adamantine; metallic to dull
Hausmannite	Mn_3O_4	Tetragonal	Pyramidal crystals	Brownish-black	Submetallic
Hematite	Fe_2O_3	Hexagonal	Tabular crystals	Steel-grey to red	Metallic to dull
Ilmenite	$FeTiO_3$	Hexagonal	Tabular crystals	Black	Metallic
Limonite	$FeO(OH).nH_2O$	Amorphous	Massive	Brown, black or yellow	Silky to dull
Magnetite	Fe_3O_4	Cubic	Octahedral crystals	Black	Metallic

STREAK	FRACTURE	CLEAVAGE	HARDNESS	SG	REACTION TO ACIDS OR H_2O
Grey	Uneven	Perfect in one direction	3.0–3.5	6.6–6.72	---
Tin-white	Uneven	Perfect in one direction	3.5	5.63–5.78	---
Silver-white	Brittle	Perfect in one direction	2.0–2.5	9.7–9.8	Soluble in concentrated HNO_3
Red	Hackly	None	2.5–3.0	8.5–9.0	Dissolves in HNO_3
Ash-grey	Conchoidal	Perfect in three directions	10	3.5–3.53	Insoluble in acids
Yellow	Uneven	None	2.5–3.0	19.3	Insoluble in HCl
Black or dark grey	---	Perfect basal	1.0–2.0	2.09–2.23	---
Grey	Hackly	Poor	4.0	7.3–7.9	Easily soluble in HCl
Grey	Hackly	None	4.0–4.5	14.0–19.0	Soluble in hot H_2SO_4
Silver-white	Hackly	None	2.5–3.0	10.1–11.1	Soluble in HNO_3
White	Conchoidal to uneven	Poor	1.5–2.5	2.07	---
Grey	---	Perfect in one direction	2.0–2.5	6.1–6.3	Soluble in hot concentrated H_2SO_4
Colourless or pale yellow	Subconchoidal	Perfect in one direction	5.0–6.5	3.9	---
---	Conchoidal	Perfect in one direction	1.5	3.87	---
White to brownish-white	Subconchoidal to uneven	Perfect in one direction	6.5	5.4–6.02	---
Colourless	Earthy	None	1.0–3.0	2.55	---
Greyish-yellow	Uneven to earthy	None	4.5	8.64–9.22	Soluble in concentrated HNO_3
Dark greenish-brown	Conchoidal	None	4.5	4.5–5.43	Decomposes in hot concentrated H_2SO_4
Colourless or greyish-green	Subconchoidal to uneven	Indistinct	5.5–6.0	4.14	---
---	Subconchoidal to uneven	Imperfect	6.0–7.0	6.99	---
Brown	Uneven	None	5.5	4.5	Insoluble in acids
Reddish-brown to brownish-black	Conchoidal	Poor	3.5–4.0	5.4–6.4	---
White	Uneven to conchoidal	None	9.0	4.0	Insoluble in acids
Brownish-red	Conchoidal to uneven	Poor	3.5–4.0	6.14	Soluble in HCl
White	Conchoidal	Good in one direction	6.5–7.0	3.4–3.5	Insoluble in acids
Black to brownish-black	Conchoidal	Indistinct	5.5–6.5	5.07–5.22	---
Brownish-yellow to yellow	Uneven	Perfect in one direction	5.0–5.5	3.4–4.3	---
Chestnut-brown	Uneven	Good basal	5.5	4.84	Soluble in hot HCl with the evolution of chlorine
Reddish-brown	Subconchoidal to uneven	None	5.0–6.0	5.26	Soluble in concentrated HCl
Black	Conchoidal to subconchoidal	None	5.0–6.0	4.75	Slowly soluble in HCl
Brown or yellow	Conchoidal to earthy	None	5.0–5.5	3.6–4.0	---
Black	Uneven	None	6.0	5.2	---

MINERAL	FORMULA	SYSTEM	HABIT	COLOUR	LUSTRE
Manganite	$MnO(OH)$	Monoclinic	Prismatic crystals	Grey or black	Submetallic
Martite	Fe_2O_3	Hexagonal	Octahedral crystals	Black	Dull to submetallic
Massicot	PbO	Orthorhombic	Massive	Yellow	Dull to greasy
Minium	Pb_3O_4	Amorphous	Massive	Shades of red	Greasy to dull
Molybdite	MoO_3	Orthorhombic	Crusts	Yellow	Silky to earthy
Periclase	MgO	Cubic	Grains	Colourless to greyish white	Vitreous
Perovskite	$CaTiO_3$	Cubic/monoclinic	Cubic crystals	Black, brown or red	Metallic adamantine
Psilomelane	$(Ba,Mn)Mn_4O_8(OH)_2$	Orthorhombic	Massive	Black	Submetallic
Pyrochlore	$(Ca,Na,Ce)(Nb,Ti,Ta)_2(O,OH,F)_7$	Cubic	Octahedral crystals	Brown or black	Vitreous to resinous
Pyrolusite	MnO_2	Tetragonal	Massive	Grey	Metallic
Quartz	SiO_2	Hexagonal	Prismatic crystals	Variable	Vitreous
Ramsdellite	MnO_2	Orthorhombic	Tabular crystals	Grey to black	Metallic
Rutile	TiO_2	Tetragonal	Prismatic or acicular crystals	Variable	Metallic adamantine
Senarmontite	Sb_2O_3	Cubic	Octahedral crystals	Colourless or greyish-white	Resinous
Spinel	$MgAl_2O_4$	Cubic	Octahedral crystals	Variable	Vitreous
Tantalite	$(Fe,Mn)(Ta,Nb)_2O_6$	Orthorhombic	Prismatic or tabular crystals	Brownish-black to black	Submetallic to subresinous
Tapiolite	$FeTa_2O_6$	Tetragonal	Prismatic crystals	Black	Subadamantine to submetallic
Tellurite	TeO_2	Orthorhombic	Acicular crystals	White	Subadamantine
Tenorite	CuO	Monoclinic	Elongated tabular crystals	Grey or black	Metallic
Thorianite	ThO_2	Cubic	Cubic crystals	Dark grey to black	Submetallic
Trevorite	$NiFe_2O_4$	Cubic	Massive	Black or brownish-black	Metallic
Uraninite	UO_2	Cubic	Octahedral crystals	Black	Greasy to submetallic
Valentinite	Sb_2O_3	Orthorhombic	Prismatic or tabular crystals	Colourless or white	Adamantine
Woodruffite	$(Zn,Mn^{2+})_2Mn_5^{4+}O_{12}.4H_2O$	Uncertain	Massive	Black	Dull
Zincite	ZnO	Hexagonal	Massive	Orangish-yellow to red	Subadamantine

HYDROXIDES

MINERAL	FORMULA	SYSTEM	HABIT	COLOUR	LUSTRE
Brucite	$Mg(OH)_2$	Hexagonal	Tabular crystals	White (often tinted)	Waxy to vitreous
Gibbsite	$Al(OH)_3$	Monoclinic	Tabular crystals	White (often tinted)	Pearly to vitreous

SULPHIDES

MINERAL	FORMULA	SYSTEM	HABIT	COLOUR	LUSTRE
Aikinite	$PbCuBiS_3$	Orthorhombic	Prismatic or acicular crystals	Greyish-black	Metallic

STREAK	FRACTURE	CLEAVAGE	HARDNESS	SG	REACTION TO ACIDS OR H_2O
Reddish-brown	Uneven	Perfect in two directions	4.0	4.2–4.4	Soluble in HCl with the evolution of chlorine
Reddish or purplish-brown	Conchoidal	—	6.0–7.0	4.8	—
Yellow	—	Imperfect	2.0	9.6	Decomposes in H_2SO_4 with the precipitation of lead sulphate
Orangish-yellow	—	None	2.5	9.0	Soluble in HCl with the evolution of chlorine
—	None	Good in one direction	1.0–2.0	4.5	—
White	Irregular	Perfect in one direction	5.5	3.56	Easily soluble in HCl
Colourless or grey	Uneven	Indistinct	5.5	4.0	Decomposes in hot H_2SO_4
Brownish-black to black	—	—	5.0–6.0	4.7	Soluble in HCl with the evolution of chlorine
Light brown to yellowish-brown	Subconchoidal to uneven	Indistinct	5.0–5.5	4.2–4.5	Slowly soluble in H_2SO_4
Black to bluish-black	Uneven	Perfect in one direction	6.0–6.5	5.0	—
—	Conchoidal	None	7.0	2.6	Insoluble in acids
Black	—	Good in two directions	3.0	4.7	—
Pale brown to black	Conchoidal to uneven	Good in one direction	6.0–6.5	4.23	Insoluble in acids
White	Uneven	Poor	2.0–2.5	5.5	Easily soluble in HCl
White	Conchoidal	None	7.8–8.0	3.55	—
Dark red to black	Subconchoidal to uneven	Good in one direction	6.0	7.95	Partially decomposes when evaporated with concentrated H_2SO_4
Brown to brownish-black	Uneven to subconchoidal	None	6.0–6.5	7.9	
—		Perfect in one direction	2.0	5.9	Easily soluble in HCl
—	Conchoidal to uneven	Poor	3.5	5.8–6.4	Easily soluble in HCl
Grey to greenish-grey	Uneven to subconchoidal	Poor	6.5	9.7	Soluble in H_2SO_4 with the evolution of hydrogen
Brown	—	Poor	5.0	5.2	Soluble with difficulty in HCl
Brownish-black to olive-green	Uneven to conchoidal	—	5.0–6.0	10.8	Slowly attacked by HCl
White	—	Perfect in one direction	2.5–3.0	5.76	Soluble in HCl
Brownish-black	Conchoidal	—	4.5	3.71	—
Orangish-yellow	Conchoidal	Perfect in one direction	4.0	5.7	Soluble in HCl
White	None	Perfect in one direction	2.5	2.39	—
White	Tough	Perfect in one direction	2.5–3.5	2.3–2.4	—
Greyish-black	Uneven	Indistinct	2.0–2.5	7.07	Decomposes in HNO_3 with the precipitation of sulphur and lead sulphide

MINERAL	FORMULA	SYSTEM	HABIT	COLOUR	LUSTRE
Alabandite	MnS	Cubic	Massive	Black	Submetallic
Argentite	Ag_2S	Cubic	Cubic crystals	Dark grey	Metallic
Arsenopyrite	$FeAsS$	Monoclinic	Prismatic crystals	Silver-white	Metallic
Bismuthinite	Bi_2S_3	Orthorhombic	Acicular crystals	Grey	Metallic
Blende	ZnS	Cubic	Tetragonal crystals	Variable	Resinous to adamantine
Bornite	Cu_5FeS_4	Cubic	Cubic crystals	Red or brown	Metallic
Boulangerite	$Pb_5Sb_4S_{11}$	Monoclinic	Prismatic crystals	Bluish-grey	Metallic
Bravoite	$(Ni,Fe)S_2$	Cubic	Cubic crystals	Grey	Metallic
Chalcocite	Cu_2S	Orthorhombic	Prismatic or tabular crystals	Blackish-grey	Metallic
Chalcopyrite	$CuFeS_2$	Tetragonal	Tetrahedral crystals	Yellow	Metallic
Cinnabar	HgS	Hexagonal	Tabular or prismatic crystals	Red	Adamantine to metallic
Covelline	CuS	Hexagonal	Massive	Indigo-blue	Submetallic to resinous
Enargite	Cu_3AsS_4	Orthorhombic	Prismatic crystals	Greyish-black or black	Metallic
Galena	PbS	Cubic	Cubic crystals	Grey	Metallic
Germanite	$Cu_3(Ge,Ga,Fe,Zn)(As,S)_4$	Cubic	Detrital masses	Dark reddish-grey	Metallic
Greenockite	CdS	Hexagonal	Pyramidal crystals	Shades of yellow	Adamantine to resinous
Jamesonite	$Pb_4FeSb_6S_{14}$	Monoclinic	Acicular crystals	Greyish-black	Metallic
Marcasite	FeS_2	Orthorhombic	Tabular crystals	Pale bronze-yellow	Metallic
Matildite	$AgBiS_2$	Orthorhombic	Massive	Black or grey	Metallic
Millerite	NiS	Hexagonal	Capillary crystals	Brassy-yellow	Metallic
Molybdenite	MoS_2	Hexagonal	Tabular crystals	Grey	Metallic
Nagyagite	$Pb_5Au(Te,Sb)_4S_{5\ 8}$	Monoclinic	Tabular crystals	Blackish-grey	Metallic
Orpiment	As_2S_3	Monoclinic	Prismatic crystals	Lemon-yellow	Pearly
Parkerite	$Ni_3(Bi,Pb)_2S_2$	Monoclinic	Platy crystals	Bronze	Metallic
Pentlandite	$(Fe,Ni)_9S_8$	Cubic	Massive	Bronze-yellow	Metallic
Polybasite	$Ag_{16}Sb_2S_{11}$	Monoclinic	Tabular crystals	Black	Metallic
Proustite	Ag_3AsS_3	Hexagonal	Prismatic crystals	Scarlet-vermilion	Adamantine
Pyargyrite	Ag_3SbS_3	Hexagonal	Prismatic crystals	Dark red	Adamantine
Pyrargyrite	Ag_3SbS_3	Hexagonal	Prismatic crystals	Black or red	Metallic adamantine
Pyrite	FeS_2	Cubic	Cubic crystals	Pale brassy-yellow	Metallic

STREAK	FRACTURE	CLEAVAGE	HARDNESS	SG	REACTION TO ACIDS OR H_2O
Green	Uneven	Perfect in one direction	3·5–4·0	4·0	Soluble in HCl with the evolution of hydrogen sulphide
Silver-grey	Subconchoidal	Poor	2·0–2·5	7·3	—
Dark greyish-black	Uneven	Good in one direction	5·5–6·0	6·07	—
Grey	None	Perfect in one direction	2·0	6·4	—
Brown or white	Conchoidal	Perfect in one direction	3·5–4·0	3·9–4·1	Soluble in HCl with the evolution of hydrogen sulphide
Pale greyish-black	Conchoidal to uneven	Poor	3·0	5·07	Soluble in HNO_3 with the precipitation of sulphur
Brownish-grey to brown	Brittle	Good in one direction	2·5–3·0	5·7–6·3	Soluble in hot HCl with the evolution of hydrogen sulphide
Grey	Conchoidal to uneven	Perfect in one direction	5·5–6·0	4·62	Soluble in hot HNO_3 with the evolution of hydrogen sulphide
Blackish-grey	Conchoidal	Indistinct	2·5–3·0	5·5–5·8	—
Greenish-black	Uneven	Indistinct	3·5–4·0	4·1–4·3	Soluble in HNO_3 with the separation of sulphur
Scarlet	Uneven	Perfect in one direction	2·0–2·5	8·09	—
Grey or black	Uneven	Perfect in one direction	1·5–2·0	4·6–4·7	—
Greyish-black	Uneven	Perfect in two directions	3·0	4·4–4·5	Soluble in aqua regia
Grey	Subconchoidal	Perfect in one direction	2·5–2·75	7·58	—
Dark grey or black	Brittle	None	4·0	4·46–4·6	Soluble in HNO_3
Orangish-yellow to brick-red	Conchoidal	Good in one direction	3·0–3·5	5·0	Soluble in concentrated HCl with the evolution of hydrogen sulphide
Greyish-black	—	Good basal	2·5	5·63	Decomposes in HNO_3 with the separation of antimony oxide and lead sulphate
Greyish-black	Uneven	Indistinct	6·0–6·5	4·9	Soluble in HNO_3
Pale grey	Uneven	None	2·5	6·9	Soluble in HNO_3 with the precipitation of sulphur
Greenish-black	Uneven	Perfect in two directions	3·0–3·5	5·5	—
Bluish-grey	None	Perfect in one direction	1·0–1·5	4·62–4·73	Decomposes in HNO_3 with the separation of molybdenum oxide
Blackish-grey	—	Perfect in one direction	1·0–1·5	7·4	Soluble in HNO_3 with a residue of gold
Pale lemon-yellow	Irregular	Perfect in one direction	1·5–2·0	3·5	Soluble in H_2SO_4
Black	—	Perfect in one direction	3·0	8·74	—
Bronze-brown	Conchoidal	None	3·5–4·0	4·6–5·0	—
Black	Uneven	Poor	2·0–3·0	6·1	Decomposes in HNO_3
Vermilion	Conchoidal to uneven	Good in one direction	2·0–2·5	5·6	Decomposes in HNO_3 with the separation of sulphur
Purplish-red	Conchoidal to uneven	Visible in one direction	2·5	5·8	Decomposes in HNO_3 with the separation of silver and antimony oxide
Red	Conchoidal	Good in one direction	2·0–3·0	5·8	Decomposes in HNO_3
Greenish-black to brownish-black	Conchoidal to uneven	Indistinct	6·0–6·5	5·0	Insoluble in HCl

MINERAL	FORMULA	SYSTEM	HABIT	COLOUR	LUSTRE
Pyrrhotite	FeS	Hexagonal	Tabular or platy crystals	Bronze-yellow to brown	Metallic
Realgar	As_4S_4	Monoclinic	Prismatic crystals	Red to orangish-yellow	Resinous to greasy
Stannite	Cu_2FeSnS_4	Tetragonal	Massive	Grey or black	Metallic
Stephanite	Ag_5SbS_4	Orthorhombic	Prismatic or tabular crystals	Black	Metallic
Stibnite	Sb_2S_3	Orthorhombic	Prismatic or acicular crystals	Grey	Metallic
Teallite	$PbSnS_2$	Orthorhombic	Tabular crystals	Greyish-black	Metallic
Tennantite	Cu_3AsS_3	Cubic	Tetragonal crystals	Grey or black	Metallic
Tetradymite	Bi_2Te_2S	Hexagonal	Massive	Pale grey	Metallic
Tetrahedrite	Cu_3SbS_3	Cubic	Tetragonal crystals	Grey or black	Metallic
Ullmannite	$NiSbS$	Cubic	Cubic crystals	Grey to silver-white	Metallic
Wurtzite	ZnS	Hexagonal	Pyramidal or prismatic crystals	Brownish-black	Resinous

SULPHATES

MINERAL	FORMULA	SYSTEM	HABIT	COLOUR	LUSTRE
Alum	$KAl(SO_4)_2.12H_2O$	Cubic	Massive	Colourless or white	Vitreous
Aluminite	$Al_2SO_4(OH)_4.7H_2O$	Monoclinic	Massive	White	Earthy
Alunite	$KAl_3(SO_4)_2(OH)_6$	Hexagonal	Rhombohedral crystals	White	Vitreous
Alunogene	$Al_2(SO_4)_3.18H_2O$	Triclinic	Crusts	Colourless or white	Vitreous or silky
Anglesite	$PbSO_4$	Orthorhombic	Tabular crystals	Colourless or white	Adamantine to resinous
Anhydrite	$CaSO_4$	Orthorhombic	Massive	Colourless or blue	Vitreous to pearly
Antlerite	$Cu_3(SO_4)(OH)_4$	Orthorhombic	Tabular or prismatic crystals	Shades of green	Vitreous
Aphthitalite	$NaKSO_4$	Hexagonal	Tabular crystals	Variable	Vitreous to resinous
Barytes	$BaSO_4$	Orthorhombic	Platy crystals	Variable	Vitreous
Brochantite	$Cu_4SO_4(OH)_6$	Monoclinic	Prismatic or acicular crystals	Shades of green	Vitreous
Celestine	$SrSO_4$	Orthorhombic	Tabular crystals	Variable	Vitreous
Chalcanthite	$CuSO_4.5H_2O$	Triclinic	Prismatic crystals	Sky-blue	Vitreous
Glauberite	$Na_2Ca(SO_4)_2$	Monoclinic	Tabular or prismatic crystals	Variable	Vitreous to waxy
Gypsum	$CaSO_4.2H_2O$	Monoclinic	Tabular or prismatic crystals	Variable	Subvitreous to pearly
Jarosite	$KFe_3(SO_4)_2OH_6$	Hexagonal	Tabular crystals	Yellow to dark brown	Subadamantine to vitreous
Langbeinite	$K_2Mg_2(SO_4)_2$	Cubic	Massive	Colourless	Vitreous
Linarite	$PbCu(SO_4)(OH)_2$	Monoclinic	Tabular crystals	Azure-blue	Vitreous
Mascagnite	$(NH_4)_2SO_4$	Orthorhombic	Crusts	Colourless or grey	Vitreous
Melanterite	$FeSO_4.7H_2O$	Monoclinic	Prismatic crystals	Green	Vitreous

STREAK	FRACTURE	CLEAVAGE	HARDNESS	SG	REACTION TO ACIDS OR H_2O
Dark greyish-black	Uneven to subconchoidal	None	3·5–4·5	4·58–4·65	Decomposes in HCl with the evolution of hydrogen sulphide
Red to orangish-yellow	Conchoidal	Good in one direction	1·5–2·0	3·56	Decomposes in HNO_3
Black	Uneven	Indistinct	4·0	4·3–4·5	Decomposes in HNO_3 with the separation of sulphur and tin oxide
Black	Subconchoidal to uneven	Imperfect	2·0–2·5	6·25	Decomposes in HNO_3 with the separation of sulphur and antimony oxide
Grey	Subconchoidal	Perfect in one direction	2·0	4·63	Soluble in HCl
Black	—	Perfect basal	1·5	6·36	Easily decomposes in hot concentrated H_2SO_4
Black or brown	Subconchoidal to uneven	None	3·7–4·5	4·6–5·0	Decomposes in HNO_3 with the separation of sulphur
Pale grey	—	Perfect in one direction	1·5–2·0	7·4	—
Black, brown or cherry-red	Subconchoidal to uneven	None	3·0–3·7	4·8–5·1	Decomposes in HNO_3 with the separation of sulphur and antimony oxide
Greyish-black	Uneven	Perfect basal	5·0–5·5	6·65	Decomposes in HNO_3
Brown	—	Visible in one direction	3·5–4·0	3·98	Soluble in HCl with the evolution of hydrogen sulphide
Colourless	Conchoidal	Poor	2·0–2·5	1·8	—
Colourless	Earthy	None	1·0–2·0	1·66	—
White	Uneven	Good in one direction	3·5–4·0	2·6–2·9	Slowly soluble in H_2SO_4
—	—	Perfect in one direction	1·5–2·0	1·77	Soluble in water
Colourless	Conchoidal	Good in one direction	2·5–3·0	6·38	
White to greyish-white	Uneven to splintery	Perfect in one direction	3·5	2·98	—
Pale green	Uneven	Good in one direction	3·5–4·0	3·9	Soluble in HCl
Variable	Conchoidal to uneven	Poor	3·0	2·65–2·70	Soluble in water
White	Uneven	Perfect in two directions	3·0–3·5	4·5	—
Pale green	Uneven to conchoidal	Perfect in one direction	3·5–4·0	3·97	—
White	Uneven	Perfect in one direction	3·3–5·0	3·97	—
Colourless	Conchoidal	Imperfect	2·5	2·28	—
White	Conchoidal	Perfect in one direction	2·5–3·0	2·75–2·85	—
White	Conchoidal	Good in two directions	2·0	2·3	Slightly soluble in water
Pale yellow	Uneven to conchoidal	Good basal	2·5–3·5	2·9–3·26	Soluble in HCl
Colourless	Conchoidal	None	3·5–4·0	2·83	Slowly soluble in water
Pale blue	Conchoidal	Perfect in one direction	2·5	5·33–5·35	Soluble in HNO_3 with the precipitation of lead sulphate
Colourless	Uneven	Good in one direction	2·0–2·5	1·77	Soluble in water
Colourless	Conchoidal	Perfect in one direction	2·0	1·9	Soluble in water

MINERAL	FORMULA	SYSTEM	HABIT	COLOUR	LUSTRE
Mirabilite	$Na_2SO_4.10H_2O$	Monoclinic	Prismatic or tabular crystals	Colourless or white	Vitreous
Natroalunite	$NaAl_3(SO_4)_4(OH)_6$	Hexagonal	Tabular crystals	White	Vitreous
Plumbojarosite	$PbFe_6(SO_4)_4(OH)_{12}$	Hexagonal	Crusts	Shades of brown	Dull
Polyhalite	$K_2MgCa_2(SO_4)_4.2H_2O$	Triclinic	Massive	Colourless or white	Vitreous
Syngenite	$K_2Ca(SO_4)_2.H_2O$	Monoclinic	Tabular or prismatic crystals	Colourless	Vitreous
Thenardite	Na_2SO_4	Orthorhombic	Tabular crystals	Colourless (often tinted)	Vitreous

CARBONATES

MINERAL	FORMULA	SYSTEM	HABIT	COLOUR	LUSTRE
Alstonite	$BaCa(CO_3)_2$	Orthorhombic	Pyramidal crystals	Colourless or white (often tinted)	Vitreous
Ancylite	$(Sr,Ca)_3(Ce,La)_4 (CO_3)_7(OH)_4.3H_2O$	Orthorhombic	Prismatic crystals	Variable	Vitreous
Ankerite	$Ca(Mn,Mg,Fe) (CO_3)_2$	Hexagonal	Rhombohedral crystals	Yellowish-brown to brown	Vitreous to pearly
Aragonite	$CaCO_3$	Orthorhombic	Prismatic crystals	Colourless or white	Vitreous
Aurichalcite	$(Zn, Cu)_5(OH)_6(CO_3)_2$	Orthorhombic	Crusts	Pale greenish-blue	Pearly
Azurite	$Cu_3(CO_3)_2(OH)_2$	Monoclinic	Prismatic crystals	Deep azure-blue	Vitreous
Bismutite	$Bi_2CO_3.H_2O$	Tetragonal	Crusts	White, grey or yellow	Vitreous to pearly
Calcite	$CaCO_3$	Hexagonal	Tabular or acicular crystals	Colourless (often tinted)	Vitreous
Cerussite	$PbCO_3$	Orthorhombic	Prismatic crystals	Colourless (often tinted)	Adamantine
Chalybite	$FeCO_3$	Hexagonal	Rhombohedral crystals	Variable	Pearly to vitreous
Dolomite	$CaMg(CO_3)_2$	Hexagonal	Rhombohedral crystals	Variable	Vitreous
Gay-lussite	$Na_2Ca(CO_3)_2.5H_2O$	Monoclinic	Prismatic crystals	Colourless or white	Vitreous
Lansfordite	$MgCO_3.5H_2O$	Monoclinic	Prismatic crystals	Colourless or white	Vitreous
Magnesite	$MgCO_3$	Hexagonal	Massive	Colourless or white	Vitreous
Malachite	$Cu_2CO_3(OH)_2$	Monoclinic	Massive	Shades of green	Silky
Nesquehonite	$MgCO_3.3H_2O$	Orthorhombic	Prismatic crystals	Colourless to white	Vitreous
Phosgenite	$Pb_2CO_3Cl_2$	Tetragonal	Prismatic or tabular crystals	White to brown	Adamantine
Rhodochrosite	$MnCO_3$	Hexagonal	Massive	Shades of pink and red	Vitreous
Shortite	$Na_2Ca_2(CO_3)_3$	Orthorhombic	Tabular or prismatic crystals	Colourless or pale yellow	Vitreous
Smithsonite	$ZnCO_3$	Hexagonal	Massive	Greyish-white to dark grey	Vitreous
Strontianite	$SrCO_3$	Orthorhombic	Prismatic or acicular crystals	Colourless or grey	Vitreous
Trona	$Na_3H(CO_3)_2.2H_2O$	Monoclinic	Massive	Colourless or grey to white	Vitreous
Witherite	$BaCO_3$	Orthorhombic	Pyramidal crystals	Colourless or white	Vitreous
Zaratite	$Ni_3CO_3(OH)_4.4H_2O$	Cubic	Massive	Emerald-green	Vitreous to greasy

STREAK	FRACTURE	CLEAVAGE	HARDNESS	SG	REACTION TO ACIDS OR H_2O
White	Conchoidal	Perfect in one direction	1·5–2·0	1·5	Easily soluble in water
White	Conchoidal	Good in one direction	3·5–4·0	2·6–2·9	Slowly soluble in H_2SO_4
Pale brown	None	Poor	1·0	3·6	Slowly soluble in HCl
—	Splintery	Perfect in one direction	3·5	2·78	Decomposes in water with the separation of gypsum
Colourless	Conchoidal	Perfect in two directions	2·5	2·6	Decomposes in water with the separation of gypsum
—	Uneven	Perfect in one direction	2·5–3·0	2·6	—
White	Uneven	Imperfect	4·0–4·5	3·7	Soluble in HCl
White	Splintery	None	4·0–4·5	3·95	Soluble in HCl with the evolution of carbon dioxide
Colourless	Subconchoidal	Perfect in one direction	3·5–4·0	3·02	—
Colourless	Subconchoidal	Good in one direction	3·5–4·0	2·9–3·0	Soluble with effervescence in HCl
—	None	Perfect basal	2·0	3·5–3·6	Soluble with effervescence in HCl
Blue	Conchoidal	Perfect in one direction	3·5–4·0	3·7–3·8	Soluble with effervescence in HCl
Grey	None	Good basal	2·5–3·5	6·1–7·7	Soluble with effervescence in HCl
White to grey	Conchoidal	Good in two directions	3·0	2·7	Soluble with effervescence in HCl
Colourless or white	Conchoidal	Good in two directions	3·0–3·5	6·55	Soluble with effervescence in HCl
White	Uneven	Perfect in two directions	3·5–4·5	3·7–3·9	Effervesces in hot HCl
White to grey	Conchoidal to subconchoidal	Perfect in two directions	3·5–4·0	2·9	Soluble with effervescence in warm HCl
Colourless or greyish-white	Conchoidal	Perfect in one direction	2·5–3·0	1·99	Easily soluble with effervescence in HCl
White	Uneven	Good in two directions	2·5	1·7	Soluble in HCl
White	Conchoidal	Perfect in one direction	3·5–4·0	3·0–3·5	Soluble with effervescence in HCl
Pale green	Irregular	Good basal	3·5–4·0	3·9–4·0	Soluble with effervescence in HCl
—	Splintery	Perfect in one direction	2·5	1·85	Easily soluble with effervescence in HCl
White	Conchoidal	Good in two directions	2·0–3·0	6·1	Soluble with effervescence in HNO_3
White	Uneven to conchoidal	Perfect in one direction	3·5–4·0	3·7	Soluble with effervescence in warm HCl
—	Conchoidal	Good in one direction	3·0	2·6	Decomposes in water with the separation of calcium carbonate
White	Uneven to subconchoidal	Good in one direction	4·0–4·5	4·43	Soluble with effervescence in HCl
—	Uneven	Good in one direction	3·5	3·75	Soluble with effervescence in HCl
—	Uneven to subconchoidal	Perfect in one direction	2·5–3·0	2·14	Soluble in water
White	Uneven	Good in one direction	3·0–3·5	4·3	Soluble with effervescence in HCl
Green	Conchoidal	—	3·5	2·57–2·69	Soluble with effervescence when heated in HCl

MINERAL	FORMULA	SYSTEM	HABIT	COLOUR	LUSTRE
PHOSPHATES					
Amblygonite	$(Li, Na)AlPO_4(F,OH)$	Triclinic	Prismatic crystals	White	Vitreous to greasy
Apatite	$Ca_5(PO_4)_3(F,Cl,OH)$	Hexagonal	Prismatic or pyramidal crystals	Variable	Vitreous to subresinous
Autunite	$Ca(UO_2)_2(PO_4)_2.$ $10-12H_2O$	Tetragonal	Platy crystals	Yellow	Pearly to vitreous
Beraunite	$Fe^2 Fe^3 (PO_4)_3$ $(OH)_5.3H_2O$	Monoclinic	Tabular crystals	Shades of red	Vitreous
Beryllonite	$NaBePO_4$	Monoclinic	Tabular or prismatic crystals	White to pale yellow	Vitreous
Evansite	$Al_3PO_4(OH)_6.6H_2O$	Amorphous	Massive	Colourless or white (often tinted)	Vitreous to resinous
Lithiophilite	$LiMn(PO_4)$	Orthorhombic	Massive	Salmon	Vitreous
Mimetite	$Pb_5(AsO_4,PO_4)Cl$	Hexagonal	Acicular crystals	Colourless, yellow or brown	Resinous
Monazite	$(Ce,La,Yt,Th)(PO_4)$	Monoclinic	Prismatic crystals	Yellow, white or brown	Waxy to resinous
Plumbogummite	$PbAl_3(PO_4)_2$ $(OH)_5.H_2O$	Hexagonal	Massive	Greyish-white to yellow	Dull to resinous
Pseudomalachite	$Cu_5(PO_4)_2(OH)_4.H_2O$	Monoclinic	Massive	Shades of green	Vitreous
Purpurite	$(Mn,Fe)PO_4$	Orthorhombic	Prismatic crystals	Reddish-purple	Satiny
Pyromorphite	$Pb_5(PO_4)_3Cl$	Hexagonal	Prismatic crystals	Green, yellow or brown	Resinous to subadamantine
Scoralite	$(Fe,Mg)Al_2(PO_4)_2$ $(OH)_2$	Monoclinic	Pyramidal crystals	Shades of blue	Vitreous
Sicklerite	$(Li, Mn, Fe)PO_4$	Orthorhombic	Massive	Yellowish-brown to dark brown	Dull
Strengite	$FePO_4.2H_2O$	Orthorhombic	Tabular crystals	Colourless, red or violet	Vitreous
Torbernite	$Cu(UO_2)_2(PO_4)_2.$ $12H_2O$	Tetragonal	Tabular crystals	Shades of green	Vitreous to subadamantine
Triphylite	$Li(Fe,Mn)PO_4$	Orthorhombic	Massive	Brownish-grey to greenish-gfey	Vitreous to subadamantine
Turquoise	$CuAl_6(PO_4)_4(OH)_8.$ $5H_2O$	Triclinic	Massive	Shades of blue and green	Vitreous to waxy
Variscite	$AlPO_4.2H_2O$	Orthorhombic	Massive	Green	Vitreous to waxy
Vivianite	$Fe_3(PO_4)_2.8H_2O$	Monoclinic	Prismatic crystals	Colourless	Vitreous to dull
Wardite	$Na_4aAl_{12}(PO_4)_8$ $(OH)_{18}.6H_2O$	Tetragonal	Pyramidal crystals	Colourless or shades of green	Vitreous
Wavellite	$Al_6(PO_4)_4(OH)_6.$ $9H_2O$	Orthorhombic	Massive	Greenish-white, green or yellow	Vitreous
Xenotime	$YtPO_4$	Tetragonal	Prismatic crystals	Brown, red or yellow	Vitreous
CHLORIDES					
Atacamite	$Cu_2Cl(OH)_3$	Orthorhombic	Prismatic or tabular crystals	Shades of green	Adamantine
Calomel	Hg_2Cl_2	Tetragonal	Tabular or prismatic crystals	Variable	Adamantine
Carnallite	$KMgCl_3.6H_2O$	Orthorhombic	Pyramidal crystals	Colourless to milky-white	Greasy to dull

STREAK	FRACTURE	CLEAVAGE	HARDNESS	SG	REACTION TO ACIDS OR H_2O
Colourless	Uneven to subconchoidal	Perfect in one direction	5·5–6·0	3·11	
White	Conchoidal to uneven	Poor	5·0	3·17–3·23	Soluble in HCl
Yellow	None	Perfect in two directions	2·0–2·5	3·1	—
Yellow to olive	—	Good in one direction	3·5–4·0	2·8–2·9	Readily soluble in HCl
White	Conchoidal	Perfect in one direction	5·5–6·0	2·81	—
White	Conchoidal	None	3·0–4·0	1·8–2·2	Easily soluble in HCl
Greyish-white	Uneven	Perfect in one direction	4·4	3·5–3·58	Soluble in HCl
White	Uneven	Indistinct	3·5–4·0	7·0	—
White	Conchoidal to uneven	Good in two directions	5·0–5·5	4·6–5·4	Slowly decomposes in HCl
Colourless or white	Uneven	None	4·5–5·0	4·0	Soluble in hot HCl
Dark green	Splintery	Good in one direction	4·5–5·0	4·35	Soluble in HCl
Reddish-purple	Uneven	Good in one direction	4·0–4·5	3·3	Easily soluble in HCl
White	Uneven	Indistinct	3·5–4·0	7·0	Soluble in HNO_3
White	Uneven	Indistinct	5·5–6·0	3·38	Slowly soluble in hot HCl
Pale yellowish-brown to brown		Good in one direction	4·0	3·2–3·4	Soluble in HCl
White	Conchoidal	Good in one direction	3·5	2·87	Soluble in HCl
Pale green		Perfect basal	2·0–2·5	3·22	Soluble in HCl
Colourless to greyish-white	Uneven to subconchoidal	Good in one direction	4·0–5·0	3·55	Soluble in HCl
White to pale green	Conchoidal to smooth	Perfect basal	5·0–6·0	2·6–2·85	Soluble with difficulty in HCl
White	Uneven to splintery	Good in one direction	3·5–4·5	2·57	
Colourless or bluish-white	Fibrous	Perfect in one direction	1·5–2·0	2·68	Easily soluble in HCl
—	—	Perfect basal	5·0	2·87	Soluble with difficulty in HCl
White	Uneven to subconchoidal	Perfect in one direction	3·25–4·0	2·36	Easily soluble in HCl
Pale brown, pale red or pale yellow	Uneven to splintery	Good in one direction	4·0–5·0	4·4–5·1	—
Apple-green	Conchoidal	Perfect in one direction	3·0–3·5	3·76	—
Pale yellowish-white	Conchoidal	Good in one direction	1·5	7·15	—
Colourless	Conchoidal	None	2·5	1·6	Soluble in water

21

MINERAL	FORMULA	SYSTEM	HABIT	COLOUR	LUSTRE
Halite	$NaCl$	Cubic	Cubic crystals	Variable	Vitreous
Sal Ammoniac	NH_4Cl	Cubic	Trapezohedral crystals	Colourless or white	Vitreous
Sylvine	KCl	Cubic	Cubic crystals	Colourless or white	Vitreous
Vanadinite	$Pb_5(VO_4)_3Cl$	Hexagonal	Prismatic crystals	Shades of red and yellow	Subresinous to subadamantine
NITRATES					
Nitratine	$NaNO_3$	Hexagonal	Massive	Colourless or white	Vitreous
Nitre	KNO_3	Orthorhombic	Crusts	Colourless or white	Vitreous
BORATES					
Borax	$Na_2B_4O_7.10H_2O$	Monoclinic	Prismatic or tabular crystals	Variable	Vitreous
Colemanite	$Ca_2B_6O_{11}.5H_2O$	Monoclinic	Prismatic crystals	Variable	Vitreous to adamantine
Ulexite	$NaCaB_5O_8.8H_2O$	Triclinic	Capillary or acicular crystals	Colourless or white	Vitreous to silky
ARSENATES					
Adamite	$Zn_2(OH)AsO_4$	Orthorhombic	Prismatic crystals	Variable	Vitreous
Annabergite	$(Ni,Co)_3(AsO_4)_2.8H_2O$	Monoclinic	Crusts	Pale apple-green	Silky to vitreous
Chalcophyllite	$Cu_{18}Al_2(AsO_4)_3(SO_4)_3(OH)_{27}.33H_2O$	Hexagonal	Tabular crystals	Shades of green	Vitreous to subadamantine
Clinoclase	$Cu_3AsO_4(OH)_3$	Monoclinic	Tabular crystals	Greenish-black to greenish-blue	Vitreous to pearly
Erythrite	$(Co,Ni)_3(AsO_4)_2.8H_2O$	Monoclinic	Prismatic to acicular crystals	Crimson to peach-red	Vitreous to pearly
Liroconite	$Cu_2Al(AsO_4)(OH)_4.4H_2O$	Monoclinic	Acicular crystals	Sky-blue to green	Vitreous
Olivenite	$Cu_2(AsO_4)(OH)$	Orthorhombic	Prismatic or acicular crystals	Olive-green to brown	Adamantine to vitreous
Pharmacolite	$CaHAsO_4.2H_2O$	Monoclinic	Massive	White to grey	Vitreous
Roselite	$Ca_2(Co,Mg)(AsO_4)_2.2H_2O$	Monoclinic	Prismatic crystals	Pink to dark red	Vitreous
Scorodite	$FeAsO_4.2H_2O$	Orthorhombic	Pyramidal or tabular crystals	Leek-green to brown	Vitreous to subadamantine
ARSENIDES					
Algodonite	Cu_6As	Hexagonal	Massive	Grey to silver-white	Metallic
Niccolite	$NiAs$	Hexagonal	Massive	Pale copper-red	Metallic
Smaltite	$CoAs_2$	Cubic	Cubic crystals	Tin-white to silver-grey	Metallic
Sperrylite	$PtAs_2$	Cubic	Cubic crystals	Tin-white	Metallic
TELLURIDES					
Altaite	$PbTe$	Cubic	Massive	Tin-white	Metallic
Calaverite	$(AuTe)_2$	Monoclinic	Bladed crystals	Brass-yellow to silver-white	Metallic
Petzite	$(Ag,Au)_2Te$	Cubic	Massive	Steel-grey to black	Metallic

STREAK	FRACTURE	CLEAVAGE	HARDNESS	SG	REACTION TO ACIDS OR H_2O
Colourless or white	Conchoidal	Perfect basal	2·0	2·2	Soluble in water
—	Conchoidal	Visible in one direction	1·5–2·0	1·53	Soluble in water
White	Uneven	Perfect basal	2·0	2·0	Soluble in water
White to yellow	Uneven to conchoidal	—	2·75–3·0	6·88	Soluble in HCl with the precipitation of lead chloride
Colourless	Conchoidal	Perfect in one direction	1·5–2·0	2·24–2·29	Easily soluble in water
Colourless or white	Subconchoidal	Perfect in one direction	2·0	2·2	Easily soluble in water
White	Conchoidal	Good in one direction	2·0–2·5	1·7	Soluble in water
White	Uneven to subconchoidal	Perfect in one direction	4·5	2·4	Soluble in HCl
—	Uneven	Perfect in one direction	2·5	2·0	Decomposes in hot water
—	Uneven	Good in two directions	3·5	4·3–4·4	—
—	None	Poor	2·5–3·0	3·0	—
Pale green	—	Perfect in one direction	2·0	2·65	—
Bluish-green	Uneven	Perfect in one direction	2·5–3·0	4·33–4·38	—
Pale shades of red	—	Perfect in one direction	1·5–2·5	2·9	—
Sky-blue to green	Uneven	Indistinct	2·0–2·5	2·9–3·0	Soluble in HNO_3
Olive-green to brown	Conchoidal	Indistinct	3·0	4·4	Soluble in HCl
—	Uneven	Perfect in one direction	2·0–2·5	2·53–2·73	Readily soluble in HCl
Red	—	Perfect in one direction	3·5	3·5–7·34	Easily soluble in HCl
—	Subconchoidal	Indistinct	3·5–4·0	3·28	Soluble in HCl
—	Subconchoidal	None	4·0	8·38	Soluble in HNO_3
Pale brownish-black	Brittle	None	5·0–5·5	7·8	Soluble in H_2SO_4
—	Conchoidal to uneven	Good in two directions	5·5	6·5	Soluble in HNO_3
Black	Conchoidal	Poor	6·0–7·0	10·58	—
—	Subconchoidal	Perfect in one direction	3·0	8·15	—
Yellowish to greenish-grey	Subconchoidal to uneven	None	2·5–3·0	9·24	Decomposes in HNO_3 with residue of gold powder
—	Subconchoidal	Good in one direction	2·5–3·0	8·7–9·02	Decomposes in HNO_3 with the separation of gold

MINERAL	FORMULA	SYSTEM	HABIT	COLOUR	LUSTRE
Sylvanite	$AgAuTe_4$	Monoclinic	Prismatic or tabular crystals	Steel-grey to silver-white	Metallic
VANADATES					
Carnotite	$K_2(UO_2)_2(VO_4)_2.3H_2O$	Orthorhombic	Platy crystals	Canary-yellow	Dull to earthy
Descloizite	$Pb(Zn,Cu)VO_4OH$	Orthorhombic	Pyramidal or prismatic crystals	Variable	Greasy
Mottramite	$Pb(Cu,Zn)VO_4OH$	Orthorhombic	Prismatic crystals	Brownish-red to blackish-brown	Greasy
Tyuyamunite	$Ca(UO_2)_2(VO_4)_2.10H_2O$	Orthorhombic	Scaly crystals	Yellow to greenish-yellow	Adamantine to waxy
TUNGSTATES					
Scheelite	$CaWO_4$	Tetragonal	Octahedral crystals	Colourless or white	Vitreous
Stolzite	$PbWO_4$	Tetragonal	Tabular crystals	Reddish-brown to brown	Resinous to subadamantine
Tungstite	H_2WO_4	Orthorhombic	Acicular crystals	Yellow or yellowish-green	Resinous to earthy
Wolframite	$(Fe,Mn)WO_4$	Monoclinic	Prismatic crystals	Dark grey, brownish-black or black	Submetallic to metallic
MOLYBDATES					
Powellite	$CaMoO_4$	Tetragonal	Pyramidal or tabular crystals	Variable	Subadamantine
Wulfenite	$PbMoO_4$	Tetragonal	Tabular crystals	Orangish-yellow to yellow	Resinous to adamantine
ANTIMONIDE					
Allemontite	$AsSb$	Hexagonal	Massive	Tin-white to reddish-grey	Metallic
FLUORIDE					
Fluorite	CaF_2	Cubic	Cubic crystals	Variable	Vitreous
SELENIDE					
Tiemannite	$HgSe$	Cubic	Tetragonal crystals	Steel-grey to blackish-grey	Metallic
SILICATES					
Actinolite	$Ca_2(Mg,Fe)_5Si_8O_{22}(OH)_2$	Monoclinic	Prismatic crystals	Green	Vitreous
Aegirine	$NaFe(Si_2O_6)$	Monoclinic	Prismatic crystals	Green	Vitreous
Aenigmatite	$Na_2Fe_5TiSi_6O_{20}$	Triclinic	Prismatic crystals	Black	Vitreous
Afwillite	$Ca_3(SiO_3.OH)_2 2H_2O$	Monoclinic	Prismatic crystals	Colourless	Vitreous
Akermanite	$Ca_2(MgSi_2O_7)$	Tetragonal	Tabular crystals	Variable	—
Albite	$NaAlSi_3O_8$	Triclinic	Tabular crystals	White (often tinted)	Vitreous
Allanite	$(Ca,Ce,La,Na)_2(Al,Fe,Be,Mn,Mg)_3(SiO_4)_3(OH)$	Monoclinic	Tabular crystals	Black to dark brown	Submetallic to resinous
Alleghanyite	$Mn_5Si_2O_8(OH)_2$	Monoclinic	Bladed crystals	Pink to brown	Vitreous to resinous
Allophane	$Al_2SiO_5.5H_2O$	Amorphous	Massive	Variable	Vitreous to subresinous

STREAK	FRACTURE	CLEAVAGE	HARDNESS	SG	REACTION TO ACIDS OR H_2O
Steel-grey to silver-white	Uneven	Perfect in one direction	1·5–2·0	8·16	Decomposes in HNO_3 with the separation of gold
	Crumbly	Good in one direction	1·0–2·0	4·1	Soluble in HCl
Variable	Subconchoidal to uneven	None	3·0–3·5	6·2	Easily soluble in HCl
Orange to brownish-red	Conchoidal	None	3·0–3·5	5·9	Easily soluble in HCl
—	—	Perfect basal	2·0	3·67–4·35	Soluble in HCl
White	Uneven to subconchoidal	Good in one direction	4·5–5·0	6·1	Decomposes in HCl leaving a residue of hydrous tungstic oxide
Colourless	Conchoidal to uneven	Imperfect	2·5–3·0	7·9–8·3	Decomposes in HCl with the separation of yellow tungstic acid
—	—	Perfect basal	2·5	5·5	—
Reddish-brown, brownish-black or black	Uneven	Perfect in one direction	4·0–4·5	7·4	Decomposes in hot concentrated HCl
—	Uneven	Indistinct	3·5–4·0	4·23	Decomposes in HCl
White	Subconchoidal to uneven	Good in one direction	2·7–3·0	6·5–7·0	Soluble in concentrated HCl
Grey	None	Perfect in one direction	3·0–4·0	6·3	—
White	Conchoidal	Perfect in one direction	4·0	3·0–3·3	—
Black	Uneven to conchoidal	None	2·5	8·19	
—	Subconchoidal to uneven	Good in two directions	5·0–6·0	2·9–3·3	Insoluble in acids
—	Uneven	Good in two directions	6·0–6·5	3·4–3·5	—
Reddish-brown	Uneven	Perfect in two directions	5·5	3·8–3·85	Partially decomposes in HCl
White	Conchoidal	Perfect in one direction	4·0	2·6	—
—	None	Indistinct	5·0–6·0	2·9	Gelatinises with HCl
Colourless	Uneven to conchoidal	Perfect in one direction	6·0–6·5	2·62–2·65	—
Grey	Uneven to subconchoidal	Poor	5·5–6·0	3·5–4·2	Gelatinises with HCl
—	Conchoidal	None	5·5	4·0	Decomposes in HCl
Colourless	Conchoidal	None	3·0	1·85–1·89	Gelatinises with HCl

MINERAL	FORMULA	SYSTEM	HABIT	COLOUR	LUSTRE
Almandine	$Fe_3Al_2Si_3O_{12}$	Cubic	Cubic crystals	Dark red	Vitreous
Analcite	$NaAlSi_2O_6.H_2O$	Cubic	Cubic crystals	Colourless or white (often tinted)	Vitreous
Andalusite	Al_2SiO_5	Orthorhombic	Prismatic crystals	Variable	Vitreous
Andesine	Plagioclase *feldspar* (50–70 per cent *albite*, 50–30 per cent *anorthite*)	Triclinic	Massive	Variable	Subvitreous to pearly
Andradite	$Ca_3Fe_2Si_3O_{12}$	Cubic	Cubic crystals	Variable	Vitreous
Anorthite	$CaAl_2Si_2O_8$	Triclinic	Prismatic or tabular crystals	Colourless, white or greenish-grey	Vitreous to pearly
Anthophyllite	$(Mg,Fe)_7Si_8O_{22}(OH)_2$	Orthorhombic	Massive	Brown	Vitreous
Apophyllite	$KCa_4Si_8O_{20}(F,OH).8H_2O$	Tetragonal	Prismatic crystals	White or grey (often tinted)	Vitreous to pearly
Astrophyllite	$(K,Na)_2(Fe^{2+},Mn)_4 TiSi_4O_{14}(OH)_2$	Triclinic	Bladed crystals	Bronze-yellow to gold yellow	Submetallic to pearly
Augite	$(Ca,Mg,Fe,Al)_2 (Al,Si)_2O_6$	Monoclinic	Prismatic crystals	Black to greenish-black	Vitreous to resinous
Axinite	$Ca_2(Mn, Fe)$, $Al_2BSi_4O_{15}OH$	Triclinic	Acicular crystals	Variable	Vitreous
Babingtonite	$Ca_2Fe^{2+}Fe^{3+}Si_5O_{14}(OH)(OH)$	Triclinic	Prismatic crystals	Black or greenish-black	Vitreous
Benitoite	$BaTiSi_3O_9$	Hexagonal	Tabular crystals	Blue or white	Vitreous
Bertrandite	$Be_4Si_2O_7(OH)_2$	Orthorhombic	Tabular crystals	Colourless or pinky-white	Vitreous to pearly
Beryl	$Be_3Al_2Si_6O_{18}$	Hexagonal	Prismatic crystals	Variable	Vitreous
Biotite	$K_2(Mg,Fe)_{4-6} (Si,Al)_8O_{20}(OH)_4$	Monoclinic	Tabular or prismatic crystals	Green to black	Pearly
Brewsterite	$(Sr,Ba)Al_2Si_6O_{16}.5H_2O$	Monoclinic	Prismatic crystals	White	Vitreous
Bytownite	Plagioclase *feldspar* (10–30 per cent *albite*, 90–70 per cent *anorthite*)	Triclinic	Massive	Variable	Vitreous to pearly
Cancrinite	c $4(NaAlSiO_4).CaCO_3.H_2O$	Hexagonal	Massive	Yellow, white or red	Subvitreous to pearly
Celsian	$BaAl_2Si_2O_8$	Monoclinic	Massive	Colourless	Greasy
Chabazite	$(Ca,Na,K)_7Al_{12} (Al,Si)_2Si_{26}O_{80}.40H_2O$	Hexagonal	Rhombohedral crystals	Variable	Vitreous
Chlorite	c $(Mg,Fe)_5Al(AlSi_3)O_{10}(OH)_8$	Monoclinic	Tabular crystals	Shades of green	Pearly
Chloritoid	$(Mg,Fe)_2Al_4Si_2O_{10}(OH)_4$	Monoclinic	Massive	Variable	Pearly
Chrysocolla	$CuSiO_3.2H_2O$	Amorphous	Massive	Shades of green and blue	Vitreous
Chrysotile	$Mg_3Si_2O_5(OH)_4$	Monoclinic	Fibrous masses	Shades of green and brown	Silky
Clinochlore	$(Mg,Fe^{2+},Al)_6 (Si,Al)_4O_{10}(OH)_8$	Monoclinic	Tabular or prismatic crystals	Variable	Pearly
Cordierite	$(Mg,Fe)_2Al_4Si_5O_{18}$	Orthorhombic	Prismatic crystals	Shades of blue	Vitreous to dull
Danburite	$CaB_2Si_2O_8$	Orthorhombic	Tetrahedral crystals	Shades of yellow	Vitreous

STREAK	FRACTURE	CLEAVAGE	HARDNESS	SG	REACTION TO ACIDS OR H_2O
White	Conchoidal to uneven	None	7·5	3·95–4·25	—
White	Subconchoidal	Poor	5·0–5·5	2·22–2·29	Gelatinises in HCl
Colourless	Subconchoidal to uneven	Good in one direction	7·5	3·16–3·2	Insoluble in acids
—	None	Perfect in one direction	5·0–6·0	2·68–2·69	—
White	Conchoidal to uneven	None	7·5	3·8–3·9	—
White	Uneven	Good in two directions	6·0–6·5	2·7	Decomposes in HCl with the separation of gelatinous silica
—	None	Perfect in one direction	5·5–6·0	2·9–3·4	Insoluble in HCl
White	Uneven	Perfect in one direction	4·5–5·0	2·3–2·5	Decomposes in HCl with the separation of silica
—	None	Perfect in one direction	3·0	3·4–3·5	Decomposes in HCl with the separation of scaly silica
White to grey or greyish-green	Uneven to conchoidal	Good in one direction	5·0–6·0	3·2–3·5	Insoluble in HCl
Colourless	Conchoidal	Good in one direction	6·5–7·0	3·27	—
—	Conchoidal	Good in two directions	5·5–6·0	3·4	Insoluble in HCl
—	Conchoidal	Poor	6·0–6·5	3·6	—
—	Flaky	Perfect basal	6·0–7·0	2·6	Insoluble in HCl
White	Conchoidal	Poor	7·5–8·0	2·6	—
Colourless	None	Perfect basal	2·5–3·0	2·7–3·1	Decomposes in H_2SO_4 leaving residue of scaly silica
—	Uneven	Perfect in one direction	5·0	2·45	—
—	Uneven	Perfect in one direction	6·0	2·72	—
Colourless	Uneven	Perfect in one direction	5·0–6·0	2·4–2·5	Effervesces in HCl
—	Conchoidal to uneven	Perfect in one direction	6·0–6·5	3·37	Insoluble in acids
White	Uneven	Poor	4·0–5·0	2·1–2·2	Decomposes in HCl with the separation of silica
Pale green	Earthy	Perfect basal	1·5–2·5	2·65–2·94	—
Colourless	Scaly	Perfect basal	6·5	3·52–3·57	Decomposes in H_2SO_4
White	Conchoidal	None	2·0–4·0	2·0–2·24	—
White	Fibrous	None	2·55	2·2	—
Colourless or greenish-white	Earthy	Perfect in one direction	2·0–2·25	2·65–2·78	Decomposes in H_2SO_4
White	Uneven	Good in one direction	7·0–7·5	2·6–2·7	—
White	Conchoidal to uneven	None	7·0	2·9–3·0	—

MINERAL	FORMULA	SYSTEM	HABIT	COLOUR	LUSTRE
Datolite	$CaBSiO_4OH$	Monoclinic	Tabular, pyramidal or prismatic crystals	Variable	Vitreous to greasy
Diopside	$MgCaSi_2O_6$	Monoclinic	Prismatic crystals	Variable	Vitreous
Dioptase	$CuSiO_2(OH)_2$	Hexagonal	Prismatic crystals	Shades of green	Vitreous
Edingtonite	$BaAl_2Si_3O_{10}.4H_2O$	Tetragonal	Massive	White, greyish-white or pink	Vitreous
Enstatite	$MgSiO_3$	Orthorhombic	Prismatic crystals	Variable	Vitreous to silky
Epidote	$Ca_2(Al,Fe)_3$ $(SiO_4)_3(OH)$	Monoclinic	Prismatic crystals	Shades of green and brown	Vitreous to pearly
Fayalite	Fe_2SiO_4	Orthorhombic	Tabular crystals	Yellow	Metallic to resinous
Forsterite	Mg_2SiO_4	Orthorhombic	Tabular crystals	Variable	Vitreous
Glaucophane	$Na_2(Mg,Fe)_3$ $Al_2Si_8O_{22}(OH)_2$	Monoclinic	Prismatic crystals	Shades of blue	Vitreous to pearly
Gmelinite	$(Na_2,Ca)Al_2Si_4O_{12}.$ $6H_2O$	Hexagonal	Rhombohedral crystals	Colourless, white (often tinted) or red	Vitreous
Grossular	$Ca_3Al_2Si_3O_{12}$	Cubic	Cubic crystals	Greenish-white to olive-green	Vitreous
Harmotome	$BaAl_2Si_6O_{16}.6H_2O$	Monoclinic	Prismatic crystals	Variable	Vitreous
Hemimorphite	$Zn_4Si_2O_7(OH)_2.H_2O$	Orthorhombic	Pyramidal crystals	White	Vitreous to adamantine
Hornblende	$(Ca,Mg,Fe,Na,Al)_{7-8}$ $(Al,Si)_8O_{22}(OH)_2$	Monoclinic	Prismatic crystals	Black or greenish-black	Vitreous
Humite	$Mg_7Si_3O_{12}(F,OH)_2$	Orthorhombic	Pyramidal crystals	White, yellow or brown	Vitreous to resinous
Idocrase	$Ca_{10}(Mg,Fe^{2+},Fe^{3+})_2$ $Al_4Si_9O_{34}(OH)_4$	Tetragonal	Prismatic crystals	Brown to green	Vitreous
Jadeite	$NaAlSi_2O_6$	Monoclinic	Massive	Shades of green and white	Subvitreous to pearly
Kaolinite	$Al_2Si_2O_5(OH)_4$	Monoclinic	Massive	White or grey	Pearly to dull
Kyanite	Al_2SiO_5	Triclinic	Bladed crystals	Blue or white	Vitreous to pearly
Labradorite	Plagioclase *feldspar* (30–50 per cent *albite*, 70–50 per cent *anorthite*)	Triclinic	Massive	Greyish-brown	Vitreous
Laumontite	$CaO.Al_2O_34SiO_2.4H_2O$	Monoclinic	Prismatic crystals	White	Vitreous
Lazurite	$Na_{4-5}Al_3Si_7O_{12}S$	Cubic	Massive	Azure-blue	Vitreous
Lepidolite	$K_2Li_3Al_4Si_7O_{21}(OH,F)_3$	Monoclinic	Aggregates	Lilac	Pearly
Leucite	$KAlSi_2O_6$	Tetragonal	Pseudocubic crystals	Grey, white or colourless	Dull
Levynite	$CaAl_2Si_3O_{10}.5H_2O$	Orthorhombic	Massive	White or greyish-green	Vitreous
Margarite	$CaAl_4Si_2O_{10}(OH)_2$	Monoclinic	Aggregates	White, violet or grey	Pearly
Melilite	$Ca_2MgSi_2O_7$	Tetragonal	Tabular crystals	White or pale yellow	Vitreous
Mesolite	$Na_2Ca_2(Al_2Si_3O_{10})_3.$ $8H_2O$	Monoclinic	Acicular crystals	White to grey	Vitreous
Microcline	$KAlSi_3O_8$	Triclinic	Prismatic crystals	White to pale yellow	Vitreous
Milarite	$K_2Ca_4Be_4Al_2Si_{24}O_{60}.$ H_2O	Hexagonal	Prismatic crystals	Colourless to pale green	Vitreous
Monticellite	$MgCaSiO_4$	Orthorhombic	Prismatic crystals	Colourless or grey	Vitreous
Muscovite	$KAl_2Si_3O_{10}(OH)_2$	Monoclinic	Tabular crystals	Colourless or white	Pearly

STREAK	FRACTURE	CLEAVAGE	HARDNESS	SG	REACTION TO ACIDS OR H$_2$O
White	Conchoidal to uneven	None	5·0–5·5	2·9–3·0	Gelatinises with HCl
White or grey	Uneven	Perfect in one direction	5·5	3·2–3·38	—
Green	Conchoidal to uneven	Perfect in one direction	5·0	3·3	Gelatinises with HCl
White	Subconchoidal to uneven	Perfect in one direction	4·0–4·5	2·7	Gelatinises with HCl
Colourless or grey	Uneven	Perfect in one direction	5·5–6·0	3·2–3·9	Insoluble in HCl
Colourless or grey	Uneven	Perfect basal	6·0–7·0	3·4–3·5	—
Colourless	Conchoidal	Good in one direction	6·5	4·0–4·14	Gelatinises with HCl
Colourless	Subconchoidal to uneven	Distinct in one direction	6·0–7·0	3·21–3·33	Decomposes in HCl with the separation of gelatinous silica
Greyish-blue	Subconchoidal to uneven	Perfect in one direction	6·0–6·5	3·1–3·11	—
—	Uneven	Good in one direction	4·5	2·04–2·17	Decomposes in HCl with the separation of silica
White	Subconchoidal to uneven	None	7·5	3·5	—
White	Uneven to subconchoidal	Good in one direction	4·5	2·44–2·5	Decomposes in HCl
White	Uneven to subconchoidal	Perfect in one direction	4·5–5·0	3·4–3·5	Gelatinises in acetic acid
—	Subconchoidal to uneven	Perfect in two directions	5·0–6·0	3·0–3·4	—
—	Subconchoidal to uneven	Good in one direction	6·0–6·5	3·1–3·2	Gelatinises with HCl
White	Subconchoidal to uneven	Indistinct	6·5	3·35–3·45	Partially decomposes in HCl
Colourless	Splintery	Good in two directions	6·5–7·0	3·33–3·35	—
—	—	Perfect basal	2·0–2·5	2·6–2·63	—
Colourless	—	Perfect in one direction	5·0–7·25	3·56–3·67	—
—	Conchoidal	Good in two directions	5·0–6·0	2·71	Decomposes with difficulty in HCl
Colourless	Uneven	Good in three directions	3·5–4·0	2·25–2·36	Gelatinises in HCl
Blue	Uneven	Indistinct	5·0–5·5	2·38–2·45	Decomposes in HCl to give gelatinous silica and hydrogen sulphide
White	None	Perfect basal	2·5–4·0	2·8–2·9	Reacts slowly with HCl
Colourless	Conchoidal	Indistinct	5·5–6·0	2·4–2·5	Soluble in HCl
—	Subconchoidal	Indistinct	4·0–4·5	2·1	Gelatinises with HCl
—	None	Perfect basal	3·5–4·5	4·0	Slowly decomposes in boiling HCl
—	Conchoidal	Good in one direction	5·0	2·9–3·1	Gelatinises with HCl
—	—	Perfect in one direction	5·0	2·2–2·4	Gelatinises with HCl
—	Uneven	Poor	6·0–6·5	2·55	—
—	Conchoidal	Indistinct	5·5–6·0	2·55–2·59	Decomposes in HCl
Colourless	Uneven	Good in one direction	5·0–5·5	3·03–3·25	Soluble in HCl
Colourless	None	Perfect basal	2·0–2·5	2·76–3·0	—

MINERAL	FORMULA	SYSTEM	HABIT	COLOUR	LUSTRE
Natrolite	$Na_2Al_2Si_3O_{10}.2H_2O$	Orthorhombic	Prismatic crystals	Colourless or white	Vitreous
Nepheline	$NaAlSiO_4$	Hexagonal	Prismatic crystals	Colourless, white or yellow	Vitreous to greasy
Nosean	$Na_8Al_6Si_6O_{24}SO_4$	Cubic	Cubic crystals	Grey, blue or brown	Subvitreous
Oligoclase	Plagioclase *feldspar* (70–90 per cent *albite*, 30–10 per cent *anorthite*)	Triclinic	Massive	White, green or red	Vitreous
Orthoclase	$KAlSi_3O_8$	Monoclinic	Prismatic or tabular crystals	White, pink, yellow or brown	Vitreous
Pectolite	$NaCa_2Si_3O_8OH$	Monoclinic	Acicular crystals	Whitish-grey	Silky
Petalite	$LiAl(Si_2O_5)_2$	Monoclinic	Massive	Colourless, white or grey	Vitreous
Phenacite	Be_2SiO_4	Orthorhombic	Lenticular crystals	Colourless, yellow or pale red	Vitreous
Phillipsite	$(Ca,Na,K)_3(Al_3Si_5O_{16}).6H_2O$	Monoclinic	Aggregates	White	Vitreous
Phlogopite	$KMg_3AlSi_3O_{10}(OH)_2$	Monoclinic	Tabular crystals	Yellowish-brown to brownish-red	Pearly
Piemontite	$Ca_2(Al,Fe,Mn)_3Si_3O_{12}OH$	Monoclinic	Prismatic crystals	Reddish-brown to reddish-black	Vitreous
Pollucite	$(Ca,Na)AlSi_2O_6.nH_2O$	Cubic	Cubic crystals	Colourless	Vitreous
Prehnite	$Ca_2Al_2Si_3O_{10}(OH)_2$	Orthorhombic	Massive	White, grey or light green	Vitreous
Pyrope	$Mg_3I_2Si_3O_{12}$	Cubic	Fragments	Deep crimson-red	Vitreous
Pyrophyllite	$Al_2Si_4O_{10}(OH)_2$	Monoclinic	Massive	Variable	Pearly to dull
Rhodonite	$MnSiO_3$	Triclinic	Tabular or prismatic crystals	Pink to grey	Vitreous
Riebeckite	$Na_2Fe_3^{2+}Fe_2^{3+}Si_8O_{22}(OH)_2$	Monoclinic	Prismatic crystals	Blue or bluish-black	Vitreous
Roepperite	$(Fe,Mn,Zn)_2SiO_4$	Orthorhombic	Massive	Yellow	Vitreous
Roscoelite	$K(V,Al)_3Si_3O_{10}(OH)_2$	Monoclinic	Scales	Clove-brown to greenish-brown	Pearly
Schorlomite	$Ca_3(Fe,Ti)_2(Si,Ti)_3O_{12}$	Cubic	Massive	Black	Vitreous
Scolecite	$CaAl_2Si_3O_{10}.3H_2O$	Monoclinic	Prismatic crystals	White	Vitreous to silky
Serpentine	$Mg_3Si_2O_5(OH)_4$	Monoclinic	Massive	Variable	Subresinous to greasy
Sillimanite	Al_2SiO_3	Orthorhombic	Elongate crystals	Variable	Vitreous
Sodalite	$Na_4Al_3Si_3O_{12}Cl$	Cubic	Cubic crystals	Grey or white (often tinted)	Vitreous
Spessartite	$Mn_3Al_2Si_3O_{12}$	Cubic	Cubic crystals	Shades of red	Vitreous
Sphene	$CaTiSiO_5$	Monoclinic	Prismatic crystals	Variable	Adamantine to resinous
Spodumene	$LiAlSi_2O_6$	Monoclinic	Prismatic crystals	Variable	Vitreous
Staurolite	$(Fe,Mg)_4Al_{18}Si_8O_{46}(OH)_2$	Monoclinic	Prismatic crystals	Shades of brown	Subvitreous
Stilbite	$NaCa_2Al_5Si_{13}O_{36}.14H_2O$	Monoclinic	Tabular crystals	White	Vitreous
Talc	$Mg_3Si_4O_{10}(OH)_2$	Monoclinic	Massive	White (often tinted)	Pearly
Tephroite	Mn_2SiO_4	Orthorhombic	Massive	Shades of red and grey	Vitreous to greasy

STREAK	FRACTURE	CLEAVAGE	HARDNESS	SG	REACTION TO ACIDS OR H_2O
White	Uneven	Perfect in one direction	5·0–5·5	2·2–2·25	Gelatinises with HCl
—	Subconchoidal	Good in one direction	5·5–6·0	2·55–2·65	Gelatinises with HCl
Variable	Uneven	Poor	5·5	2·25–2·46	Gelatinises with HCl
—	Conchoidal to uneven	Perfect in one direction	6·0–7·0	2·65–2·67	Unaffected by acids
—	Conchoidal	Good in two directions	6·0	2·6	Unaffected by acids
White	Uneven	Perfect in two directions	5·0	2·68–2·78	Decomposes in HCl with the separation of silica
Colourless	Conchoidal	Perfect in one direction	6·0–6·5	2·39–2·46	Unaffected by acids
—	Conchoidal	Good in one direction	7·5–8·0	2·97–3·0	—
Colourless	Uneven	Good in two directions	4·0–4·5	2·2	Gelatinises with HCl
—	None	Perfect basal	2·5–3·0	2·78–2·85	Decomposes in H_2SO_4 with the separation of silica
Red	Uneven	Perfect in one direction	6·5	3·4	Unaffected by acids
Colourless	Conchoidal	Poor	6·5	2·9	Slowly decomposes in HCl with the separation of silica
Colourless	Uneven	Good in one direction	6·0–6·5	2·8–2·95	Decomposes slowly in HCl
Dark red	Conchoidal	None	7·5	3·7	Gelatinises in HCl
Variable	Uneven	Perfect basal	1·0–2·0	2·8–2·9	Partially decomposes in H_2SO_4
White	Conchoidal to uneven	Perfect in two directions	5·5–6·0	3·5–3·7	Slightly affected by HCl
—	—	Perfect in two directions	4·0	3·43	—
Yellow to reddish-grey	—	Good in two directions	5·5–6·0	4·0	Gelatinises in HCl
—	None	Perfect basal	1·0–2·0	2·92–2·94	—
Greyish-black	Conchoidal	None	7·0–7·5	3·81–3·88	Gelatinises with HCl
—	—	Good in one direction	5·0–5·5	2·16–2·4	Gelatinises with HCl
Variable	Conchoidal to splintery	Poor	2·5–4·0	2·5–2·65	—
Colourless	Uneven	Perfect in one direction	6·0–7·0	3·23–3·24	—
Colourless	Conchoidal to uneven	Good in two directions	5·5–6·0	2·14–2·3	Decomposes in HCl with the separation of gelatinous silica
White	Subconchoidal	None	7·0–7·5	4·15–4·27	—
White	Subconchoidal	Good in two directions	5·0–5·5	3·4–3·56	Decomposes in H_2SO_4
White	Uneven to subconchoidal	Perfect in one direction	6·5–7·0	3·13–3·2	—
Colourless or grey	Subconchoidal	Good in one direction	7·0–7·5	3·65–3·75	—
Colourless	Uneven	Perfect in one direction	3·5–4·0	2·09	Decomposes in HCl
White	None	Perfect basal	1·0	2·7–2·8	—
Pale grey	Subconchoidal	Good in two directions	5·5–6·0	4·0–4·12	Gelatinises in HCl

MINERAL	FORMULA	SYSTEM	HABIT	COLOUR	LUSTRE
Thomsonite	$NaCa_2Al_5Si_5O_{20}.6H_2O$	Orthorhombic	Massive	White	Vitreous to pearly
Thorite	$ThSiO_4$	Tetragonal	Prismatic or pyramidal crystals	Brownish-yellow to black	Vitreous to resinous
Topaz	$Al_2SiO_4(OH,F)_2$	Orthorhombic	Prismatic crystals	Yellow or white (often tinted)	Vitreous
Tourmaline	$(Na,Ca)(Li,Mg,Fe^{2+}Al)_3(Al,Fe^{3+})_6B_3Si_6O_{27}(O,OH,F)_4$	Hexagonal	Prismatic or acicular crystals	Black (often tinted)	Vitreous to resinous
Tremolite	$Ca_2Mg_5Si_8O_{22}(OH)_2$	Monoclinic	Bladed crystals	White to dark grey	Vitreous
Uranophane	$Ca(UO_2)_2Si_2O_7.6H_2O$	Orthorhombic	Acicular crystals	Yellow	Vitreous
Uvarovite	$Ca_3Cr_2Si_3O_{12}$	Cubic	Cubic crystals	Emerald-green	Vitreous
Willemite	Zn_2SiO_4	Hexagonal	Prismatic crystals	White or greenish-yellow	Vitreous to resinous
Wollastonite	$CaSiO_3$	Monoclinic	Tabular crystals	White	Vitreous
Zinnwaldite	$K_2(Li,Fe,Al)_6(Si,Al)_8O_{20}(F,OH)_4$	Monoclinic	Tabular crystals	Pale violet, yellow or brown	Pearly
Zircon	$ZrSiO_4$	Tetragonal	Prismatic crystals	Colourless (often tinted)	Adamantine
Zoisite	$Ca_2Al_3Si_3O_{12}OH$	Orthorhombic	Prismatic crystals	Variable	Vitreous

STREAK	FRACTURE	CLEAVAGE	HARDNESS	SG	REACTION TO ACIDS OR H_2O
Colourless	Uneven to subconchoidal	Perfect in one direction	5·0–5·5	2·3–2·4	Gelatinises in HCl
Pale orange to dark brown	Conchoidal	Good in one direction	4·5–5·0	5·4	Gelatinises with HCl
Colourless	Subconchoidal to uneven	Perfect in one direction	8·0	3·4–3·65	Partially decomposes in H_2SO_4
Colourless	Subconchoidal to uneven	Poor	7·0–7·5	2·98–3·2	Unaffected by acids
—	Subconchoidal to uneven	Perfect in one direction	5·0–6·0	2·9–3·2	—
—	None	Poor	2·0–3·0	3·81–3·9	Soluble in warm HCl with the separation of silica
Greenish-white	Subconchoidal to uneven	None	7·5	3·42	—
Colourless	Conchoidal to uneven	Poor	5·5	3·89–4·18	Gelatinises in HCl
White	Uneven	Perfect in one direction	4·5–5·0	2·8–2·9	Decomposes in HCl with the separation of silica
—	None	Perfect basal	2·5–3·0	2·62–3·2	—
Colourless	Conchoidal	Poor	7·5	4·68–4·7	—
Colourless	Uneven to subconchoidal	Perfect in one direction	6·0–6·5	3·25–3·37	—

A to Z ENCYCLOPEDIA

AA or APHROLITH

Type of *basalt* lava flow with a clinker-like tumultuous surface.
Occurrence: USA—Hawaiian Islands.

AASBY DIABASE

Igneous rock of the *gabbro* clan. A hypabyssal rock. Contains *biotite, ilmenite* and *apatite*, plus *labradorite, augite* and *olivine*.
Occurrence: Sweden—Aasby.
A type of *olivine-dolerite*.
See *diabase*.

ABSAROKITE

Igneous rock of the *diorite* clan. A porphyritic extrusive rock. Contains phenocrysts of *olivine* and *augite* in a groundmass of *labradorite* with *orthoclase* rims, *olivine, augite* and some *leucite*.
Occurrence: USA—Wyoming (Absaroka Range in the Yellowstone National Park).

ABYSSAL

Formed at depth. Abyssal sediments (deposited below about 2km in oceans) comprise various *oozes* and *clays*.

ACANTHITE

$4[Ag_2S]$ Silver sulphide. Monoclinic. Prismatic to long prismatic crystals. Iron-black. Metallic lustre. Black streak. Uneven fracture. Indistinct cleavage. Hardness $2 \cdot 0 – 2 \cdot 5$. SG $7 \cdot 2 – 7 \cdot 3$. Chemical identification as for *argentite*.

Occurrence: USA—Colorado (Georgetown, Rice). Czechoslovakia—Bohemia (Joachimstal). Germany—Saxony (Himmelfurst Mine near Freiberg). Mexico—Chihuahua. With *argentite*.

ACCESSORY

Relevant but present in too small a concentration for classification purposes.

ACETIC ACID

CH_3COOH, corrosive and useful for testing when undiluted.

ACHONDRITE

Type of *meteorite*.

ACHROÏTE

Colourless gemstone variety of *tourmaline*.
Occurrence: Italy—Island of Elba.

ACICULAR

See *crystal*.

ACID

A substance which in itself, or in solution, has a pH value below $7 \cdot 0$. Such indicators as litmus are affected (litmus will turn red). Often used for testing rocks and minerals, the resulting chemical action assisting with identification.
Acidic igneous rocks have a light colour and contain a high proportion of *quartz* (over 10 per cent) and of silica (over 70 per cent).
See *alkali*.

ACMITE

Brown or brownish-green variety of *aegirine*. Forms long, prismatic crystals with sharp terminations.
Occurrence: In association with *aegirine*.

ACTINOLITE

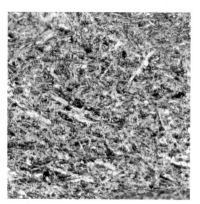

$Ca_2(Mg, Fe)_5Si_8O_{22}(OH)_2$ Calcium magnesium silicate varying to calcium magnesium iron silicate. Monoclinic. Long, slender or blade-like prismatic crystals; can be columnar, fibrous, radiating, compact or granular. Green. Vitreous lustre. Subconchoidal to uneven fracture. Excellent prismatic cleavage at 56 degrees. Hardness $5 \cdot 0 – 6 \cdot 0$. SG $2 \cdot 9 – 3 \cdot 3$. Pleochroic in shades of yellow and green. Insoluble in acid. Thin splinters fuse to a black or white *glass*. Iron-rich varieties fuse more easily.
Occurrence: Worldwide in metamorphosed *limestone, gneiss, serpentine, schist* and *granite*. Varieties are *actinolite asbestos, ferrotremolite* and *nephrite*.

ACTINOLITE ASBESTOS or AMIANTHUS

Fibrous form of *actinolite*. Long, parallel, flexible, easily separated fibres.

Occurrence: USA—New Jersey, North Carolina, Pennsylvania.

ADAMANTINE

Diamond-like in hardness or lustre.

ADAMELLITE

Igneous rock of the *granite* clan. A coarse-grained, often porphyritic, plutonic rock. Contains *biotite, hornblende* and *quartz,* with *orthoclase* and *plagioclase* in a ratio ranging between 35:65–65:35. If porphyritic, *orthoclase* phenocrysts.

Occurrence: Worldwide. Notable examples are England—Westmorland (Shap Fell) and Austria—the Tyrol (Adamello Complex).

ADAMITE

$4[Zn_2(OH)AsO_4]$ Basic zinc arsenate.

Above and left: adamite. *Below:* adularia.

Orthorhombic. Drusy crusts of short-prismatic or horizontally elongated crystals. Light yellow, greenish, rose or violet. Vitreous lustre. Uneven fracture. Domal cleavage. Hardness 3·5. SG 4·3–4·4. Often brilliantly fluorescent yellow-green. Reluctantly fuses with slight decrepitation. Loses fluorescence on first heating, whitens and becomes opaque giving slight arsenical smell. Becomes less fusible as water bubbles away.
Occurrence: USA—Nevada. France—Cap Garonne. Greece—Laurium. Chile—Chañarcillo. Mexico—Durango (Ojuela Mine). Usually lining cavities in *limonite*.

ADAMSITE

Variety of *muscovite*. A greenish-black *mica*.
Occurrence: USA—Vermont (Derby). In a *micaceous schist*.

ADINOLE

Low grade metamorphic rock. A mosaic-textured *shale* or *slate* metamorphosed by intrusions of *dolerite*. Consists of an intergrowth of *albite*, or *albite* and *quartz* with interstitial *iron* and *chlorite* ores.
Occurrence: England—Cornwall.

ADOBE

Sedimentary rock of doubtful mode of formation. A deposit similar to *loess*.
Occurrence: USA—California, Colorado, Mississippi, Texas.

ADSORBENT

The property of attracting substances such as atmospheric gases and moisture to form a layer on the surface. Differs from absorbent in that there is no penetration to the interior.

ADULARIA or VALENCIANITE

Variety of *orthoclase*. Pure or nearly pure potassium silicate. Prismatic, basal plane and hemiorthodome crystals. Opalescent, transparent or slightly cloudy. Pearly lustre.
Occurrence: Switzerland—Adular Mountains. Peru—Valencia Silver Mine.

AEGIRINE

$NaFe(Si_2O_6)$ Sodium iron silicate. Monoclinic. Short, prismatic crystals with blunt terminations; also acicular or fibrous crystals. Green. Vitreous lustre. Uneven fracture. Prismatic cleavage. Hardness 6·0–6·5. SG 3·4–3·5. Easily fuses to shiny, black magnetic bead giving yellow sodium flame.
Occurrence: USA—Arkansas (Magnet Cove in the Ozark Mountains), Colorado

(Colorado Springs), New Jersey (Beemerville). Canada—near Montreal. Greenland—Kangerdluarsuk. Norway—Langesund Fiord. Brazil. In soda-rich igneous rocks.
Varieties are *acmite* and *aegirine-augite*.

AEGIRINE-AUGITE

Green variety of *aegirine*. A rock-making *augite* consisting of a solid solution series containing less than 75 per cent *aegirine*. Shows characteristics intermediate between *aegirine* and *augite*.
Occurrence: Norway—Alnö Island. Japan—Hokkaidô. Northern Nigeria. Sudan. In soda-rich igneous rocks such as *syenites*, *nepheline-syenites* and *phonolites*.

AENIGMATITE

$Na_4(Fe^{2+},Fe^{3+},Ti)_{13}Si_{12}O_{42}$. Iron titanium sodium and aluminium silicate. Triclinic. Prismatic crystals. Black. Vitreous lustre. Reddish-brown streak. Prismatic, uneven fracture. Distinct cleavage. Hardness 5·55. SG 3·8–3·85. Brittle. Fuses to brownish-black *glass*. Partially decomposes in acid.
occurrence: Greenland—Kangerdluarsuk. Norway—Langesund Fiord. East Africa. In soda-rich *trachytes*.
Varieties are *cossyrite* and *rhonite*.

AEOLIAN

Wind-blown.

AEOLIAN DEPOSITS

Sedimentary, wind-deposited rocks. Contain mainly well-rounded *quartz* grains featuring millet-seed pitting. Characterised by large scale cross-bedded structures such as sand dunes.
See *loess*.

AFWILLITE

$4[Ca_3(SiO_3OH)_2 \cdot 2H_2O]$ Hydrated calcium silicate. Monoclinic. Prismatic crystals. Colourless. Vitreous lustre. White streak. Conchoidal fracture. Perfect cleavage parallel to one crystal plane. Hardness 4·0. SG 2·6. Brittle crystals which appear translucent if mineral is corroded. Responds to tests for alkali, e.g. turns litmus blue.
Occurrence: South Africa—Kimberley. In pipes of *kimberlite*.

AGALMATOLITE

Synonym of *talc*.

AGARIC MINERAL

A white earthy variety of *calcite* found in caverns.

AGATE

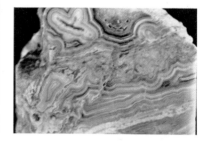

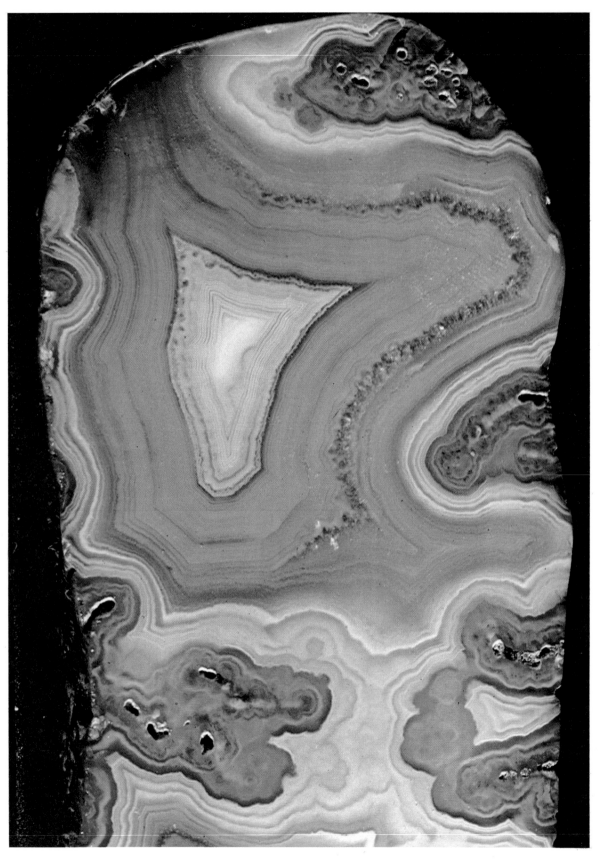

40

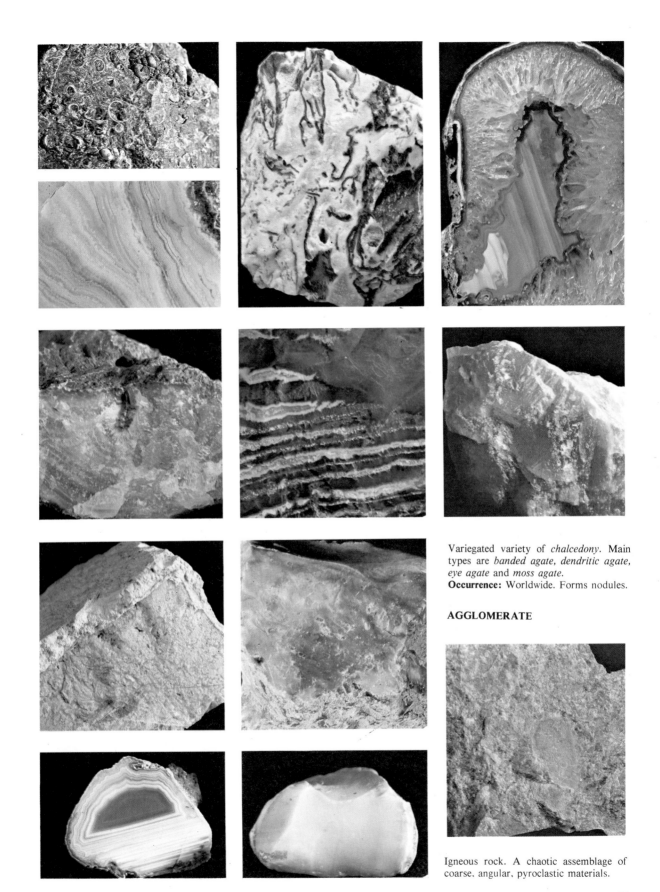

Variegated variety of *chalcedony*. Main types are *banded agate*, *dendritic agate*, *eye agate* and *moss agate*.
Occurrence: Worldwide. Forms nodules.

AGGLOMERATE

Igneous rock. A chaotic assemblage of coarse, angular, pyroclastic materials.

AGGREGATE

A mass of units or parts somewhat loosely associated one with the other.

AIKINITE or BELONITE

$PbCuBiS_3$ Lead copper bismuth sulphide. Orthorhombic. Prismatic to acicular and striated crystals; also massive. Blackish lead-grey, tarnishing brown or copper-red, sometimes with yellowish-green coating. Metallic lustre. Greyish-black streak. Uneven fracture. Indistinct cleavage. Hardness 2·0–2·5. SG 7·07. Decomposes in nitric acid with precipitation of sulphur and lead sulphide.
Occurrence: USA—Idaho, North Carolina, Utah. Australia—Tasmania (Dundas). France—Gard, Isere. Mexico. USSR—Ural Mountains (Beresovsk district). In *quartz* in veins with *gold* and *iron* ores.

AILSYTE

Igneous rock of the *granite* clan. A greenish-grey, fine-grained equigranular plutonic to hypabyssal rock. Contains *feldspar, quartz, riebeckite* and *zircon*. Surface spotted with irregular, light-blue blotches.
Occurrence: Scotland—Firth of Clyde (Island of Ailsa Craig).

AJOITE

$Cu_6Al_2Si_{10}O_{29} \cdot 5\frac{1}{2}H_2O$ Hydrous copper aluminium silicate. Platy crystals of uncertain system; also massive. Bluish-green. SG 2·96.
Occurrence: USA—Arizona (Ajo in Pima County).

AKENOBEITE

Igneous rock of the *diorite* clan. A leucocratic hypabyssal rock. Consists of a random aggregation of thick, tabular crystals of *feldspar* with a fine *quartz* grained interstitial aggregate. Contains *orthoclase* in excess of *oligoclase* and very little ferromagnesian mineral. *Quartz, chlorite, biotite* and *epidote* may occur.

Occurrence: Japan—Tajima (Akénobé district).

ÅKERITE

Igneous rock of the *syenite* clan. A well-developed granular plutonic rock with rectangular development of the *feldspars*. Contains *orthoclase, oligoclase*, often abundant *biotite*, and a little *quartz*. *Olivine* usually absent.
Occurrence: Norway—Oslo.

ÅKERMANITE

$Ca_2(MgSi_2O_7)$ Calcium magnesium silicate. Tetragonal. Thin, tabular crystals. Colourless, greyish, green or brown. No fracture. Indistinct cleavage. Hardness 5·0–6·0. SG 2·944. Gelatinises in hydrochloric acid.
Occurrence: Sweden.
Formed from certain slags on cooling. Isomorphous with *gehlenite* and *melilite*.

ALABANDITE

4(MnS) Manganese sulphide. Cubic. Granular or massive. Iron-black, tarnishing brown on exposure. Submetallic lustre. Green streak. Uneven fracture. Perfect cleavage. Hardness 3·5–4·0. SG 4·0±. Brittle. Dissolves in hydrochloric acid with evolution of hydrogen sulphide.
Occurrence: Worldwide. As primary material in vein deposits with sulphides and manganese spars.

ALABASTER

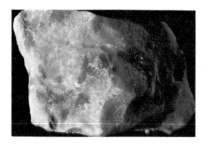

Variety of *gypsum*. A white, or delicately shaded, fine-grained massive structure. Alternatively variety of *calcite* (*onyx marble*).

ALALITE

Gemstone variety of *diopside* occurring in broad, right-angled prisms. Colourless to faint greenish or clear green. Usually striated longitudinally.
Occurrence: Italy—Piedmont (Ala Valley).

ALASKITE

Igneous rock of the *granite* clan. A light-coloured, hypautomorphic-granular, xenomorphic plutonic rock. Contains alkali-*feldspars* and *quartz* with small amounts of ferromagnesian minerals.
Occurrence: USA—Alaska, Nevada.

ALBERTITE

Variety of *bitumen*. Mixture of different hydrocarbons, partly oxygenated. Jet-black. Brilliant, pitch-like lustre. Conchoidal fracture. Hardness 1·0–2·0. SG 1·097. Only partially soluble in turpentine.
Occurrence: Scotland. Canada—Nova Scotia. In irregular fissures in rocks of subcarboniferous age.
Variety is *impsonite*.

ALBITE

4[NaAlSi$_3$O$_8$] Sodium aluminium silicate. Triclinic. Tabular or elongated crystals; often massive and either lamellar or granular. Normally white; occasionally bluish, grey, reddish, greenish or green. Vitreous lustre; cleavage sometimes pearly. Colourless streak. Uneven to conchoidal fracture. Perfect cleavage along one axis. Hardness 6·0–6·5. SG 2·62–2·65. Brittle. Bead test fuses to colourless or white *glass*, imparting intense yellow to flame.
Occurrence: Worldwide. Constituent of many crystalline rocks.
Varieties are *amelia albite*, *cleavelandite*, *pericline* and *peristerite*.

ALBITITE

Igneous rock of the *syenite* clan. A light-coloured, fairly coarse, granular hypabyssal rock irregular in form. Contains mostly *albite*, with minute amounts of ferromagnesian minerals, varying with locality.
Occurrence: USA—California (Meadow Valley in Plumas County). Australia—Kangaroo Island.

ALBITOPHYRE

Igneous rock of the *syenite* clan. A leucocratic, fine-grained porphyritic hypabyssal rock. Contains phenocrysts of *albite* in a groundmass of *chalcedony, quartz* and calcium compounds.
Occurrence: France.

ALBORANITE

Igneous rock of the *gabbro* clan. A mottled hyalopilitic extrusive rock. Contains phenocrysts of *bytownite-anorthite* in a groundmass of *augite, hypersthene*, iron ores, *apatite* and interstitial *glass*.
Occurrence: Spain—Island of Alboran.

ALEXANDRITE

Variety of *chrysoberyl* (*of Werner*). Gemstone, often with very large, twin, six-sided or six-rayed crystals. Emerald green colour changes to columbine-red with transmitted light. SG 3·6.
Occurrence: USSR—Ural Mountains. Burma. Sri Lanka.

ALEXOITE

Igneous rock of the *ultrabasic* clan. Aporphyritic plutonic rock. Contains largely *olivine* pseudomorphed by fibres of *serpentine* and iron-nickel ore.
Occurrence: Canada—Ontario (Alexo Mine).

ALGAL LIMESTONE

Creamy-coloured sedimentary rock. Formed in a shallow marine environment as a breakdown product of stony, calcareous algae. Consists of dome-shaped or columnar masses of laminated structure, or sand-grain sized needles of *aragonite*.
Occurrence: Bahamas—Andros Island (dome-shaped and columnar masses). Canada—Saskatchewan. France—the Pyrenees. South Sea Islands—Nullipore Limestones (sand-grain sized needles).

ALGARVITE

Igneous rock of the *ultrabasic* clan. A dark-coloured, medium-grained plutonic rock. Contains *diopside, aegirine-augite* or *aegirine* with *biotite* and small amounts of *nepheline* and *garnet*.
Occurrence: Portugal—Province of Algarve.

ALGERIAN ONYX

Old usage for stalagmitic variety of *calcite*.

ALGODONITE

Cu$_6$As Copper arsenide. Hexagonal. Minute crystals; often massive and granular. Steel-grey; silver-white on polished surface. Bright, metallic lustre, becoming dull on exposure. Subconchoidal fracture. No cleavage. Hardness 4·0. SG 8·38. Bead test on charcoal gives arsenical fumes and a malleable, metallic globule which on treatment with soda gives pure *copper*. Not soluble in hydrochloric acid; dissolves in nitric acid.

Occurrence: USA—Lake Superior region. USSR—Chili.
Isomorphous with *domeykite*.

ALKALI

Substance which alone or in solution has a pH value greater than 7·0 and will turn litmus blue (see acid). Occasionally used in analysis.
Alkaline igneous rocks have *feldspar* components mainly based on sodium and/or potassium.

ALLAGITE

Derivative of *rhodonite*. Result of the tendency of manganese protoxide and other protoxides present to unite with carbonic acid in alkaline carbonated waters (to form manganese carbonate) and penetrate the silicate. Dull green or reddish-brown colour.
Occurrence: Germany—Harz Mountains (near Elbingerode, Schebenholze). Isomorphous with *photicite*.

ALLALINITE

Igneous rock of the *gabbro* clan. A plutonic rock. Contains *uralite, talc, saussurite* and altered *olivine*.
Occurrence: Switzerland—Allalin.

ALLANITE or ORTHITE

(Ca, Ce, La, Na)$_2$(Al, Fe, Be, Mn, Mg)$_3$(SiO$_4$)$_3$(OH) Very variable silicate with rare-earth minerals, thorium, cerium,

disprosium, lanthanum, yttrium and erbium oxides, often comprising up to 20 per cent of mineral weight. Monoclinic. Flat, tabular crystals; also long and slender; also massive and granular. Black to dark brown. Submetallic, pitchy or resinous lustre; occasionally vitreous. Grey streak; sometimes greenish or brownish. Uneven or subconchoidal fracture. Poor cleavage. Hardness 5·5–6·0. SG 3·5–4·2. Brittle. *Borax* bead fuses easily and swells up to a dark, magnetic *glass*. Gelatinises with hydrochloric acid.
Occurrence: Scotland. USA—New York State (New York). Canada. Denmark—Greenland. Norway. Madagascar. In albitic and common feldspathic *granite*, *gneiss*, *syenite*, *zircon-syenite* and *porphyry*. Radioactive.

ALLEGHANYITE

$2[Mn_5Si_2O_8(OH)_2]$ Manganese silicate, sometimes with appreciable titanium or *iron*. Monoclinic. Grains (without titanium) or blade-like crystals (with titanium). Bright pink (without titanium) or clovebrown (with titanium). Vitreous to resinous lustre. Conchoidal fracture. No cleavage. Hardness 5·5. SG 4·02. Fuses quietly to black enamel. Readily decomposes in 1:1 hydrochloric acid, leaving silica skeleton.
Occurrence: Wales (with titanium). USA—Virginia (Allegheny Mountains) (without titanium).

ALLEMONTITE

AsSb Arsenic antimonide. Hexagonal. In-

distinct branching crystals; also fibrous, reniform masses, curved lamellar or fine granular. Tin-white or reddish-grey, tarnishing grey or brownish-black. Metallic lustre may be glossy or dull. Grey streak. No fracture. Perfect cleavage in one direction. Hardness 3·0–4·0. SG 6·277. When heated with charcoal emits *arsenic* and *antimony* fumes and fuses to a metallic globule, which burns away leaving a white coating of arsenic oxide.
Occurrence: Canada—British Columbia. France—Allemont. In veins with *arsenic*, *arsenolite*, *antimony* or *kermesite*.

ALLIVALITE

Igneous rock of the *ultrabasic* clan. A pale-coloured layered plutonic intrusion with dark or rusty spots in the groundmass. Contains *olivine* and *anorthite* in all proportions with *chromite* as an accessory.
Occurrence: Scotland—Isle of Rhum (Allival).

ALLOCHETITE

Igneous rock of the *syenite* clan. A greyish-green porphyritic hypabyssal rock. Contains phenocrysts of *labradorite*, *orthoclase*, *titanaugite*, *nepheline*, *magnetite* and *apatite* in a dense groundmass of *augite*, *biotite*, *magnetite*, *hornblende*, *nepheline* and *orthoclase*. Dark constituents of the groundmass form a microlitic felt.
Occurrence: Austria—South Tyrol (Allochet Valley).

ALLOCHROÏTE

Synonym of *aplome*.

ALLOPHANE

$Al_2SiO_5.5H_2O$ Hydrous aluminium silicate. Amorphous. Pale sky-blue, sometimes greenish to deep green, brown, yellow or colourless. Vitreous to sub-resinous lustre; bright and waxy internally. Colourless streak. Imperfectly conchoidal and

shining fracture. No cleavage. Hardness 3·0. SG 1·85–1·89. Brittle. Gelantinises in dilute hydrochloric acid.
Occurrence: England—London (near Woolwich). USA—Massachusetts (near Richmond), Pennsylvania (Friedensville Zinc Mines). France—near Lyons. Germany. Often incrusting fissures or cavities in mines. Incrustations thin and *hyalite*-like; sometimes stalactitic.

ALLUVIAL

Transported by streams or floods, and then deposited as sediment. Often used interchangeably with fluvial.

ALLUVIAL FAN

Mass of sediment deposited at a point where there is a decrease in the gradient of flow.

ALLUVIUM

Sedimentary rock. Consists of coarse detrital materials, such as silt, sands and gravels, laid down by a stream, interbedded with finer-textured, often laminated silts and muds laid down on the flood-plain.
Occurrence: Worldwide, with different associated minerals. England—Cornwall (alluvial *cassiterite*). USA—Alaska, California (auriferous gravel and sand deposits yielding *gold*). In valleys of mature, meandering streams.

ALMANDINE

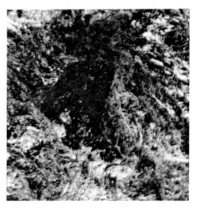

$8[Fe_3Al_2Si_3O_{12}]$ Iron aluminosilicate. Member of *garnet* family. Cubic. Usually dodecahedral or trapezohedral crystals. Very dark, almost violet-red. Vitreous lustre. White streak. Conchoidal to uneven fracture. Cleavage non-existent, but occasional partings. Hardness 7·5. SG 3·95–4·25. F 3·0. Fuses to black magnetic globule. Gemstone.
Occurrence: Scotland (not gemstone quality). USA—New York (Adirondack Mountains), Idaho (Emerald Creek, Latch County, Lewiston, Nog Perce County).

Canada—Baffin Island. India. Madagascar.
Varieties are *common garnet* and *precious garnet*.

ALMANDINE SPINEL

Bluish or violet-red gem variety of *spinel*.

ALNOÏTE

Igneous rock of the *gabbro* clan. A dark, bluish-black to greyish-black, porphyritic hypabyssal rock. Consists of phenocrysts of *mica* in a groundmass of *olivine, augite, melilite, nepheline* and *garnet*.
Occurrence: Sweden—Island of Alnö, Westnorrland Coast. In *nepheline-syenite*. A type of *lamprophyre*.

ALPHITITE

Sedimentary rock. Term used for *clays* consisting largely of *rock-flour*, such as those washed and laid down from glacial debris.

ALSBACHITE

Igneous rock of the *granite* clan. A porphyritic hypabyssal rock. Consists of a groundmass of *quartz, orthoclase* and almost colourless *mica* flakes containing phenocrysts of *quartz, feldspar*, occasional larger *micas* and rose-red *garnets*.
Occurrence: Germany—Alsbach (Melibocus).

ALSTONITE, BARYTOCALCITE (of Johnston) or BROMLITE

$2[BaCa(CO_8)_2]$ Orthorhombic. Dihexagonal dipyramid crystals with strongly horizontally striated faces. Colourless to snow-white; also greyish, pale cream, pink or pale rose-red (bleaching on exposure). Vitreous lustre. White streak. Uneven fracture. Imperfect cleavage in one direction. Hardness 4·0–4·5. SG 3·707. Dissolves in dilute hydrochloric acid.
Occurrence: England — Cumberland Brownley Hill Lead Mine near Alston), Durham (near New Brancepath), Northumberland (Fallowfield Lead Mine near Hexham). With *calcite, barite* and *witherite* as a low-temperature hydrothermal deposit. Isomorphous with *barytocalcite (of Brooke)*.

ALTAITE

$4[PbTe]$ Lead telluride. Cubic. Usually massive; rarely in cubes or cubo-octahedrons. Tin-white with a yellowish tinge tarnishing to bronze-yellow. Metallic lustre. Subconchoidal fracture. Perfect cleavage. Hardness 3·0. SG 8·15. Sectile mineral. F 1·5. Gives fumes of *tellurium* oxide in open tube, forming a white sublimate.

Occurrence: USA—Colorado (Red Cloud Mine in Boulder County), North Carolina (Kings Mountain Mine in Gaston County). Canada—British Columbia. Western Australia—Kalgoorlie. Chile—Coquimbo Province. USSR—Siberia (Altai Mountains). In veins with native *gold* and other tellurides and sulphides.

ALTERATION

Process caused by the extremely hot chemical-carrying fluids sometimes associated with igneous activity, in which certain forms of minerals are converted to others. In serpentinisation, for instance, *serpentines* may be produced from *olivines*.

ALUM or POTASH ALUM

$4[KAl(SO_4)_2 . 12H_2O]$ Hydrated potassium aluminium sulphate. Natural alum usually massive, with columnar or granular structure; also stalactitic or as mealy coating. Colourless and transparent, or white. Vitreous lustre. Conchoidal fracture. Traces of cleavage in one plane. Hardness 2·0–2·5. SG 1·757. Sweetish and astringent taste. Melts in its water of crystallisation at 91 degrees C.
Occurrence: England — Yorkshire (Whitby). Scotland—near Glasgow (Campsey, Hurlet). USA—California, Nevada, Tennessee (Sevier County). Germany—*brown coal* deposits throughout country. Italy—Mount Vesuvius. As an efflorescence or crevice filling in *argillaceous rocks* and *brown coals* containing disseminated *pyrite* or *marcasite*.

ALUMINITE or HYDRARGILLITE

$Al_2SO_4(OH)_4 . 7H_2O$ Hydrous basic aluminium sulphate. Monoclinic. Usually forms compact, reniform masses. White. Dull, earthy lustre. Earthy fracture. No cleavage. Hardness 1·0–2·0. SG 1·66. Adheres to tongue. Meagre to touch. Forms hepatic mass with soda on charcoal in blowpipe flame.
Occurrence: England—Sussex (New-

haven). USA—Missouri (Joplin). France—Auteuil. Germany—Bohemia (Kuchelbad). India—Punjab (Salt Range). With beds of *clay* in the tertiary and post-tertiary formations.

ALUMOBERESOFITE

$8[Mg,Fe)(Cr,Al)_2O_4]$ Variety of *chromite* and member of *spinel* group. Magnesium substitutes for iron and aluminium substitutes for chromium in the crystal lattice.
Occurrence: USSR—Ural Mountains.

ALUMOCHROMITE

$8[Fe(Cr,Al)_2O_4]$ Variety of *chromite* and member of *spinel* group. Aluminium substitutes for chromium in the crystal lattice.
Occurrence: USSR—Leningrad.

ALUM-SHALE

Sedimentary rock. A *shale* impregnated with *alum*.
Occurrence: England—Yorkshire.

ALUNITE

$KAl_3(SO_4)_2(OH)_6$ Hydrous potassium aluminium sulphate. Member of *alunite group*. Hexagonal. Rhombohedral crystals resembling cubes; massive form has fibrous, granular or impalpable texture. White, sometimes greyish or reddish. Vitreous lustre; basal plane somewhat pearly. White streak. Uneven, flat conchoidal fracture; massive varieties, splintery, sometimes earthy. Distinct cleavage in one direction. Hardness 3·5–4·0. SG 2·6–2·9. Brittle. Insoluble in water; practically insoluble in hydrochloric or nitric acids; slowly soluble in dilute sulphuric acid; readily soluble in nitric acid after ignition.
Occurrence: USA—Colorado (Hinsdale County, Mineral County), Nevada (near Beatty in Nye County). Australia—New South Wales (Bullah Delah). Hungary—near Beregszasz. Italy—Tolfa (near Civita Vecchia). Spain—near Almeria. In rocks that have been 'alunitised' by solfataric action; sometimes formed by action of sulphuric acid from oxidation of *pyrite* in aluminous rocks.

ALUNITE Group

Members of series are isostructural and conform to general formula $A'B'''(SO_4)_2(OH)_6$ where $A = Na,K,Pb,NH_4$ or Ag and $B = Al$ or Fe. Either tabular or pseudocubic on a rhombohedron. Small, rare, imperfect crystals. Group comprises *alunite, natroalunite, jarosite, ammoniojarosite, natrojarosite, argentiojarosite, carphosiderite, beaverite* and *plumbojarosite*.

ALUNOGENE

$Al_2(SO_4)_3$. $18H_2O$ Hydrated aluminium sulphate. Triclinic. Small, rare and prismatic crystals usually delicate fibrous masses or crusts or as efflorescence; also massive, fibrous. Single crystals colourless and transparent; aggregates white or tinged yellowish or reddish by impurities. Vitreous to silky lustre. Perfect cleavage in one direction. Hardness 1·5–2·0. SG 1·77. Acidic, sharp taste. Water soluble.
Occurrence: USA—Arizona, California, New California. New Mexico, New York. Canada—British Columbia, Nova Scotia. Czechoslovakia—Bohemia. France— Haute Saône, Rhône district. Italy— Mount Vesuvius. Chile. Peru. USSR— Kamchatka (Avacha volcano). West Indies—Lesser Antilles (Martinique). In neighbourhood of volcanoes and *shales*, especially *alum shales*, where *pyrite* is decomposing.

ALURGITE

c $K_2(Mg, Al)_{4-5}(Al,Si)_8O_{20}(OH)_4$. Variety of *lepidomelane*. A *mica* with composition between *muscovite* and *phlogopite*. Monoclinic. Massive, consisting of scales rarely having a hexagonal outline. Purple to cochineal-red. Pearly to vitreous lustre. Rose-red streak. No fracture. Eminent basal cleavage. Hardness 2·25–3·0. SG 2·985–3·0.

Occurrence: Italy—Piedmont (St Marcel). With manganese ores.

ALUSHTITE

Clay mineral related to *kaolinite*. A hydrated aluminium silicate containing 13·7 per cent water and a little magnesia.
Occurrence: USSR—Crimea (especially near Alushta). Forms bluish or greenish crusts, nests and veins in *quartz* veins in black *clay-slates*.

AMALGAM

Varying proportions of mercury and silver. Formula uncertain. Cubic. Rhombdodecahedral crystals; also massive. Silver-white. Metallic lustre. Silver-white streak. Uneven to conchoidal fracture. Brittle. Hardness 3·0–3·5. SG 10·5–14·0. When heated in closed tube gives a residue of silver. Soluble in nitric acid.
Occurrence: Spain—Almaden. With *cinnabar* or *silver*.
Variety is *arquerite*.

AMATRICE

Trade name for a gemstone variety of *variscite*. Consists of differing proportions of *variscite*, *utahlite* and *wardite* in a matrix of *quartz* and *chalcedony*.
Occurrence: USA—Nevada (Ely), Utah.

AMAZONITE or AMAZON STONE

Gemstone variety of *microcline*. Verdigris-green. Hardness 6·5. SG 2·56–2·58. Often coated with *albite* crystals in parallel position.

Occurrence: USA—Colorado (near Crystal Peak, Pike's Peak). Canada— Ontario (Parry Sound, Renfrew Sound), Quebec (Kipewa). South Africa—North West Cape Province. South West Africa. Brazil. Madagascar.

AMAZON JADE

Synonym of *microcline*.

AMAZON STONE

Synonym of *amazonite*.

AMBER

A resin containing considerable succinic acid and very variable carbon: hydrogen: oxygen ratios. Traces of sulphur to one per cent. Yellow, sometimes reddish, brownish or whitish; often clouded, sometimes fluorescent. Resinous lustre. White streak. Conchoidal fracture. No cleavage. Hardness 2·0–2·5. SG 1·05–1·096. Tasteless. Negatively electrified on friction. Softens at 150 degrees C, melts at 250–300 degrees C.

Occurrence: Sea amber washed up on coasts of England (Essex, Norfolk and Suffolk), Denmark, Sweden and Prussian Baltic coast. Pit amber mined in various Baltic locations. Worldwide occurrence of material similar to amber, but containing no succinic acid.

Amber comes from certain pines which flourished in Oligocene times.

Varieties are *roumanite* and *simetite*.

AMBLYGONITE or HEBRONITE

$2[(Li,Na)AlPO_4(F,OH)]$ with proportion of lithium greater than sodium and proportion of fluorine greater than hydroxide. Basic aluminium sodium lithium phosphate. Triclinic. Small, rough, short prismatic crystals; also large cleavable masses; columnar, compact. Usually white to milky or creamy-white. Vitreous to greasy lustre; pearly on well-developed cleavages. Uneven to subconchoidal fracture. Perfect cleavage. Hardness 5·5–6·0. SG 3·11. Brittle. Easily fuses with intumescence and cools to opaque white bead. Colours a flame-red (presence of lithium).

Occurrence: USA—South Dakota (Custer district). Czechoslovakia. Germany—Saxony. Italy—Island of Elba. Chiefly in lithium and phosphate-rich *granite-pegmatites*; often in crystals of enormous size.

Isomorphous with *montebrasite* and *natro-montebrasite*.

AMBONITE

Collective name for various igneous rocks of the *diorite* clan. Porphyritic extrusive rocks. Three main types, *bronzite-*, *mica-* and *hornblende-andesites* and *dacites*, mostly containing *cordierite* and *garnet*. Phenocrysts are set in a fine-grained lithoidal or microcrystalline groundmass.

Occurrence: Dutch East Indies—Island of Ambon.

AMELIA ALBITE

Exceptionally pure variety of *albite*. Forms clear, glassy crystals of *cleavelandite* habit.

Occurrence: USA—Virginia (Amelia County). In *pegmatite*.

AMESITE

Variety of *corundophilite*. Foliated, hexagonal plates, resembling Tyrolean green *talc*. Apple-green. Pearly lustre on cleavage face. Hardness 2·5–3·0.

Occurrence: USA—Massachusetts (Chester). With *diaspore*.

AMETHYST

Clear purple or bluish-violet variety of *quartz*.

Occurrence: England—Cornwall. Ireland. USA—California, Maine, Montana, North Carolina, Pennsylvania, Texas, Virginia, Wyoming (Yellowstone National

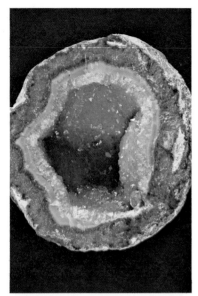

AMIANTHUS

Old name for *actinolite asbestos*.

AMMONIOJAROSITE

$(NH_4)Fe_3(SO_4)_2(OH)_6$ Basic ammonium ferric iron sulphate. Member of *alunite group*. Hexagonal. Crystals form small lumps and irregular flattened nodules composed of microscopic tabular grains. Light ochrous-yellow. Dull and waxy to earthy lustre. SG 3·112. Darkens in a flame and gives off sulphur dioxide at red heat.
Occurrence: USA—Southern Utah (Kaibab Fault) Czechoslovakia—Bohemia (Valachov).

AMORPHOUS

No regularity of structure (see crystal), even on the submicroscopic scale. An example is *glass*, in fact a supercooled liquid.

AMOSITE, ANTHOPHYLLITE ASBESTOS or MONTASITE

$4[(Fe,Mg,Al)_7(Si,Al)_8O_{22}(OH)_2]$ Variety of *anthophyllite*. *An* ortho-*amphibole*. An *amphibole-asbestos* occurring as long fibres and forming one type of commercial *asbestos*.
Occurrence: South Africa—Transvaal (Lydenburg and Pietersburg districts).

AMPHIBOLE Group

Magnesium iron calcium silicates, sometimes sodium, (rarely potassium) with or without aluminium. Variable composition can be represented by the formula $X_{7-8}(Si_4O_{11})_2(OH)_2$ where X mainly includes combinations of the above-named minerals. The hydroxyl group is always present. Commonly bladed crystals terminated by three faces. Prism angle 124 degrees; cleavages at 124 degrees. Subdivided according to crystal system; orthorhombic—*anthophyllite*; monoclinic—*cummingtonite-grünerite* series, *tremolite-actinolite* series, *horn-blende* series, alkali-*amphibole* series (*arfvedsonite, glaucophane, riebeckite*); and triclinic—*cossyrite*.

AMPHIBOLE-MAGNETITE

Medium grade contact metamorphic rock. Granulose texture. Contains *grünerite*, ferruginous silicates and *magnetite*. Often banded structure. Formed by contact metamorphism of ferruginous *cherts*.

AMPHIBOLITE

Medium to high grade regional metamorphism of basic igneous rocks and some impure calcareous sediments. Essentially *hornblende* and *plagioclase*. Foliated due to alignment of *amphibole* prisms. Segrega-tion banding sometimes developed.
Occurrence: Scotland—North West Highlands. Canada—Laurentian Shield. Finland and Sweden—Fennoscandian Shield.

AMYGDALOID, -AL

Igneous rocks containing cavities, 'vugs', often filled with secondary minerals.

AMYGDALOIDAL BASALT

Igneous rock of the *gabbro* clan. An extrusive rock, comprising a form of vesicular, *olivine*-free *basalt*. Vesicles contain secondary minerals such as *quartz, calcite* or *zeolites*.
Occurrence: England—Derbyshire.

ANABOHITSITE

Igneous rock of the *ultrabasic* clan. A plutonic rock. Contains *olivine, hornblende*-bearing *pyroxenite, hypersthene* and *c* 30 per cent *ilmenite* and *magnetite*.
Occurrence: Madagascar—Anabohitsy. Periphery of *troctolite*.

ANALBITE (of Winchell)

$4[(Na,K)AlSi_3O_8]$ with proportion of sodium to potassium at least 9 : 1. Variety of *anorthoclase*. Member of *feldspar* family.

Park). South Africa. Rhodesia. Brazil. Uruguay. India. Sri Lanka. Lines interior walls of hollow cavities.

ANALCIME

Synonym of *analcite*.

ANALCIMITE or ANALCITITE

Igneous rock of the *gabbro* clan. An extrusive, *olivine* and *feldspar*-free rock. Groundmass contains mainly *analcite* and *augite*.
Occurrence: Italy—Scano.

ANALCIMITE-TINGUAITE

Igneous rock of the *syenite* clan. An olive-green, compact, aphanitic hypabyssal rock. Silky lustre; a few white phenocrysts. Contains *aegirine* and *feldspar* in clear, colourless groundmass of *nepheline* and *analcite*.

Occurrence: USA—Massachusetts (Pickard's Point).

ANALCITE or ANALCIME

$16[NaAlSi_2O_6 \cdot H_2O]$ Hydrated sodium aluminium silicate. Cubic. Trapezohedrons or cubes; massive form granular or compact with concentric structure. Colourless or white; occasionally greyish, greenish, yellowish or reddish-white. Vitreous lustre. White streak. Subconchoidal fracture. Cubic cleavage, in traces. Hardness 5·0–5·5. SG 2·22–2·29. Brittle. F 2·5. Fuses to colourless *glass*. Gelatinises in dilute hydrochloric acid.
Occurrence: Scotland—Dunbartonshire and Stirlingshire (Kilpatrick Hills). Northern Ireland—County Antrim. USA— New Jersey. Canada—Nova Scotia. Greenland. Iceland. Italy—Cyclopean Islands. Norway. In trap rock cavities with other *zeolites*.

Member of *zeolite* family. Varieties are *cluthalite*, *eudnophite*, *potash-analcime*, *picranalcime*, *euthalite*.

Above and below: analcite.

ANALCITITE

Synonym of *analcimite*.

ANAMESITE

Igneous rock of the *gabbro* clan. A fine-grained extrusive rock. Texture intermediate between dense *basalt* and coarse *dolerite*.

ANAPAITE

$Ca_2Fe^{2+}(PO_4)_2 . 4H_sO$ Hydrated calcium iron phosphate. Triclinic. Tabular crystals; also aggregates. Greenish to greenish white. Vitreous lustre. White streak. Perfect cleavage in one direction. Hardness 3·5. SG 2·81. Easily soluble in hydrochloric acid.
Occurrence: USA—California (Kings County). USSR—Black Sea coast (Sheljesny Rog on the Taman Peninsula).

ANATASE, DAUPHINITE, HYDRO-TITANITE, OCTAHEDRITE, OISANITE, WISERINE or XANTHI-TANE

$4[TiO_2]$ Titanium dioxide. Tetragonal. Acute pyramidal crystals, often highly modified; also tabular. Usually various shades of brown, passing into indigo-blue and black; also greenish, blue-green, pale lilac or slate-grey. Adamantine or metallic adamantine lustre, sometimes glossy. Colourless to pale yellow streak. Subconchoidal fracture. Perfect cleavage in one direction. Hardness 5·0–6·5. SG 3·90. Brittle. Chemical identification as for *rutile*.
Occurrence: England—Cornwall (Liskeard, Tintagei Cliffs), Devonshire (Tavistock). USA—Colorado (Gunnison County), Massachusetts (Sommerville). Canada—Nova Scotia (Sherbrooke). French and Swiss Alpine regions. Brazil.

In vein or crevice deposits in *gneiss* or *schist*.
Polymorphous with *brookite* and *rutile*.

ANCHORITE

Igneous rock of the *diorite* clan. A nodular and veined plutonic rock, the face being variegated with dark patches of ferromagnesian minerals and light-coloured silicate minerals such as *feldspar* and *quartz*.
Occurrence: England—Warwickshire (Anchor Inn near Nuneaton).

ANCYLITE

$(Sr,Ca)_3(Ce,La)_4(CO_3)_7(OH)_4.H_2O_3$ Basic hydrated strontium calcium cerium group of rare earths carbonate. Orthorhombic. Pseudo-octahedral or short prismatic. Pale yellow with orange tinge; can be yellowish-brown to brown or grey. Vitreous lustre on faces; greasy on fracture surfaces. White streak. Splintery and rather tough fracture. No cleavage. Hardness 4·0–4·5. SG 3·95. Brittle. Loses water when heated with charcoal. Dissolves in hydrochloric acid with the evolution of carbon dioxide.
Occurrence: Greenland—Julianehaab district (Narsarsuk). Groups and crusts of small, rounded crystals in *nepheline-syenite*.
Variety is *calcio-ancylite*.

ANDALUSITE

$4[Al_2SiO_5]$ Aluminium silicate. Orthorhombic. Usually coarse, nearly square prismatic; massive form imperfectly columnar or radiated and granular. Whitish, rose-red, flesh-red, violet, pearl-grey, reddish-brown or olive-green. Vitreous, often weak lustre. Colourless streak. Subconchoidal, uneven fracture. Distinct, sometimes perfect cleavage. Hardness 7·5. SG 3·16–3·2. Brittle. Unaffected by acids.
Occurrence: England—Cumberland. Scotland—Argyllshire (near Balahulish). Ireland—Killiney Bay. Germany. Spain—Andalusia. Most common in *argillaceous schist*.
Varieties are *chiastolite* and *manganandalusite*. Isostructural with *kyanite* and *sillimanite*.

ANDEN-DIORITE

Igneous rock of the *diorite* clan. A plutonic rock. Variety of quartziferous *diorite* comprising mainly *augite*.
Occurrence: Argentina—the Andes.

ANDESINE or PSEUDO-ALBITE

Plagioclase feldspar containing 50–70 per cent *albite*. Triclinic. Crystals rare, usually cleavable or granular massive. White, grey, greenish, yellowish or flesh-red. Subvitreous to pearly lustre. No fracture. Perfect cleavage in one direction. Hardness 5·0–6·0. SG 2·68–2·69. Fuses if in thin splinters.
Occurrence: USA—Maine (Sanford). France—Department of Var (l'Esterel). Iceland—Vapnefiord. As primary mineral in intermediate igneous rocks such as *diorites* and *andesites*. Varieties are *oligoclase-andesine* and potash-*andesine*.

ANDESINITE

Igneous rock of the *diorite* clan. A granular hypabyssal rock. Composed entirely of *andesine*.

ANDESITE

Igneous rock of the *diorite* clan. A holocrystalline porphyritic extrusive rock. Contains phenocrysts of sodic *plagioclase* and one or more *biopyriboles* in groundmass of the same minerals.
Occurrence: USA—Rocky Mountains, Sierra Nevada Mountains. Argentina—

the Andes. Indonesia—Java, Sumatra. Japan. Aleutian Islands. In volcanic regions.

ANDESITE-PORPHYRY

Igneous rock of the *diorite* clan. A light-coloured, glassy textured hypabyssal rock. Contains *porphyry* in the groundmass and crystal fragments of *plagioclase, augite, hornblende* and a little *magnetite*.
Occurrence: Indonesia—Java (Eerste Punt).

ANDESITE-TUFF

Igneous rock of the *diorite* clan. A light coloured, glassy textured porphyritic hypabyssal rock. Contains phenocrysts of *plagioclase, augite* and *hornblende* in a glassy ground mass of *andesite* composition.
Occurrence: USA—California (Stillwater in Shasta County).

ANDRADITE, CALDERITE or MELANITE

$8[Ca_3Fe_2Si_3O_{12}]$ Calcium iron silicate. Member of the *garnet* family. Cubic. Small, lustrous, dodecahedral crystals; granular massive, compact, lamellar, desseminated. Various shades of yellow, green and brown ranging to greyish-black and black. Vitreous lustre. White streak. Conchoidal to uneven fracture. No cleavage, but occasional partings. Hardness 7·5. SG 3·8–3·9. Darkens in blowpipe flame, then fuses to black magnetic globule.
Occurrence: USA—Arizona (Stanley

Above and below: andradite.

Buttes). France—Pyrenees (near Barèges). Italy—Mont Somma, Mount Vesuvius. Found coating seams with *magnetite, epidote, feldspars, nephelite* and *leucite*. Gemstone varieties are *topazolite, uralian emerald* and *demantoid*. Non-gemstone varieties are *aplome, calderite (of Piddington), colophonite, polyadelphite* and *pyreneite*.

ANDREWSITE

$(Cu,Fe^{2+})_3Fe^{3+}_6(PO_4)_4(OH)_{12}$ with ferrous to copper $c2:3$. Basic copper ferric iron phosphate with ferrous iron substituting for copper. Orthorhombic. Botryoidal aggregates with radial fibrous structure. Dark green to bluish-green. Somewhat silky lustre. Cleavage in two directions parallel to fibre length. Hardness 4·0. SG 3·475.
Occurrence: England—Cornwall (West Phoenix Mine near Liskeard). Associated with *limonite, dufrenite, cuprite* and *chalcosiderite*.

ANEMOUSITE

Synonym of *labradorite*.

ANGLESITE or LEAD VITRIOL

$4[PbSO_4]$ Lead sulphate. Orthorhombic. Thin to thick tabular crystals, rhomboidal in outline; commonly massive, granular to compact. Colourless to white. Adamantine lustre, inclining to resinous and vitreous. Colourless streak. Conchoidal fracture. Good cleavage in one direction. Hardness 2·5–3·0. SG 6·38. Brittle. Often fluoresces yellow in ultra-violet light.

Occurrence: Worldwide.
Usually formed by oxidation of *galena*.
Variety is *barytoanglesite*.

ANGLESOBARYTES

Synonym of *hokutolite*.

ANHYDRITE

$4[CaSO_4]$ Calcium sulphate. Orthorhombic. Crystals rare, rectangular pinacoidal or elongated parallel to domes; usually massive, granular, fibrous or in contorted concretionary forms. Colourless to bluish or violet; transparent in perfect crystals. Vitreous to pearly lustre. White or greyish-white streak. Uneven to splintery fracture. Perfect cleavage. Hardness 3·5. SG 2·98. Brittle. F 3·0. Fuses to white enamel.
Occurrence: USA—Louisiana, New York (Balmat), Texas (oil drillings). Germany—Magdeburg district (Stassfurt Mine). Poland—Salt Mines near Kraków. Switzerland—Bex-les-Bains (in perfect crystals). In sedimentary terrains with *gypsum, limestone, dolomite* and salt beds. Anhydrous form of *gypsum*.
Varieties are *tripestone* and *vulpinite*.

ANHYDROUS

Without water content.

ANION

A negatively charged ion; the ion in an electrolysed solution that migrates to the anode.

ANKARAMITE

Ultrabasic igneous rock. A dark-coloured hypabyssal rock. Contains phenocrysts up to 10mm in diameter of *olivine* and *pyroxene* in a compact groundmass of *augite, titanite* and a little *biotite*.
Occurrence: Madagascar—Ampasindava (Ankaramy).

ANKARATRITE

Igneous rock of the *gabbro* clan. A dark-coloured extrusive rock. Contains phenocrysts of *olivine*, ferromagnesian constituents, *nepheline* and *biotite*, *ilmenite* and often *perofskite* in a groundmass of titaniferous *augite*.
Occurrence: Madagascar—Mount Ankaratra.

ANKERITE

$Ca(Mn,Mg,Fe)(CO_3)_2$ with more than 10 per cent ferrous carbonate and proportion of iron greater than magnesium. Calcium iron carbonate, with magnesium replacing iron in the lattice. Hexagonal. Crystals commonly rhombohedral; massive, coarse to fine granular. Yellowish-brown or brown, turning dark brown on weathering; pink or rose in varieties with much manganese. Vitreous to pearly lustre. Subconchoidal fracture. Perfect cleavage in one direction. Hardness 3·5–4·0. SG $c3·02$. Brittle. When heated in blowpipe flame, darkens in colour and becomes magnetic.
Occurrence: England—Northumberland (*coal* fields), Yorkshire (Northern Pennines). USA—New York (Antwerp Iron Mine in Jefferson County), Idaho (Shoshone County).
Same mode of occurrence as *dolomite*.
Member of *dolomite* group.

ANNABERGITE or NICKEL BLOOM

$(Ni,Co)_3(AsO_4)_2.8H_2O$ Hydrous nickel arsenate. Monoclinic. Small, slender capillary needles form earthy crusts and films. Light apple-green. Silky or vitreous lustre. Never solid enough for fracture. Cleavage

Above: ankerite. *Below:* annabergite.

side pinacoid, usually not visible. Hardness 2·5–3·0. SG 3·0. Earthy mineral. Can be reduced to magnetic, metallic bead by strong heating in reducing flame. *Borax* bead reddish-brown in oxidising flame, opaque grey in reducing flame.
Occurrence: USA—Nevada (Humboldt County). Canada—Ontario (Cobalt). Germany—Saxony. Greece—Laurium. Near surfaces of cobalt nickel *silver arsenic* sulphide veins.

ANNITE (of Dana)

Variety of *lepidomelane*.
Occurrence: USA—Massachusetts (Cape Ann). In *granite* with *orthoclase*, *albite* and *zircon*.

ANNITE (of Winchell)

Synonym of *hydroxyl-annite*.

ANOMITE

Variety of *biotite*. Characterised by different optical properties to *biotite*. SG 2·87.
Occurrence: USA—New York (Greenwood Furnace in Orange County). Austria—Steinegg. Sweden—Island of Alno, Westernorrland. USSR—Siberia (Lake Baikal). In *gneiss*, *basalt* or *limestone*.

ANOPHORITE

$2[Na_3Mg_3Fe^{2+}.Fe^{3+}(Ti,Si)_8O_{22}(OH_2)]$ Var-

iety of alkali-*hornblende*, resembling *catophorite* in its pleochroism, but differing in chemical composition. Typical *magnesioarfvedsonite*, except for high titanium content.
Occurrence: Germany—Baden (Katzenbuckel). In *shonkinite*.

ANORTHITE

8[CaAl$_2$Si$_2$O$_8$] Calcium aluminium silicate. Triclinic. Prismatic or tabular complex crystals; compact, cleavable, lamellar. Colourless, white, grey or greenish. Vitreous to pearly lustre. White streak. Uneven fracture. Basal, brachypinacoidal cleavage. Hardness 6·0–6·5. SG 2·7. Brittle. Decomposes in hydrochloric acid with separation of gelatinous silica.
Occurrence: USA—New Jersey (marble quarry near Franklin). Italy—Mount Etna, Mount Vesuvius. Japan—Miyake. In basic igneous rocks or crystalline *limestones*. End member of *plagioclase feldspar* group. Variety is *indialite*.

ANORTHITE-BASALT

Igneous rock of the *gabbro* clan. An extrusive rock. Contains nine per cent *quartz*, small amounts of *augite* and *magnetite* in a colourless base with *anorthite*.
Occurrence: Japan—crater of Mount Fujiyama.

ANORTHITE ROCK

Igneous rock of the *gabbro* clan. A light grey to colourless, faint greenish tinge, coarse-textured plutonic rock. Contains mainly *anorthite*.
Occurrence: USA—Lake Superior region. Variety of *anorthosite*.

ANORTHITFELS

Synonym of *calciclasite*.

ANORTHITISSITE

Igneous rock of the *gabbro* clan. A hypabyssal rock. Contains 70 per cent *hornblende*, 22 per cent *anorthite*, eight per cent ore and accessory *apatite*.
Occurrence: USSR—Middle Ural Mountains (Koswinski Mountains).
Similar to *harrisite*.

ANORTHITITE

Synonym of *calciclasite*.

ANORTHOCLASE

4[(Na,K)AlSi$_3$O$_8$] Sodium potassium *feldspar* with proportion of sodium usually greater than potassium. Variety of *microcline*. Crystals often rhombohedral.
Occurrence: USA—Wyoming County (Obsidian Cliff in Yellowstone National Park). Italy—Pantellaria. Southern Norway. Kenya—Mount Kilimanjaro. Member of *feldspar* family. Variety is *analbite (of Winchell)*.

ANORTHOCLASITE

Igneous rock of the *syenite* clan. A leucocratic plutonic rock of pure *anorthoclase*.

ANORTHOSITE

Igneous rock of the *gabbro* clan. A whitish to dark grey, hypautomorphic-granular plutonic rock. Composed essentially of *labradorite* or *bytownite*, with accessory *orthopyroxene* and *clinopyroxene*.
Occurrence: USA—New York (Adirondack Mountains). Canada—Laurentian Shield. Norway—Bergen (Ekersund-Soggendal region).
Variety is *anorthite rock*.

ANTHOPHYLLITE

4[(Mg,Fe)$_7$Si$_8$O$_{22}$(OH)$_2$] Hydrated magnesium iron silicate. Orthorhombic. Crystals rare; usually in embedded masses with fibrous structure. Brown, sometimes with greyish or greenish tints. Vitreous lustre. No fracture. Prismatic cleavage. Hardness 5·5–6·0 (usually splinters and appears softer). SG 2·9–3·4. Insoluble in

acid. Fuses with some difficulty to black magnetic *glass*.
Occurrence: England—Cornwall (Lizard Point). Scotland. USA—North Dakota (Franklin County), Pennsylvania (Delaware County). Greenland. Norway—Kongsberg. Rare mineral of metamorphic rocks, usually associated with ore minerals.
Varieties are *amosite*, *kupfferite* and *gedrite*.

ANTHOPHYLLITE ASBESTOS

Synonym of *amosite*.

ANTHRACITE

Sedimentary rock. A humic *coal*. Contains very low percentage of volatiles. Massive, banded structure. Black. Sub-metallic lustre. Conchoidal fracture. Clean to handle. Ignites only at high temperature.
Occurrence: South Wales—Pembrokeshire. USA—Pennsylvania.

ANTHRACONITE or STINKSTEIN

A coal-black *marble* or *limestone* yielding a fetid odour when struck or rubbed. Hard and compact. Colouration and odour from inclusions of bituminous or carbonaceous matter.
Occurrence: Germany. Mediterranean.

ANTHRAXOLITE

Coal-like and lustrous variety of *bitumen* with high percentage of carbon and low

percentage of volatiles. Hardness 3·0–4·0. SG c2·0.
Occurrence: Canada—Ontario, Quebec. Coaly body of *asphaltite* group.

ANTICLINAL

See anticline.

ANTICLINE

A fold in the form of an arch.

ANTIGORITE

16[$Mg_3Si_2O_5(OH)_4$] Variety of *serpentine*. Monoclinic. Thin lamellar, easily separated into translucent or sub-translucent folia. Brownish-green in reflected light; leek-green in transmitted light. Hardness 2·5 along fibre and 4·5 across fibre. SG 2·622. Smooth to touch, but not greasy.
Occurrence: South Africa—Transvaal (Congo Vaal Mine). Italy—Piedmont (Antigorio Valley). Produced under stress in dislocation-metamorphism.
Varieties are *picrolite* and *williamsite*.

ANTIMONITE

Synonym of *stibnite*.

ANTIMONY

2[Sb] Native antimony. Hexagonal.

Pseudocubic or thick tabular crystals, generally massive and lamellar; also radiated, botryoidal or reniform. Tin-white. Metallic lustre. Grey streak. Uneven fracture. Very brittle. Perfect, easy cleavage in one direction. Hardness 3·0–3·5. SG 6·61–6·72. F 1·0. When heated with charcoal gives coating of oxide which tinges reducing flame bluish-green.
Occurrence: USA—California (Kern County, Riverside County). Canada—New Brunswick, Quebec. Australia—New South Wales, Queensland. Czechoslovakia—Bohemia. France—Isère. Germany—Harz Mountains. Italy—Sardinia. Sweden—Sala. Borneo. Chile. In veins mainly with *silver*, *antimony* and *arsenic* ores, often with *stibnite* and *allemontite*.

ANTIMONY ROCK

Igneous rock of the *diorite* clan. A light grey, occasionally spotted with pink, medium to fine-grained plutonic rock. Contains mainly *albite* with *biotite*, *nepheline*, *microperthite* and *calcite*.
Occurrence: Canada—Ontario (Goodaham).

ANTIPERTHITE

Member of *feldspar* group. An intergrowth of *orthoclase* and *plagioclase* with *plagioclase* dominant.
Occurrence: Norway—Finmarken (Seiland). USSR—Kola Peninsula (Volchia Tundra). India—Madras (Pallavaram). In *oligoclase* and *andesine* layer of *plagioclase*. In rocks of *granulite* facies; occasionally in volcanic rocks.

ANTLERITE

$Cu_3(SO_4)(OH)_4$ Basic copper sulphate. Orthorhombic. Small tabular to short-prismatic crystals; also soft, fibrous masses resembling *malachite*. Bright to dark green. Vitreous lustre. Pale green streak. Uneven fracture. Pinacoidal cleavage. Hardness 3·5–4·0. SG 3·9. Dissolves in hydrochloric acid without effervescence; chips of calcite added to resulting solution

cause precipitation of white needles of calcium sulphate.
Occurrence: USA—Alaska (Kennecott), Arizona (Antler Mine), Nevada (near Black Mountain). Chile—Chuquicamata. In secondary, weathered zone of *copper* ore deposits.
Chemical composition same as *brochantite*.

ANTOZONITE

Variety of *fluorite*. Dark violet-blue.
Occurrence: Germany—Bavaria (Wölsendorf).

ANTRIMOLITE

Variety of *mesolite*. A fibrous stalactitic formation, with the fibres radiating from the centre. Hardness 3·5–4·0. SG 2·096.
Occurrence: Northern Ireland—County Antrim (Ballintoy). As a form of *thomsonite* or as a mesotype on *calcite*.

ANYOLITE

Gem variety of *zoisite*. Chrome-rich with inclusion of black *amphibole* which acts as matrix for large, hexagonal *ruby* crystals.
Occurrence: Tanzania.

APACHE TEARS

Gemstone variety of *obsidian*. Consists of

glassy, pebble-like solid cores of unaltered *glass* from decomposed *obsidian*. When cut, transparent cores produce stones of a grey or light grey colour. Occasional presence of silky striations give a cat's eye effect.

Occurrence: USA—South Western states, especially New Mexico.

APACHITE

Igneous rock of the *syenite* clan. A grey, weathering in tones of red, medium to fine-grained extrusive rock. Contains *aenigmatite* and minerals of the *hornblende* group equal to the amount of *pyroxene*.

Occurrence: USA—West Texas (Apache Mountains, especially near Fort Davis). May form sheets.

APATITE Series

Ca$_5$(PO$_4$)$_3$(F,Cl,OH) Calcium phosphate fluorphosphate or chlorphosphate. Hexagonal. Crystals common, consisting of combinations of prism and pyramid, with or without the basal plane; also mammillated, concretionary and massive. Usually pale sea-green or bluish-green, yellowish-green or yellow. Vitreous or subresinous lustre. Conchoidal and uneven fracture. White streak. Very poor cleavage, parallel to basal plane. Hardness 5·0. SG 3·17–3·23. Brittle. Soluble in hydrochloric acid (carbonate varieties with slight effervescence); on addition of sulphuric acid the resulting solution gives a precipitate of calcium sulphate.

Occurrence: Primary constituent in igneous rocks, but only in accessory amounts. Present in small quantities in most metamorphic rocks, especially crystalline *limestones*.

Series comprises *carbonate-apatite*, *hydroxyl-apatite*, *chlor-apatite* and *fluor-apatite*.

Varieties are *asparagus stone, collophane, hydroxyl-apatite, manganapatite, moroxite, osteolite, phosphorite, saamite* and *staffelite*.

APHANITIC

Very fine-grained. Often applied to igneous rocks, the grains of which are microscopic.

APHRITE

A lamellar variety of *calcite*.

APHROLITH

Synonym of *aa*.

APHROSIDERITE

2[(Fe^{2+},Mg,Al)$_6$(Si,Al)$_4$O$_{10}$(OH)$_8$] with silicon = 2·5–2·8. Basic iron magnesium aluminium silicate. Variety of *ripidolite*. Massive. Dark olive-green. Soft. SG 2·8–3·0.

Occurrence: Austria—Upper Styria. Germany—Nassau (Gelegenheit Mine near Weilburg, near Muttershausen). Hexagonal fine scales with *hematite*.

APHTHITALITE

2[NaKSO$_4$] Sodium potassium sulphate with ratio of sodium and potassium from 3:1 to 1:3. Hexagonal. Thin to thick tabular crystals with pronounced trigonal development; also as bladed aggregates, imperfect mammillary and in crusts; also massive. White, rarely colourless; also grey, blue or greenish; sometimes reddish due to included iron oxide. Vitreous to resinous lustre. Conchoidal to uneven fracture. Brittle. Fair cleavage in one direction, poor in second. Hardness 3·0. SG 2·656–2·71. Saline and bitter taste. Soluble in water and acids. Easily fusible.

Occurrence: USA—California (San Bernadino County), Hawaii, New Mexico (Eddy County). Germany—Stassfurt (Souglashall), Westeregeln. Italy—Mount Etna, Mount Vesuvius. Spain—Galicia (Kalusz). As incrustation in fumaroles of volcanoes and as constituent of oceanic and terrestrial salt deposits.

Variety is *glaserite*.

APLITE, APLOGRANITE, GRANITE APLITE or HAPLITE

Igneous rock of the *granite* clan. An off-white, cream, yellowish, reddish or grey, saccharoidal hypabyssal rock. Contains *quartz* and potash-*feldspar* with accessory *muscovite, lepidolite, biotite, hornblende, diopside, apatite, zircon* and *iron* ore.

Occurrence: Scotland—Ben Nevis, Glencoe. Australia—New South Wales. Norway—Oslo.

APLOGRANITE

Synonym of *aplite*.

APLOME or ALLOCHROÏTE

8[(Ca,Mn)$_3$Fe$_2$Si$_3$O$_{12}$] Calcium manganese iron silicate. Variety of *andradite*. Dodecahedral crystals with faces striated parallel to shorter diagonal. Dark brown, yellowish-green or brownish-green.

Occurrence: Germany—Black Forest, Saxony. USSR—Siberia (on the Lena River).

APOPHYLLITE

2[KCa$_4$Si$_8$O$_{22}$(F,OH).8H$_2$O] Basic hydrated calcium magnesium fluorsilicate. Tetragonal. Square prismatic crystals resembling cubes; also acute pyramidal, massive and lamellar. White or greyish; occasionally greenish, yellowish, rose-red or flesh-red. Vitreous lustre; pearly on basal plane. White streak. Uneven fracture. Highly perfect cleavage in one direction. Hardness 4·5–5·0. SG 2·3–2·5. Brittle. F 1·5. Decomposes in hydrochloric acid with the separation of slimy silica.

Occurrence: Scotland—Fifeshire. USA—Colorado (Table Mountain near Golden). New Jersey (Bergen Hill). Canada—Nova Scotia (Basin of Mines). Australia. Europe. India. As secondary mineral in *basalt* and related rocks in association with various *zeolites*.

APORHYOLITE

Igneous rock of the *granite* clan. A holocrystalline extrusive rock. An alkali-lime *rhyolite* containing *magnetite* and *hematite*.

Occurrence: USA—Maine (Fox Islands), Pennsylvania (South Mountains).

APPINITE

Group name for a series of rocks of the *syenite* or *diorite* clan. Dark-coloured plutonic rocks rich in *hornblende*.
Occurrence: Scotland—Glen Fyne (Garabal Hill complex). Channel Islands—North Guernsey, South East Jersey.

AQUAMARINE

Gemstone variety of *beryl*. Hexagonal. Large crystals of flawless clarity striated parallel to prism edge; sometimes tapering form due to erosion. Clear sky-blue,

Above: aragonite
Below: aquamarine

bluish-green or clear green. SG 2·68–2·71.
Occurrence: Northern Ireland—Mourne Mountains. USA. South West Africa—Rössing. Brazil. USSR—Ural Mountains. India. Madagascar. Tanzania.

AQUA REGIA

Mixture of one part of concentrated nitric acid with four of concentrated hydrochloric acid; extremely corrosive, attacking *gold* and many other generally inert substances.

ARAGONITE or KTYPÉITE

4[$CaCO_3$] Calcium carbonate; contains traces of strontium, lead and zinc. Orthorhombic. Single, short to long prismatic crystals; sometimes columnar aggregates and crusts of straight or divergent fibres. Colourless to white but often tinted. Vitreous lustre, inclined to resinous on fracture surface. Colourless streak. Subconchoidal fracture. Distinct cleavage in one direction. Hardness 3·5–4·0. SG 2·9–3·0. Brittle. Dissolves with effervescence in cold dilute hydrochloric acid.
Occurrence: Worldwide.
Polymorphous with *calcite* and *vaterite*. Varieties are *flos-ferri* and *tarnowitzite*. Member of *aragonite group*.

ARAGONITE Group

A series of orthorhombic crystals of

pseudo-hexagonal symmetry. Partially solid solutions extend between members of the group at ordinary temperatures. Series comprises *aragonite*, *witherite*, *strontianite*, *cerussite* and *alstonite*.

ARAPAHITE

Igneous rock of the *gabbro* clan. A black, aphanitic extrusive rock. Contains 50 per cent *magnetite* and equal amounts of *bytownite* and *augite* with a small amount of *apatite*. Very porous.
Occurrence: USA—Colorado (Pole Mountain in North Park).

ARBORESCENT

Branch-like, tree-like.

ARDUINITE or MORDENITE

$(Ca,Na_2,K_2)_4Al_8Si_{40}O_{96}.28H_2O$ A red, radially fibrous member of the *zeolite* family.
Occurrence: Italy—Venice (Val dei Zuccanti).

ARENACEOUS

Clastic sedimentary rock, grain size between 2mm and 0·0625mm.

ARENACEOUS ROCKS

Sedimentary sands and *sandstones* forming fragmental and detrital deposits. Medium-textured, showing linear, cross-bedded or deformational structure. Consists mostly of *quartz* but can contain appreciable amounts of *feldspar* and white *mica*.
Occurrence: Worldwide. Laid down in shallow marine, fresh water, glacial or terrestrial environments.

ARENDALITE

A variety of *epidote* found in Arendal in Norway.

ARFVEDSONITE

$2[(NaCa)_{2.5}(Fe^{2+},Fe^{3+}Mg)_5Si_8O_{22}(OH)_2]$

Slightly basic sodium calcium and chiefly ferrous iron metasilicate. Monoclinic. Long prismatic crystals; often tabular or in prismatic aggregates. Pure black; deep green in thin scales. Vitreous lustre. Deep bluish-grey streak. Uneven fracture. Perfect prismatic cleavage. Hardness 6·0. SG 3·44–3·45. Brittle. F 2·0. Fuses with intumescence to black magnetic globule.
Occurrence: Greenland—Kangerdluarsuk. Norway—Langesund Fiord. Member of the *amphibole* group.
Varieties are *kataphorite* and *magnesioarfvedsonite*.

ARGENTINE

Silvery.

ARGENTITE (α form) or SILVER GLANCE

Ag_2S Silver sulphide. Cubic. Crystals usually cubic, but faces often too distorted and branching to be recognisable; commonly massive. Dark lead-grey. Metallic lustre, usually tarnishing dull black. Shining streak. Subconchoidal fracture. Poor cleavage. Hardness 2·0–2·5. SG 7·3. Sectile. Fuses into bead on charcoal which gives *silver* button in oxidising flame. When button is added to hydrochloric acid, deposit of silver chloride formed with evolution of hydrogen.
Occurrence: England—Cornwall (Liskeard). USA—Colorado (Leadville), Montana (Butte). Czechoslovakia— Kremnitz, Schemnitz. Germany—Saxony. Bolivia. Chile. Mexico. Peru. Usually in veins. Most common primary ore of *silver*.
Acanthite is β form of *argentite*

ARGENTOJAROSITE

$AgFe_3(SO_4)_2(OH)_6$ Basic silver ferric iron sulphate. Hexagonal. Very fine-grained masses and coatings of micaceous crystals, or flattened scales with hexagonal outlines. Yellow to brown. Brilliant lustre. No fracture. Good cleavage in one direction. SG 3·66. Borax bead gives bottle-green colour in reducing flame (indicating the presence of *iron*).
Occurrence: USA—Utah (Tintic Standard Mine near Dividend). Secondary mineral found with *anglesite*, *barite* and *quartz*. Member of the *alunite group*.

ARGILLACEOUS ROCKS

Sedimentary rocks. Consist of *clays* and muds forming fragmental or detrital deposits. Fine-grained texture. Layered structure or as minute, flaky crystals. Formed from insoluble decomposed products of chemically weathered rocks deposited from water suspension. Contain mainly *quartz*, *feldspar* and *mica* with chemically precipitated organic matter and various colloids.
Occurrence: Worldwide.

ARGILLITE

Sedimentary rock of *argillaceous* type. Compact texture. Consists of *quartz*, *mica* and *feldspar* cemented with silica.
Occurrence: Worldwide.

ARIZONA RUBY

Synonym of *pyrope*.

ARIZONITE

$Fe_2O_3.3TiO_2$ Ferric metatitanate. Probably monoclinic. Irregular masses. Dark steel-grey. Metallic to sub-metallic lustre. Subconchoidal fracture. Consists of mixture of *hematite*, *ilmenite*, *anatase* and *rutile*.
Occurrence: USA—Arizona (near Hackberry). In *pegmatite* veins.

ARIZONITE ROCK

Igneous rock of the *granite* clan. A leucocratic, homogeneous, fine-grained hypabyssal rock. Contains mainly *quartz* with *orthoclase*, *anorthite* and *albite* plus associated *zircon*, *muscovite* and *garnet*.
Occurrence: Germany—Saxony (Granulitgebirge). Norway—Stavanger (Island of Sjölyst).
Related to *aplite*.

ARKANSITE

Variety of *brookite*. Forms stout brown to iron-black (often dull on the surface) crystals. Altered by paramorphism to *rutile*.

Occurrence: USA—Arkansas (Magnet Cove in the Ozark Mountains). With *schorlomite*, *rutile* and *brookite*.

ARKITE

Igneous rock of the *syenite* clan. A leucocratic, granular porphyritic hypabyssal rock. Contains phenocrysts of *leucite* in holocrystalline groundmass of *nepheline*, *garnet* and *pyroxene*.
Occurrence: USA—Arkansas (Magnet Cove in the Ozark Mountains).

ARKOSE

Sedimentary rock. A coarse-grained *sandstone* or *grit* formed from rapid disintegration of *granite* or *gneiss* with no alteration by weathering. Contains a minimum of 20 per cent *feldspar*.
Occurrence: Scotland—Ross and Cromarty (Torridon). USA—California (coast ranges).

ARQUERITE

A *silver*-rich variety of *amalgam* found in Chile.

ARSENIC

2[As] Native arsenic. Hexagonal. Natural crystals rare, acicular; usually granular massive, in concentric layers. Tin-white. Nearly metallic lustre in fresh fracture. Tin-white streak. Uneven and fine granular fracture. Perfect cleavage in one direction. Hardness 3·5. SG 5·63–5·78. Brittle. Flame volatilises in blowpipe without fusion, giving garlic odour of arsenic trihydride.
Occurrence: USA—Arizona (Santa Cruz County), Louisiana (Winnfield Salt Dome). Canada—Ontario (Lake Superior region), Quebec (Montreal). Western Australia—Kalgoorlie. Czechoslovakia—Bohemia (Joachimstal). France—Alsace. Germany—Saxony. Italy. Rumania. Chile. Japan. In hydrothermal veins, most commonly with *silver*, cobalt or nickel ores.

ARSENOLITE

16[As$_2$O$_3$] Arsenic oxide. Cubic. Crystals usually minute octahedra or capillary. White, occasionally with bluish, yellowish or reddish tinge. Conchoidal fracture. Perfect cleavage in one direction. Hardness 1·5. SG 3·87. When heated in closed tube, sublimes and condenses above in small octahedra.
Occurrence: USA—Kansas (Ellis County), Nevada (Ophir Mine). Canada—British Columbia (Watson Creek), Ontario (Sudbury district). Czechoslovakia. France (Aveyron, Loire, Saône-et-Loire). Germany—Baden, Saxony. Hungary. Italy. Peru. Secondary mineral from oxidation of *arsenopyrite*, native *arsenic* or other *arsenic* minerals.

ARSENOPYRITE or MISPICKEL

FeAsS Iron arsenide sulphide. Monoclinic.

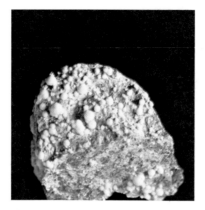

Prismatic or short prismatic columnar crystals, either straight or divergent; can be granular or compact. Silver-white, inclining to steel-grey. Metallic lustre. Dark greyish-black streak. Uneven fracture. Distinct cleavage in one direction. Hardness 5·5–6·0. SG 6·07. Brittle. When heated in closed tube, gives sublimate of *arsenic* as bright grey crystals near heated end and brilliant black amorphous deposit further away.
Occurrence: Worldwide. In diverse types of deposits; most commonly high-temperature deposited ore veins with other ores such as *silver*, nickel, cobalt and other *arsenic* ores.
Cobalt-rich variety is *danaite*.

ARSOITE

Igneous rock of the *syenite* clan. A grey, porphyritic extrusive rock. Contains phenocrysts of *sanidine, andesine, diopside* and *olivine* in a groundmass of *sanidine, oligoclase, diopside, magnetite* and *sodalite*.
Occurrence: Italy—Ischia.

ARTINITE

$2[Mg_2CO_3(OH)_2 . 3H_2O]$ Basic hydrated magnesium carbonate. Monoclinic. Acicular crystals; also massive. White. Vitreous to silky lustre. White streak. Brittle fracture. Perfect cleavage in one direction. Hardness 2·5. SG 2·02. Easily soluble

with effervescence in hydrochloric acid.
Occurrence: USA—Nevada (Luning), New Jersey (Hoboken).

ASBESTOS

Fibrous form of *amphibole* or *serpentine*. The long, fine, flexible fibres are easily separated by the fingers. White to greenish and brownish colour. Heat and (except for *chrysotile*) acid resistant.

Types of asbestos are *actinolite asbestos* or *amianthus*—form to which the name asbestos was originally given; *chrysotile*—fibrous *serpentine; amosite*—fibrous *anthophyllite;* and *crocidolite*—fibrous soda-*amphibole*. Mountain leather, mountain cork and mountain wood are commercial varieties of asbestos, varying in compactness and matting of fibre.

ASBOLANE, ASBOLITE, BLACK COBALT or BLACK COBALT OCHRE

Type of *wad* containing cobalt oxide, often with nickel oxide and cuprous oxide. Grades into *heterogenite* and other cobalt oxide ores.
Occurrence: Germany—Thuringia (Kamsdorf). Poland—Silesia. Congo—Katanga.

ASBOLITE

Synonym of *asbolane*.

ASCHAFFITE

Igneous rock of the *diorite* clan. A hypabyssal rock. Contains *quartz* and *feldspar*.
Occurrence: Germany—Bavaria (Aschaffenburg).

ASHCROFTINE or KALITHOMSONITE

$40[NaKCaAl_4Ai_5O_{18},8H_2O]$ Hydrated sodium potassium calcium and aluminium silicate. Member of the *zeolite* family. Tetragonal. Fine, pink crystalline powder.
Occurrence: Greenland—Narsarsuk. Uganda.

ASPARAGUS STONE

A translucent greenish-yellow variety of *apatite*.

ASPHALT

Sedimentary deposit. Solid variety of *bitumen*. Mixture of various higher hydrocarbons. Soft, pitchy substance, or solid

with conchoidal fracture. Black or brown.
Occurrence: England—Derbyshire, Leicestershire, Shropshire (near Shrewsbury). USA—California (near Santa Barbara). Canada—Alberta. Cuba. Trinidad—Pitch Lake. Venezuela. Middle East—Dead Sea area. Penetrates *sandstones* and *dolomites* in such locations as Kentucky and Oklahoma in the USA and Neufchatel in France to provide natural paving material.

ASPHALTITE

Group-term for the purer *bitumens* such as *albertite*, *grahamite*, *anthraxolite* and *uintaite*, to distinguish them from bituminous sands and *limestones* which commercially are classified as *asphalt*.

ASPHALTUM

Mineral pitch. Amorphous. Massive. Brownish black to black. Pitchy lustre. SG 1–2.
Occurrence: England—Derbyshire (Matlock), Shropshire (Haughmond Hill). USA—California (Santa Barbara).

ASTITE

High grade contact metamorphic rock. A variety of *hornfels*. Contains chiefly *andalusite* and *mica*.
Occurrence: Italy—Italian Alps (Cima d'Asta).

ASTROPHYLLITE

$4[(K,Na)_2(Fe^{2+},Mn)_4TiSi_4O_{14}(OH)_2]$ Potassium, sodium, manganese and ferrous iron titanate silicate. Triclinic. Elongated crystals; sometimes lengthened into thin strips or blades or arranged in stellate group. Bronze-yellow to gold-yellow. Submetallic to pearly lustre. No fracture. Perfect cleavage in one direction. Hardness 3·0. SG 3·3–3·4. Decomposes in hydrochloric acid with the separation of thin scales of silica.
Occurrence: USA—Colorado (Pike's Peak). Greenland—Kangerdluarsuk. Norway—Langesund Fiord.

ATACAMITE

$4[Cu_2Cl(OH)_3]$ Basic copper chloride. Orthorhombic. Slender prismatic, frequently tabular crystals; massive fibrous or granular to compact. Various shades of bright green grading to dark emerald-green and blackish-green. Adamantine lustre.

Apple-green streak. Conchoidal fracture. Perfect cleavage in one direction. Hardness 3·0–3·5. SG 3·76. Brittle. When heated in closed tube, much water is given off and grey sublimate is formed.
Occurrence: England—Cornwall (sparingly). USA—California (Boleo, El Toro). Australia—Cornwall, Moonta and Wallaroo Mines. Italy—Mount Etna, Mount Vesuvius. Chile—Atacama Desert. A secondary mineral formed by oxidation of other *copper* minerals, especially under saline conditions.

ATAXITE

A type of *meteorite* consisting of a fine intergrowth of *kamacite* and *taenite*.

ATLANTITE

Igneous rock of the *gabbro* clan. A dark-coloured extrusive rock. Contains phenocrysts of predominantly *augite* and *olivine* in a groundmass of *titanaugite* with *olivine* and *labradorite*.
Occurrence: Malawi.

ATTAPULGITE

Synonym of *palygorskite*.

AUGEN

German word for 'eyes' referring to eye-like structures, inclusions or crystals in a rock or mineral.

AUGEN-SCHIST or MYLONITE GNEISS

Low grade, light-coloured regional metamorphic rock. Contains 'eyes' of *feldspar* simulating phenocrysts, parallel *mica* flakes or aggregates of flakes, and irregular lenticular banding and foliation due to layers and patches of different grain size and composition.
Occurrence: Canada—Laurentian Shield. Finland and Sweden—Fennoscandian Shield.

AUGITE

(Ca,Mg,Fe,Al)$_2$(Al,Si)$_2$O$_6$ Calcium magnesium iron and aluminium silicate. Member of the *pyroxene* family. Variable composition; transitions occur to the other monoclinic *pyroxenes*. Monoclinic. Usually prismatic crystals, often short and thick or coarsely lamellar; coarse or fine granular. Black or greenish-black. Vitreous to resinous lustre. White to grey and greyish-green streak. Uneven to conchoidal fracture. Good prismatic cleavage. Hardness 5·0–6·0. SG 3·2–3·5. Brittle. F 3·0. Insoluble in hydrochloric acid.

Occurrence: USA—Maine, Massachusetts, Vermont. Canada—Ontario and Quebec. Germany—the Eifel. Italy—Mount Etna, Piedmont, Mount Vesuvius. Azores Island. Cape Verde Islands. Worldwide occurrence as short prismatic crystals in many volcanic rocks and such metamorphic rocks as *pyroxene-gneisses*.

Varieties are *diallage*, *fassaite* (*of Werner*), *omphasite* and *titanaugite*.

AUGITITE

Igneous rock of the *ultrabasic* clan. A dark grey or black extrusive rock. Contains *augite* with small amounts of *amphibole*, accessory *magnetite* or *ilmenite* and *apatite* occurring as phenocrysts in a base of glassy *analcite*. Gelatinises with hydrochloric acid.
Occurrence: Cape Verde Islands.

AUGITOPHYRE

Igneous rock of the *gabbro* clan. A porphyritic hypabyssal rock. Type of *diabase* with phenocrysts of *augite*.
Occurrence: Greece—South of Levetsova.

AURICHALCITE

(Zn,Cu)$_5$(OH)$_6$(CO$_3$)$_2$ Basic zinc copper carbonate. Crystals never well-defined; usually in crusts of thin, fragile scales. Pale greenish-blue. Pearly lustre. No fracture. Micaceous, flexible cleavage. Hardness 2·0. SG 3·5–3·6. Fragile. Colours flame green. Soluble with effervescence in hydrochloric acid, with evolution of carbon dioxide, to give green solution which turns blue on addition of ammonia.
Occurrence: England—Cumberland, Derbyshire (Matlock). Scotland—Lanarkshire (Leadhills). USA—Arizona, New Mexico, Utah. South West Africa—Tsumeb. France. Greece. Italy. USSR. Congo—Mindouli. In secondary, weathered zone of *copper* zinc ore deposits.

AURIFEROUS

Gold-bearing.

AUSTINITE

4[CaZn(AsO$_4$)(OH)] Basic calcium zinc arsenate. Orthorhombic. Acicular or bladed crystals; also massive. Colourless or white. Subadamantine to silky lustre. Good cleavage in one direction. Hardness 4·0–4·5. SG 4·13.
Occurrence: USA—Utah (Gold Hill in Tooele County). Bolivia—Sica Sica Province (Lilli Mine near Lomitos).

AUSTRALITES

Type of *tektites* found in Tasmania, Australia.

AUTOMORPHIC

A rock in which the mineral crystals are fully developed.

AUTUNITE

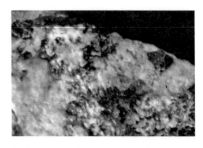

Ca(UO$_2$)$_2$(PO$_4$)$_2$.10–12 H$_2$O Hydrous calcium uranium phosphate. Tetragonal. Scattered square plates. Lemon-yellow to greenish-yellow. Pearly to vitreous lustre. Yellowish streak. No fracture. Perfect basal and prismatic cleavage. Hardness 2·0–2·5. SG 3·1. Shows intense fluorescence which serves to identify the mineral.
Occurrence: England—Cornwall (Redruth). USA—Colorado Plateau, North Carolina (Mount Pine), Utah (Marysville), Washington (Near Spokane). Australia—Mount Painter. France—Saône-et-Loire (Autin).

Late secondary uranium mineral formed as alteration product of uranium-containing minerals. Magnesium enriched form is *novacekite*.

AVENTURINE or AVENTURINE FELDSPAR

Synonym of *oligoclase*.

AVENTURINE QUARTZ

Variety of *quartz* exhibiting a schiller. Contains small crystals of *mica* or *iron*.
Occurrence: USSR—Siberia. India. Tanzania. (Variety has crystals of green chrome *mica* and is used for jewellery.) Germany. Spain—near Almeida. USSR. (Iron-coloured types. Creamy-white to reddish-brown colour.)

AVEZAKITE

Igneous rock of the *ultrabasic* clan. A porphyritic, cataclastic hypabyssal rock. An *ilmenite*-rich, *pyroxene*-bearing *hornblendite* with characteristics intermediate between *pyroxenite* and *hornblendite*.
Occurrence: France—the Pyrenees (South West of Lannemezan at Avezac-Prat).

AVIOLITE

Metamorphic rock. A variety of *hornfels*. Contains essentially *mica* and *cordierite*.
Occurrence: Italy—the Alps (Monte Aviolo).

AVOGADRITE

4[(K,Cs)BF$_4$] Potassium cesium fluoroborate. Orthorhombic. Minute tabular to platy crystals, or dense crusts. Colourless to white; yellowish or reddish when impure. SG 2·617. Bitter taste. Easily fuses. Very soluble in water.

Occurrence: Italy—Mount Vesuvius. Mixed with *sassolite* and other salts.

AXINITE, HYALITE or OISANITE

$2[Ca_2(Mn,Fe)Al_2BSi_4O_{15}OH]$ Aluminium calcium borosilicate with varying amounts of iron and manganese. Triclinic. Thin and very sharp-edged crystals; sometimes massive or lamellar. Clove-brown, plum-blue or pearly-grey. Highly vitreous lustre. Colourless streak. Conchoidal fracture. Distinct cleavage. Hardness 6·5–7·0. SG 3·27. Brittle. Readily fuses with intumescence and colours flame pale-green.

Occurrence: England—Cornwall (Botallack Mine near St Just in Penwith), Devonshire (Brent Tor). USA—Maine (Phippsburg), New York (Cold Spring), Pennsylvania (Bethlehem). Europe. Chile. Mexico. A mineral of contact metamorphism where lime-rich rocks are in contact with certain igneous rocks.

Variety is *manganaxinite*.

AZURITE or CHESSYLITE

$2CuCO_3.Cu(OH)_2$ Basic hydrated copper carbonate. Monoclinic. Modified prismatic crystals; usually massive or earthy. Deep azure-blue. Vitreous lustre, verging on adamantine. Blue streak. Conchoidal fracture.

Perfect cleavage in one direction, but interrupted. Hardness 3·5–4·0. SG 3·7–3·8. Brittle. Chemical identification as for *malachite* from which it is distinguished by azure-blue colour.

Occurrence: England—Cornwall (Redruth). USA—Arizona, New Mexico. Australia—New South Wales (Broken Hill), Queensland. South West Africa—Tsumeb. France—Chessy. Greece—Laurium. Italy—Sardinia. USSR—Siberia. Congo—Katanga. With other oxidised *copper* minerals in the zone of weathering of *copper* lodes and deposits.

BABINGTONITE

$Ca_2Fe^{2+}Fe^{3+}Si_5O_{14}(OH)$ Hydrous calcium iron silicate. Triclinic. Always in crystals, usually small, very brilliant and equidimensional with striated faces. Black or greenish-black. Vitreous lustre. Conchoidal fracture. Two pinacoidal cleavage. Hardness 5·5–6·0. SG 3·4. Brittle. Insoluble in hydrochloric acid. Easily fuses to black magnetic globule.

Occurrence: England—Devonshire. Scotland—Sutherland. USA—Massachusetts, New Jersey (Great Notch, Patterson). Italy—Baveno. Norway—Arendal. A late, hydrothermally deposited secondary mineral usually associated with *zeolites*.

BADDELEYITE

ZrO_2 Zirconium dioxide. Monoclinic.

Usually flattened and short to long prismatic crystals with striated faces. Colourless to yellow, green, reddish or greenish-brown, brown or iron-black. Greasy to vitreous lustre. White to brownish-white streak. Subconchoidal to uneven fracture. Nearly perfect cleavage in one direction. Hardness 6·5. SG 5·4–6·02. Brittle. In blowpipe flame glows, turns white and is nearly or quite infusible.

Occurrence: USA—Montana (near Bozeman), Texas (Davis Mountains). Italy—Monte Somma. Brazil. Sri Lanka. Congo—Kilo, Nedi.

BAHIAITE

Igneous rock of the *ultrabasic* clan. A glistening brownish-black holocrystalline-granular plutonic rock. A variety of *hypersthenite* containing *orthopyroxene* and *amphibole* with smaller amounts of *olivine* and *pleonaste*.

Occurrence: Brazil—Bahia (near Furnaca).

BAKERITE

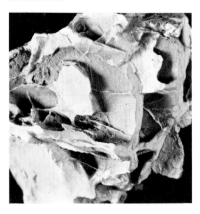

Variety of *datolite*.

Occurrence: USA—California (Borax Mines in Death Valley area). In cryptocrystalline masses deposited by hydrothermal solutions.

Chemically similar to *howlite*.

BALAS RUBY

Rose-red gemstone variety of *spinel*.
Occurrence: North India.

BALDITE

Igneous rock of the *gabbro* clan. Dark stone, greyish-black, dark green or greenish-grey, dense to fine-grained *analcite-basalt* occurring in a hypabyssal mode. Contains *pyroxene* in groundmass of colourless *analcite* with accessory *augite* and *iron* ore.
Occurrence: Scotland—East Lothian (near Kidlaw). USA—Colorado (Cripple Creek), Montana (Highwood and Littlewood Mountains). Australia—New South Wales.

BALLSTONE

Sedimentary rock. A type of *limestone*. Irregular non-bedded ball shaped concretions. Consists of colonies of corals in matrix of calcareous mud.
Occurrence: England—Shropshire (Wenlock).

BANAKITE

Igneous rock of the *diorite* clan. An extrusive rock similar to *absarokite*. Contains less *olivine* and *augite* than *absarokite*;

Below: Ballstone

may contain *quartz* but no *olivine*.
Occurrence: USA—Wyoming (Yellowstone National Park).

BANALSITE

$4[Na_2BaAlSi_4O_{16}]$ Barium aluminosilicate with sodium as dominant alkali. Orthorhombic. Compact and coarsely crystalline. Pure white. Pearly lustre on cleavage surfaces. No fracture. Prismatic cleavage. Hardness 6·0. SG 3·065. Decomposes in 1:1 hydrochloric acid to leave a colourless residue.
Occurrence: Wales—Benallt Mine. With *tephroite* and *alleghanyite*.

BAND

Stripe of different colour and thus of different composition in a rock. Banded rocks exhibit such a pattern, usually uniformly.

BANDED AGATE

Variegated variety of *chalcedony*. Irregularly clouded with various colours.
Occurrence: Worldwide. Forms nodules.

BANKET

Low grade metamorphosed sedimentary rock laid down in fluviatile environment. Greenish colour. Consists of conglomerate of vein-*quartz* pebbles in quartzitic matrix with traces of *chloritoid*, scales of *chlorite*

and *gold* and *iron* minerals.
Occurrence: South Africa—Transvaal (Witwatersrand area).

BARITE

Synonym of *barytes*.

BARKEVIKITE

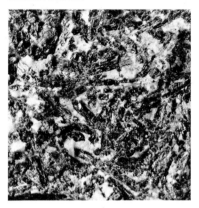

$c2(NaCa_2(Mg,Fe^{2+},Fe^{3+},Al)_5(Si,Al)_8O_{23}(OH))$ Sodium calcium magnesium iron and aluminium silicate. A clino-*amphibole*. Monoclinic. Short or long prismatic crystals. Deep velvet-black. Vitreous lustre. Uneven fracture. Prismatic, perfect cleavage. Hardness 4. SG 3·428. Brittle. Easily fuses, but less easily than *arfvedsonite*.

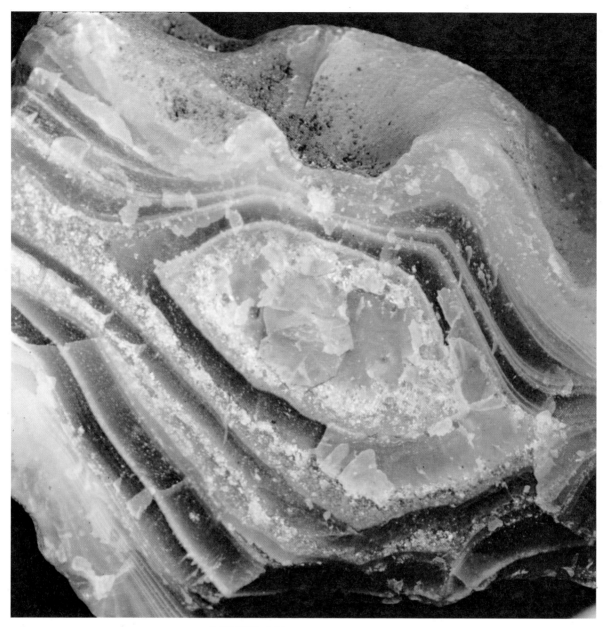

Above: banded agate

Occurrence: Norway—Christiania and Langesund Fiords. A constituent of *augite-syenite*. Related to *arfvedsonite*.

BARROISITE

Synonym of *carinthine*.

BARSHAWITE

Igneous rock of the *ultrabasic* clan. A pinkish, even-grained hypabyssal rock. Contains *barkevikite* and *titanaugite* in a groundmass of *andesine* with accessory *nepheline*, *orthoclase* and *analcite*.
Occurrence: Scotland—Renfrewshire (Barshaw near Paisley).

BARYTES, BARITE or HEAVY SPAR

$4[BaSO_4]$ Barium sulphate. Ortho-rhombic. Common-prism, basal pinacoid and dome crystals; massive, coarsely lamellar, granular, compact, columnar. Colourless to ·white; often tinged with yellow, red or brown. Vitreous lustre, approaching resinous or pearly. White streak. Uneven fracture. Perfect cleavage parallel to basal plane and prism. Hardness 3·0–3·5. SG 4·5. Brittle. F 3·0. Colours flame green.
Occurrence: England. Scotland. USA.

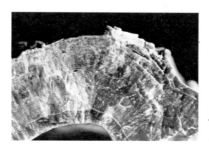

Canada. Europe. In lead and zinc veins with *galena*, *blende*, *fluorite* and *quartz*. Variety is *hokutolite*.

BARYTOANGLESITE

(Pb,Ba)SO$_4$ Variety of *anglesite* with *c* 8

per cent barium sulphate replacing the lead sulphate.
Occurrence: Germany—Oldenburg (Borbeck near Essen).

BARYTOCALCITE (of Brooke)

BaCO$_3$. CaCO$_3$ Barium calcium carbonate. Monoclinic. Prismatic crystals with vertically striated faces; also massive. White, greyish, greenish or yellowish. Vitreous to resinous lustre. White streak. Uneven to subconchoidal fracture. Perfect cleavage in one direction. Hardness 4·0. SG 3·64–3·66. Brittle. Soluble in dilute hydrochloric acid. Colours flame yellowish-green.
Occurrence: England—Cumberland (Alston Moor).
Isomorphous with *alstonite*.

BARYTOCALCITE (of Johnston)

Synonym of *alstonite*.

BASAL CONGLOMERATE

Sedimentary rock. A marine *conglomerate* with graded or small-scale cross-bedded structures resulting from sea invasion of ancient land area. Contains spherical *quartz* grains with sand and pebbles.
Occurrence: England—Lake District (lowest beds of carbonaceous *limestone*), North Pennines, Shropshire (Cambrian *conglomerate*). North West Scotland.

BASALT

Igneous rock of the *gabbro* clan. A dark-coloured, variable texture extrusive rock. Contains essentially basic *plagioclase* and a ferromagnesian mineral, usually *augite* but may be *hypersthene* or *hornblende*. *Quartz* or *orthoclase* may be present in small amounts.
Occurrence: Worldwide.

BASALT-TUFF

Igneous rock of the *gabbro* clan. A brown, *sandstone*-like textured pyroclastic rock. Consists of devitrified basaltic *glass* with fragments of *augite*, *olivine* and *plagiolase*, *zeolites* and *calcite*.
Occurrence: Italy—Sicily (Palagonia).

BASANITE, LYDIAN STONE or TOUCHSTONE

Igneous rock of the *gabbro* clan. A dark-coloured, fine-grained extrusive rock. Contains *augite* and *olivine* with *analcite*, *nepheline* or *leucite*. Similar to *tephrite* but contains *olivine*. Very hard. Used for testing precious metals by their streak.
Occurrence: England—Derbyshire (Calton Hill). Eastern Mediterranean—especially Greece and Italy.
A type of flinty *jasper*.

BASIC

Refers to igneous rocks with a low free *quartz* content. See acid, alkali.

BASSANITE or VIBERTITE

CaSO$_4$ · $\frac{1}{2}$H$_2$O Semi-hydrated calcium sulphate. Consists of microscopic needles in parallel formation. White. SG 2·69–2·76. Converted to *anhydrite* when heated to redness.
Occurrence: Italy—Mount Vesuvius. In cavities of *leucite-tephrite* blocks or *gibbsite*.

BASTITE

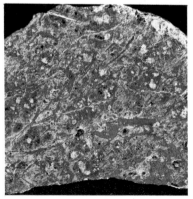

c(Mg,Fe)SiO$_3$ · 4/5H$_2$O Mg : Fe *c*4) Magnesium iron silicate. An altered *enstatite*. Leek-green to olive and pistachio-green or

brown. Bronze-like or metalloidal lustre on chief cleavage face. Perfect cleavage in one direction. Becomes brown in blowpipe flame. Borax bead shows green in reducing flame, yellow in hot oxidising flame.
Occurrence: Germany—Black Forest (Todtmoos), Harz Mountains (Baste near Harzburg). In foliated form in granular extrusive rocks.
Variety of *serpentinite*.

BATUKITE

Igneous rock of the *ultrabasic* clan. A dark-coloured aphanitic porphyritic extrusive lava. Contains phenocrysts of *augite* and some *olivine* in groundmass of *augite*, *magnetite* and *leucite*.
Occurrence: Indonesia—Celebes (Batuku).

BAUMONTITE (of Jackson)

Synonym of *chrysocolla*.

BAUXITE

$Al_2O_3 \cdot 2H_2O$ Mixture of hydrated aluminium oxides. Amorphous. Round, concretionary disseminated grains; massive, oölitic, earthy, clay-like. Whitish, greyish, ochre-yellow, brown or red. Dull lustre. Earthy fracture. No cleavage. Hardness 1·0–3·0. SG 2·55.
Occurrence: USA—Alabama, Arkansas, Georgia, Wyoming (Yellowstone National Park). France—near Arles (Baux). Germany—Nassau. Brazil. Jamaica. USSR. India. Africa. Existence of independent bauxite molecule questionable. Probably a sedimentary deposit consisting of a mixture of aluminium minerals such as *gibbsite*, *boehmite* and *diaspore*. Bauxite deposits formed under oxidising conditions in areas of low or moderate topographic relief, but with minimum erosion, where temperature consistently exceeds 25 degrees C and rainfall exceeds evaporation.

BAVALITE

$2[(Fe^{2+}.Mg,Al)_6(Si,Al)_4O_{10}(OH)_8]$ Basic magnesium aluminium ferrous iron silicate. Variety of *daphnite*. Near *chamosite* in composition. Greenish-black, bluish or greyish. Hardness 4·0. SG 3·99. Oölitic mineral. Fuses with difficulty to black magnetic scoria.

Occurrence: France—Brittany. Forms beds in old, schistose rocks. Member of *chlorite group*.

BAVENITE

$4[Ca_4(Be,Al)_4Si_9(O,OH)_{28}]$ Basic calcium beryllium aluminium silicate. Orthorhombic. Radiating tufts of white needles, or pale brown radiating aggregates of platy fibres, spherules or crystals.
Occurrence: USA—California (Himalaya Mine near San Diego). Switzerland—Baveno, around Graubünden. USSR—Central Ural Mountains (Emerald Mines). In drusy cavities in *granite*. Pseudomorphous with *beryl*.

BAYLDONITE

$(Pb,Cu)_7(AsO_4)_4(OH)_2 \cdot H_2O$ Hydrated basic copper lead arsenate. Monoclinic. Massive. Shades of green. Resinous lustre. Pale green streak. Subconchoidal fracture. Hardness 4·5. SG 5·35. Soluble with difficulty in hydrochloric acid.
Occurrence: England—Cornwall (Penberthy Croft Mine near Saint Hilary).

BEAD TEST

Many analytical procedures involve the heating of a bead of the test substance in a flame, either alone or with a suitable chemical reagent. The behaviour of the sample, and the resulting colour of the flame, assist in its identification.

BEAUMONTITE (of Jackson)

Variety of *chrysocolla*.
Occurrence: France—Chessy.

BEAUMONTITE (of Lévy)

Barium-bearing variety of *heulandite* with characteristics identical to *heulandite*.
Occurrence: USA—Maryland (Jones's Falls near Baltimore). Found with *haydenite* on a *syenite schist*. Forms minute crystals.

BEAVERITE

$Pb(Cu,Fe,Al)_3(SO_4)_2(OH)_6$ Basic lead copper ferric iron and aluminium sulphate. Hexagonal. Earthy and friable masses of microscopic, yellow hexagonal plates. SG 4·36. Insoluble in water. Soluble in hydrochloric acid.
Occurrence: USA—Nevada (Bass Mine in Yellow Pine district), Utah (near Frisco in Alta district). With *plumbojarosite* in oxidised part of lead-*copper* ores in arid regions.
Member of the *alunite group*.

BEBEDOURITE

Igneous rock of the *ultrabasic* clan. A dark-coloured, medium to coarse-grained plutonic rock. Contains essentially *biotite* and *pyroxene*.
Occurrence: USA—Arkanas (Magnet Cove in the Ozark Mountains), Montana (Libby). Brazil—Minas Gerais (Bebedours in Salitre Mountains).
Cromaltite is a *garnet-bebedourite*.

BED

Homogeneous rock layer. In International terminology a bed is the smallest stratigraphic unit.

BEEF

Term used to describe fibrous *calcite* similar in habit to *satin spar*.

BEERBACHITE

Igneous rock of the *gabbro* clan. A dark-coloured, saccharoidal, panautomorphic-granular hyabyssal rock. Contains *diallage* and *plagioclase* with accessory *magnetite*, *apatite* and *biotite*.
Occurrence: Germany—Odenwald (Frankenstein, Nieder Beerbach).

BEIDELLITE

$R_{0.33}Al_2(Si,Al)_4O_{10}(OH)_2 \cdot nH_2O$ (R = sodium, calcium, magnesium, potassium or combinations of these). Hydrous aluminium metasilicate. Variety of *montmorillonite*. Orthorhombic. Differs from *montmorillonite* in that aluminium replaces silica and magnesium does not replace aluminium in the lattice.
Occurrence: USA—Colorado (Beidell in Saguache County). Forms a *rock-flour*.

BEKINKINITE

Feldspathoid igneous rock. A plutonic rock. Contains *barkevikite*, *nepheline* and *olivine*.
Occurrence: Madagascar—Mount Bekinkina.
Similar to *theralite*.

BELOEILITE

Igneous rock of the *syenite* clan. A granular plutonic rock. Contains *sodalite*, potash-*feldspar* and *pyroxenes* with accessory *nepheline*.
Occurrence: Canada—Quebec (St Hilaire Mountain).

BELONITE

Synonym of *aikinite*.
Term also used to describe acicular, colourless and transparent crystals (probably *feldspar*) in such glassy volcanic rocks as *obsidian*.

BELUGITE

Igneous rock intermediate in composition between the *gabbro* and *diorite* clans; the *feldspar* present may be *andesine* or *labradorite*. A plutonic rock.
Occurrence: USA—Alaska (Beluga River).

BEMENTITE (of Köenig)

cMn$_8$Si$_7$O$_{17}$(OH)$_{10}$ Variable formula. Manganese silicate with small amounts of magnesium oxide, ferrous oxide and zinc oxide. *Micaceous* structure forming radiated stellate masses with small foliated structure resembling *pyrophyllite*. Pale greyish-yellow. Pearly lustre. No fracture. Perfect cleavage. SG 2·981. Soft. Readily fuses to black *glass*. Dissolves in hot hydrochloric acid without gelatinisation.
Occurrence: USA—New Jersey (Franklin Furnace), Washington (Crescent Mine in Clallam County). Associated with *calcite*.
A manganese *serpentine* and member of the *friedelite* family.

BEMENTITE (old usage)

Slender, prismatic, transparent or nearly colourless crystals of *danburite* covered with or enclosing fine scaly *chlorite* and enclosing needles of *tourmaline*.
Occurrence: Switzerland—South of Disentis.

BENITOITE

BaTiSi$_3$O$_9$ Barium titanium silicate. Hexagonal. Good tabular crystals up to 37mm across. Unevenly coloured blue and white; sometimes pink. Vitreous lustre. Conchoidal fracture. Poor pyramidal cleavage. Hardness 6·0–6·5. SG 3·6. Has very distinctive crystals which eliminate need for identifying tests since only one occurrence. Valuable gemstone when transparent or of good colour and free from flaws.
Occurrence: USA—California (Diablo Range of the California Mountains in San Benito County). In a *natrolite* dyke in green *schist* with *serpentine* and *neptunite*.

BENTONITE

Sedimentary rock of *argillaceous* type. Formed from devitrification of volcanic ash laid down in sea water. Light green or pale greenish-yellow. Contains *montmorillonite* in extremely small crystals with *orthoclase*, *plagioclase* and *biotite*. Has fracture like hard wax. Strongly adsorbent. When moistened, swells to many times its own volume. Thin sections show relics of *pumice* structure.
Occurrence: North America.
Some *fuller's earths* consisting principally of *montmorillonite*, such as those found in Bedfordshire, England, should be reclassified as *bentonite*.

BERAUNITE

Fe^{2+}Fe^{3+}(PO$_4$)$_3$(OH)$_5$.3H$_2$O Basic hydrated ferrous ferric iron phosphate. Monoclinic. Tabular crystals, striated and more or less elongated; usually radiated foliated globules and crusts. Reddish-brown to dark hyacinth-red and blood-red. Vitreous lustre, inclining to pearly on cleavage surfaces. Yellow to olive-drab streak. Good cleavage in one direction. Hardness 3·5–4·0. SG 2·8–2·9. Easily fuses to black bead. Readily soluble in hydrochloric acid.
Occurrence: Ireland—County Cork (Roury Glen). USA—Arkansas, New Hampshire, New Jersey (Middletown), Pennsylvania (Cumberland County, Northampton County). Czechoslovakia

Bohemia. Germany—Bavaria, Hesse, Saxony. Sweden—Kirunavaara. USSR—Kerch Peninsula. As secondary mineral in *iron* ore deposits and as alteration product of *triphylite* or other primary phosphates in *pegmatite*.

BERESITE

Igneous rock of the *granite* clan. A porphyritic hypabyssal rock. Contains *quartz* and *mica* with a little *pyrite*. Usually much kaolinised. Similar in composition to *greisen*.
Occurrence: USA—Nevada (Belmont). USSR—Ural Mountains (Beresovsk).

BERESOFITE (of Shepard)

Synonym of *crocoite*.

BERESOFITE (of Simpson) or BERESOVSKITE

8[(Fe,Mg)Cr$_2$O$_4$] Iron magnesium chromate with ratio of iron to magnesium c 3 : 1. Variety of *chromite* with magnesium replacing iron in the lattice. Composition grades into *magnesiochromite*.
Occurrence: USSR—Ural Mountains (Beresovsk).

BERESOVSKITE

Synonym of *beresofite* (*of Simpson*).

BERGALITE or BERGALITH

Feldspathoid igneous rock. A black, pitch-like porphyritic hypabyssal rock. Contains phenocrysts of *hauyne*, *apatite*, *perovskite*, *melilite* and *magnetite* in a groundmass of the same material plus *nepheline*, *biotite* and brown interstitial *glass*.
Occurrence: Germany—Baden (Oberbergen in the Kaiserstuhl).
A type of *lamprophyre*.

BERINGITE

Igneous rock of the *diorite* clan. A dark-coloured hypabyssal rock. Contains mainly *barkevikite* with *sodaclase* and *orthoclase*.
Occurrence: USSR—Kamchatka. Bering Sea—Bering Island.

BERMUDITE

Igneous rock of the *gabbro* clan. An extrusive rock of lamprophyric character. Contains small *biotite* crystals, accessory iron ores and *apatite* in a groundmass of *analcite*, *nepheline* and *sanidine*.
Occurrence: Bermuda Islands.
Effusive equivalent of *ouachitite*.

BERONDRITE

Igneous rock of the *gabbro* clan. A dark-coloured extrusive rock. Contains elongated crystals of brown *hornblende* with itaniferous *augite*.

Occurrence: Madagascar—Berondra Valley.

A type of *theralite*.

BERTHIERINE

$(Fe^{2+}, Fe^{3+}, Mg, Al)_{2 \cdot 8-2 \cdot 9}(Si,Al)_2O_5(OH)$. Variety of *chamosite*. Near *chamosite* in structure. Bluish-grey, black or greenish-black. Dark greenish-grey streak. Hardness 2·5. Fuses with difficulty to black magnetic globule.

Occurrence: France—Bourgogne, Champagne, Hayanges, Lorraine and Moselle. In beds of *iron* ore.

Structure allied to *kaolinite* and *antigorite*. Much *chamosite* of this type should be reclassified as berthierine.

BERTRANDITE

$4[Be_4Si_2O_7(OH)_2]$ Basic beryllium orthosilicate. Orthorhombic. Usually tabular crystals. Colourless to slightly yellow. Vitreous lustre; pearly on basal plane. No fracture. Perfect cleavage. Hardness 6·0–7·0. SG 2·59–2·6. Infusible in blowpipe flame, but becomes opaque. Insoluble in acids.

Occurrence: USA—Colorado (Chaffee County), Maine (Stoneham), Virginia (Amelia Court House). Czechoslovakia—Bohemia (Pisek). France—Nantes (la Villeder). In cavities implanted on *quartz* or *feldspar*.

BERYL

$Be_3Al_2Si_6O_{18}$. Beryllium aluminium sili-

cate. Hexagonal. Usually forms large, prismatic hexagons; also massive, granular, columnar. White, blue, yellow, green, pink. Vitreous lustre. White streak. Conchoidal fracture. Poor basal cleavage. Hardness 7·5–8·0 SG 2·6. Glows whitely in blowpipe flame and does not decrepitate too violently. Fuses with great difficulty to white *glass*.

Occurrence: Northern Ireland—Mountains of Mourne. USA—California, North Carolina, New England, New Mexico, South Dakota. Common accessory mineral in *pegmatite* veins all over the world. Occurs almost exclusively in *pegmatite*. Gemstone varieties are *aquamarine, heliodor, emerald, vorobyevite* and *goshenite*.

BERYLLONITE

$12[NaBePO_4]$ Sodium beryllium phosphate. Monoclinic. Tabular to short prismatic crystals; faces frequently dull or roughened. Snow-white or pale yellowish. Brilliant, vitreous lustre. White streak. Conchoidal fracture. Perfect cleavage in one direction. Hardness 5·5–6·0. SG 2·81. Brittle. F 3·0. Decrepitates in flame and fuses to cloudy *glass*.

Occurrence: USA—Maine, Newry, Stoneham. In *pegmatite*.

BERZELIANITE

$4[Cu_2Se]$ Copper selenide. Cubic. Den-

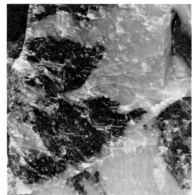

dritic Crusts. Silver white (Rarnishes black). Metallic lustre. Shining white streak. Hardness 1–2. SG 6·71. Soluble in concentrated nitric acid.

Occurrence: Germany—Harz (Lehrbach). Sweden—Shrikerum.

BERZELIITE (of Kühn)

$8[(Ca,Na)_3(Mg,Mn)_2(AsO_4)_3]$ Magnesium, divalent manganese, calcium and sodium arsenate with proportion of magnesium greater than manganese. Cubic. crystals rare, as trapezohedra with small, modifying faces; usually massive or as rounded grains. Yellow, honey-yellow to orange-yellow and yellowish-red. Resinous lustre. Nearly white to orange-yellow streak. Subconchoidal to uneven fracture. Brittle. No cleavage. Hardness 4·5–5·0. SG c4·08. Fuses in flame to grey or brown bead. Easily soluble in nitric acid or hydrochloric acid.

Occurrence: Sweden—Örebro (Grythytte parish), Wermland (Långban, Nordmark). In *limestone* skarn or secondary veinlets. Member of *garnetoid* group.

BETAFITE

$[(U,Ca)(Nb,Ti)_3O_9 \cdot nH_2O$ Uranium calcium niobium (columbium) tantalum and titanium oxide or niobate (columbate)-tantalate. Usually octahedral crystals; sometimes elongated. Greenish-brown. Waxy to vitreous to submetallic lustre. Conchoidal fracture. No cleavage. Hard-

ness 4·0–5·5. SG 3·7–5·0. Brittle. Very radioactive. Fuses with difficulty to black slag.
Occurrence: Norway—near Tangen. USSR—near Lake Baikal. Madagascar. In *granite pegmatites*.

BEUDANTITE (of Lévy)

$PbFe_3AsO_4SO_4(OH)_6$ Basic lead ferric iron sulphate-arsenate. Hexagonal. Rhombohedral, often pseudocubic crystals. Black, dark green or brown. Vitreous to resinous lustre. Greyish-yellow or greenish streak. Cleavage easy in one direction. Hardness 3·5–4·5. SG 4·0–4·3. When heated before blowpipe fuses to grey slag. Soluble in hydrochloric acid.
Occurrence: Western Australia—near Mount McGrath. France—Haute-Vienne. Germany—Rhineland (Horhausen). Greece—Laurium.

BIDALOTITE

Synonym of *gedrite*.

BIELENITE

Igneous rock of the *ultrabasic* clan. A blackish-grey, fine-grained rock, probably extrusive in origin. Contains *pyroxene* and *olivine* with accessory *chromite* and *magnetite*.
Occurrence: Czechoslovakia—Moravia (Biele Valley). Loose fragments.

BIGWOODITE

Igneous rock of the *syenite* clan. A dark-coloured, medium-grained plutonic rock. Contains *microcline, microcline-microperthite, sodaclase* and *hornblende*.
Occurrence: Canada—Ontario (Sudbury district).

BIKITAITE

$LiAlSi_2O_6 \cdot H_2O$ Hydrated lithium aluminium silicate. Variety of *spodumene*. Monoclinic. Does not form crystals. Colourless to white. White streak. No cleavage. Hardness 6·0. SG 2·29–2·34.

Occurrence: Rhodesia—Bikita. In association with *eucryptite* and *quartz* in *pegmatite*.

BILLITONITE

General term for the *obsidianites* of the Malay Peninsula.

BINARY GRANITE

Igneous rock of the *granite* clan. A light-coloured, fine to coarse-grained plutonic rock. Contains two *micas-biotite* and *muscovite* with abundant *quartz*, usually *garnet* and various accessory minerals.
Occurrence: Britain, Ireland, North America. Europe.

BINDHEIMITE

$8[Pb_2Sb_2O_6(O,OH)]$ Lead pyroantimonate. Cubic. Cryptocrystalline masses. Yellow, brown, white or green. Resinous to dull lustre. White or yellow streak. Earthy to conchoidal fracture. Hardness 4·0–4·5. SG 4·6–5·6. Dissolves in hydrochloric acid leaving a residue of lead chloride.
Occurrence: England—Cornwall (Bodannon Mine near St Endellion). USA—Idaho (the Coeur d'Alene district of Shoshone County), Nevada (Montezuma Mine in Humboldt County).

BIOPYRIBOLE

Group name for *biotite* and/or *pyroxene* and/or *amphibole* which may be present in a rock.

BIOTITE

$K(Mg,Fe)_3AlSi_3O_{10}(OH)_2$ Hydrous magnesium iron aluminium and potassium silicate. Member of the *mica* family. Monoclinic. Tabular or short prismatic crystals; often in disseminated scales or massive aggregates of cleavable scales. Usually green to black. Splendent lustre is more or less pearly on cleavage surface; sometimes submetallic when black. Colourless streak. No fracture. Basal, highly perfect cleavage.

Hardness 2·5–3·0. SG 2·7–3·1. Completely decomposes in sulphuric acid, leaving silica in thin scales.
Occurrence: Worldwide as common constituent of igneous and contact metamorphic rocks. Notable occurrences of crystals are: USA—New England. Germany—Black Forest. Hungary. Italy—Mount Vesuvius.
Varieties are *anomite/eastonite* (*of Winchell*) *hydrobiotite* and *siderophyllite*.

BIOTITITE or GLIMMERITE

Igneous rock of the *ultrabasic* clan. A black, granular plutonic rock. Contains mainly *biotite* with chiefly *apatite* as accessory.
Occurrence: USA—Montana (near Libby). Italy—South East of Rome (Villa Senni). Rumania—Banat (Valley of Magurii).

BIRD'S EYE SLATE

Quarryman's term for *slate* crowded with squeezed concretions of a lenticular shape.
Occurrence: Channel Islands.

BIRKREMITE

Igneous rock of the *granite* clan. A leucocratic plutonic rock. Consists of alkali-*feldspars* with *quartz* and *hypersthene*.
Occurrence: Norway—Birkrem. Forms laccoliths.
Variety of *kalialaskite*.

BIRNESSITE

δMnO_2 Form of manganese oxide. Hexagonal. Very fine crystals. Black. Dull lustre. Hardness 1·5. SG 2·9–3·0. Identification as for other manganese minerals.
Occurrence: Scotland—Aberdeenshire (Birness). In manganese pan.
Probably formed by air oxidation of manganese minerals in alkaline conditions.

BISCHOFITE (of Fischer)

Synonym of *plumbogummite*.

BISCHOFITE (of Ochsenius)

$MgCl_2 \cdot 6H_2O$ Hydrated magnesium chloride. Monoclinic. Crystalline granular and foliated; sometimes fibrous. Colourless to white. Vitreous to dull lustre. Conchoidal to uneven fracture. No cleavage. Hardness 1·0–2·0. SG 1·604. Deliquescent. Stinging to bitter taste. Melting point 117 degrees C. Very water soluble (25gm/100ml).

Occurrence: Germany—Anhalt, Magdeburg and Saxony salt deposits. In saline deposits with *halite* and *carnallite*.

BISMITE

Synonym of *bismuth ochre*.

BISMUTH

Bi Native bismuth, sometimes with traces of *sulphur, arsenic* and *tellurium*. Hexagonal. Rhombohedral crystals resemble cubes; usually massive, foliated or granular. Silver-white with faint tinge of red. Metallic lustre easily tarnishes. Silver-white streak. Brittle fracture when cold, malleable when heated. Perfect cleavage parallel to the basal pinacoid. Hardness 2·0–2·5. SG 9·7–9·8. Sectile. When heated on charcoal, fuses and volatilises to form an orange-red incrustation. Dissolves in fairly concentrated nitric acid; on addition of water solution becomes milky.

Occurrence: England—Cornwall, Devonshire. USA—Colorado, Connecticut, South Carolina. Canada—Ontario (Cobalt district). Australia. Europe. Bolivia. In veins with ores of tin, *silver*, cobalt and nickel.

Below: bismuth

BISMUTH GLANCE or BISMUTHIN-ITE

Bi_2S_3 Bismuth sulphide. Orthorhombic. Small, needle-like crystals; usually massive. Lead-grey, but tarnish common. Metallic lustre. Lead-grey streak. No fracture. Perfect cleavage in one direction. Hardness 2·0. SG 6·4–6·5. Somewhat sectile. When heated on charcoal with potassium iodide and sulphur, gives yellow and bright red incrustations.
Occurrence: England—Cornwall, Cumberland, Devonshire. USA—Colorado, Connecticut, Minnesota, Pennsylvania, Utah. Europe. Bolivia. In veins with *copper*, lead, tin and other ores.

BISMUTHINITE

Synonym of *bismuth glance*.

BISMUTH OCHRE or BISMITE

Bi_2O_3 Bismuth trioxide. Monoclinic. Earthy, pulverulent. Yellow, greenish-yellow or greyish-green. Subadamantine to dull lustre. Greyish to yellow streak. Uneven to earthy fracture. No cleavage. Hardness 4·5. SG 8·64–9·22. Chemical identification as for native *bismuth*.
Occurrence: England—Cornwall. USA—New Mexico, North Carolina. Australia—Tasmania. Czechoslovakia. France. Germany. Bolivia. Mexico. USSR. Alteration

product, usually impure and hydrated, occurring on *bismuth* and *bismuth glance*.

BISMUTITE

Bi_2CO_5 . H_2O Basic bismuth carbonate. Tetragonal. Fibrous or earthy crusts. White, greyish or yellowish. Vitreous to pearly lustre in coarse, fibrous material. Grey streak. No fracture. Distinct basal cleavage. Hardness $c2·5$–$3·5$ depending on state of aggregation. SG 6·1–7·7. Readily fuses. Soluble with effervescence in acids.
Occurrence: England—mainly Cornwall. USA. Australia—New South Wales. South Africa—Cape Province. France. Germany. Bolivia. Brazil. Peru. Alteration product of native *bismuth* and *bismuth glance* found in oxidised portions of veins carrying *bismuth*.

BITTER SPAR

An *iron*-bearing variety of *dolomite* which turns brown on exposure.

BITUMEN

Sedimentary deposits. Group name for natural *asphalts*, mineral waxes and related substances consisting of mixtures of various higher hydrocarbons. Generally black or dark brown and soluble to various degrees in organic liquids. Characteristic pitch odour. Burns with a smoky flame.
Occurrence: Worldwide in surface pools, veins or impregnations occupying the pore space of sands and other rocks. Varieties are *albertite*, *anthraxolite*, *asphalt*, *asphaltites*, *elaterite*, *grahamite*, and *ozokerite*.

BITUMINOUS COAL or HOUSEHOLD COAL

A *humic coal*. Fixed carbon content varies inversely with the oxygen content. Laminated or banded structure from interbedding of *fusain*, *vitrain*, *durain* and *clarain*. Generally well-jointed, breaking into rectangular lumps with joint faces covered by *ankerite* or *pyrite*.

Occurrence: Widely distributed except in South America, Oceania and Africa where relatively scarce.

BIXBYITE

$16[Mn,Fe)_2O_3]$ Manganese ferric iron oxide. Cubic. Crystals cubic; occasionally deeply pitted. Black. Metallic to submetallic lustre. Black streak. Irregular fracture. Traces of cleavage in one direction. Hardness 6·0–6·5. SG 4·945. F 4·0. Dissolves with difficulty in hydrochloric acid, releasing characteristic chlorine fumes.
Occurrence: USA—Utah. South Africa—Transvaal (Posmasburg). Rhodesia. Sweden—near Murjek (Langbau). Argentina—Patagonia (Girona). India—Central Provinces. In cavities with differing associated minerals.
Massive variety is *sitaparite*.

BJEREZITE

Igneous rock of the *syenite* clan. A dark-coloured, compact, porphyritic hypabyssal rock. Contains phenocrysts of *nepheline*, *pyroxene* with *aegirine* borders, *andesine* and *orthoclase* in fine-grained groundmass of *pyroxene*, brown *mica*, *andesine*, potash-*feldspar*, *nepheline* and some *zeolites*.
Occurrence: USSR—Siberia (along the Bjerez River).

BLACK-BAND IRONSTONE

Variety of *clay ironstone*. Contains sufficient carbonaceous matter to allow for calcining without addition of fuel.

BLACK COBALT or BLACK COBALT OCHRE

Synonym of *asbolane*.

BLACK DIAMOND

Synonym of *framesite bort*.

BLACK JACK

Synonym of *blende*.

BLACK LEAD

Synonym of *cerussite*.

BLACK MANGANESE

Synonym of *hausmannite*.

BLADED

See *crystal*.

BLAE

Scottish term for greyish-blue carbonaceous shale. Contains low content of bituminous matter. Brittle. Weathers to a crumbling mass which passes into soft *clay*.
Occurrence: Scotland—the Lothians. Associated with *oil shales*.

BLAIRMORITE

Igneous rock of the *syenite* clan. A porphyritic extrusive rock. Contains flesh-red *analcite* phenocrysts in dark green groundmass of *analcite*, alkali-*feldspar* and alkali-*pyroxenes*. Total *analcite* content is greater than 70 per cent.
Occurrence: Canada—Alberta (Blairmore in Crowsnest Pass).

BLANFORDITE

$8[(Na,Ca,Mn,Fe^{2+},Fe^{3+},Al)(Si,Al)O_3]$ A *pyroxene*. Manganoan *aegirine-augite*. Monoclinic. Brown to deep lavender.
Occurrence: India—Bhandard district (Chikla). With manganese ores.

BLEBBY

In borax bead test, descriptive term meaning full of bubbles.

BLEISCHWEIF

Synonym of *galena*.

BLENDE, BLACK JACK, SPHALE-RITE or ZINC BLENDE

ZnS Zinc sulphide, nearly always with iron as mixture of zinc and iron sulphides. Cubic. Tetragonal or dodecahedral crystals, often distorted resembling rhombo-hedra; also cleavable masses, coarse to fine granular, fibrous, concretionary. Brown, black, yellow; can be red, green, white or nearly colourless. Resinous to adamantine lustre. Brownish to light yellow and white streak. Conchoidal fracture. Perfect cleavage in one direction. Hardness 3·5–4·0. SG 3·9–4·1. Brittle. Dissolves in dilute hydrochloric acid with evolution of hydrogen sulphide.
Occurrence: Worldwide. Most common zinc mineral, usually occurring with *galena*.
Varieties are *brunckite*, *marmatite* and *cleiophane*.

BLÖDITE or BLOEDITE

$2[Na_2Mg(SO_4)_2 . 4H_2O]$ Hydrated sodium magnesium sulphate. Monoclinic. Short prismatic, often highly modified crystals; also massive, granular or compact. Colourless, sometimes bluish-green or reddish due to minute inclusions. Vitreous lustre. Conchoidal fracture. No cleavage. Brittle. Hardness 2·5–3·0. SG 2·25. Faintly saline and bitter taste. Easily soluble in cold water.
Occurrence: USA—California (Obispo County), New Mexico (Torrance County). Austria—Hallstatt, along Ischl River. Italy—Sicily. Spain—Galicia. Chile—Atacama Desert, Chuquicamata. USSR—Turkmeniya, mouth of Volga. India—Punjab. In salt deposits of oceanic origin with *halite*, *kainite*, *carnallite* and other salts.

BLOEDITE

Synonym of *blödite*.

BLOOD STONE

Synonym of *heliotrope*.

BLUE ASBESTOS

Synonym of *crocidolite*.

BLUE GROUND

Sedimentary deposit. Slaty-blue or bluish-

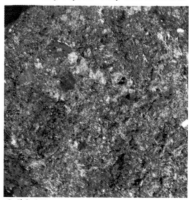

Above: blue john

green *kimberlite-breccia* of *diamond* pipes.
Occurrence: South Africa—Kimberley. Beneath a superficial oxidised covering known as *yellow ground.*

BLUE JOHN

Variety of *fluorite* containing film-like inclusions of petroleum. Banded colourless, white, blue and purple. Ornamental, used for vases or similar objects.
Occurrence: England—Derbyshire (Treak Cliff in Kinderscout district).

BLUE MUD

Sedimentary deposit. Variety of deep-sea mud. Bluish-grey. Contains calcium carbonate (from trace, up to 35 per cent), organic matter of carbonaceous nature and finely divided iron sulphide. Sometimes contains visible fragments of rock or shells. Often gives distinct smell of hydrogen sulphide from products of decomposing organic matter.
Occurrence: Worldwide.

BLUE VITRIOL

Synonym of *chalcanthite.*

BOEHMITE

$4[AlO \cdot OH]$ Basic aluminium oxide. Orthorhombic. Microscopic, lenticular crystals, usually dissociated or in pisolitic aggregates. Cleavage observable in one direction. SG 3·01–3·06.
Occurrence: Scotland—Ayrshire. USA—Georgia (Linwood–Barton district). France. Germany—Harz Mountains (Stangenrod). Hungary—Baratka. Usually distributed in *bauxite.*
Member of *lepidocrocite* family. Dimorphous with *diaspore.*

BOGHEAD COAL or TORBANITE

Sedimentary rock. A type of *coal* formed from oil-bearing algae mixed with small quantities of detrital sand. Dark brown or black. Soft. Rough, dull, nearly conchoidal fracture. Unlaminated structure with traces of bedding. Yields much oil on distillation.
Occurrence: Scotland—West Lothian

(Torbane Hill). USA—Alaska, Pennsylvania. Australia—New South Wales. France. USSR.

BOG IRON-ORE

Variety of *limonite*. A precipitate of iron carbonate in loose, porous earthy form.
Occurrence: USA—Alabama, Missouri, Ohio, Tennessee. Cuba. In swampy or low-lying ground, often impregnating and enveloping fragments of wood, leaves, mosses and other flora.

BOGUSTITE

Igneous rock of the *gabbro* clan. A mesotype, a granular plutonic rock. Contains basic *plagioclase* and *hornblende* or *augite* with interstitial *analcite* and accessory *apatite*, *magnetite* and *biotite*.
Occurrence: Poland—Silesia.

BOJITE or HORNBLENDE GABBRO

Igneous rock of the *gabbro* clan. A brownish-black to brownish-green plutonic rock. Contains *augite*, *hornblende* and *labradorite-bytownite* with *biotite* and *quartz* as most common accessory minerals.
Occurrence: USA—Minnesota (St Louis River near Duluth). Germany—Bavaria (Oberrötzdorf, Pfaffenreuth).

BONE BED

Sedimentary rock. Contains abundant vertebrate remains and thus highly phosphatic. Recognisable fish-scales or bones act as nuclei for secondary concentration of amorphous calcium phosphate to form nodules.
Occurrence: South West England (Rhaetic bone bed). Wales—Borderland (Ludlow bone bed). USA—Florida (pebble-phosphates).

BONE TURQUOISE

Synonym of *odontolite*.

BONINITE

Igneous rock of the *diorite* clan. A porphyritic extrusive rock. A *glass*-rich, nearly *feldspar*-free, *olivine-bronzite-andesite* with phenocrysts of *olivine*, *bronzite* and *augite*. Similar to *sanukite*, but contains *olivine*.
Occurrence: Japan—Bonin Islands (Peel Islands).

BONITE

Synonym of *phillipsite*.

BORACITE

$Mg_3B_7O_{13}Cl$ Magnesium borate and chloride. Cubic. Cubic or octahedral crystals; also columnar and massive, or granular. Colourless, white, yellow, greenish or greyish. Vitreous lustre. White streak. Uneven, conchoidal fracture. Brittle. Imperfect cleavage in two directions. Hardness 7·0. SG 2·95. Insoluble in water.

Soluble in hot hydrochloric acid.
Occurrence: Germany—Stassfurt.

BORAX

$Na_2B_4O_7 \cdot 10H_2O$ Hydrous sodium borate. Monoclinic. Usually large and well-formed prismatic to tabular crystals that tend to dehydrate, then crumble. White, colourless, yellowish, greyish, bluish or greenish. Vitreous lustre. White streak. Conchoidal fracture. Good cleavage in one direction. Hardness 2·0–2·5. SG 1·7. Brittle. When heated with charcoal, swells and fuses easily to clear *glass* sphere that clings to charcoal and colours the flame yellow. Water soluble. Sweetish and astringent taste.
Occurrence: USA—California, Nevada, New Mexico. USSR. India. Iran. Tibet. As single crystals or in crusts of various minerals in salt lake beds or as efflorescence on the soil in arid regions.

BORNITE, ERUBESCITE, PEACOCK ORE or PHILLIPSITE (of Beudant)

$8[Cu_5FeS_4]$ Cuprous iron sulphide. Cubic. Cubic or dodecahedral crystals with rough or curved faces; also massive, granular, compact. Copper-red or pinchbeck-brown on fresh fracture tarnishing iridescent purple on exposure to moist atmosphere. Metallic lustre. Pale greyish-black streak.

Small conchoidal, uneven fracture. Traces of cleavage in one direction. Hardness 3·0. SG 5·06–5·08. Brittle. F 2·0. Soluble in nitric acid with whitish precipitate of sulphur. Gives faint sulphur sublimate when heated in closed tube and sulphur fumes when heated in open tube.
Occurrence: England—Cornwall (Redruth). USA. Canada—British Columbia, Quebec. Australia. South Africa—Namaqualand (O'okiep), Transvaal (Messina). Austria. Germany. Italy. Chile. Peru. Madagascar. An important *copper* ore deposit.

BOROLANITE

Igneous rock of the *syenite* clan. A mesotype, crystalline granular hypabyssal rock. Contains *orthoclase, melanite* and *nepheline* or its alteration products *muscovite* and *natrolite* with accessory *biotite* and *pyroxene*. Pseudoporphyritic texture due to aggregates of *orthoclase* and *nepheline*.
Occurrence: Scotland—Sutherlandshire (Assynt). Norway—North of Tveitasen.

BORT

Variety of *diamond*. A grey to black, granular to cryptocrystalline stone. No distinct cleavage. Also badly coloured or flawed *diamond* without gem properties.

BOSTONITE

Igneous rock of the *syenite* clan. Light yellow to grey or greyish-green, aplitic hypabyssal rock characterised by rough, unoriented laths of *feldspar*. Contains alkali-*feldspar, biotite,* and *iron*-rich *amphibole* with accessory *apatite, zircon* and *iron* ores. *Quartz*-bearing varieties common.
Occurrence: USA—Massachusetts (near Boston), Vermont (Lake Champlain Valley). Australia—New South Wales (Nandewar Mountains). Brazil.

BOTRYOIDAL

With round grape-like masses, each generally consisting of radiating needles.

BOULANGERITE

$Pb_5Sb_4S_{11}$ Lead antimony sulphide. Monoclinic. Long prismatic to fibrous, deeply striated crystals; also solid feathering masses. Bluish lead-grey. Metallic lustre. Brownish-grey to brown streak. Brittle fracture. Good cleavage parallel to elongation; absent across fibres which are flexible. Hardness 2·5–3·0. SG 5·7–6·3. Brittle. Splintery. F 1·0. Decrepitates in flame, then melts to flat, bubbly mass which clings

Left: bornite

to charcoal. Soluble in hot hydrochloric acid with evolution of hydrogen sulphide.
Occurrence: USA—California, Colorado, Idaho, Montana, Nevada, Washington. Europe. Mexico. Peru.
Medium to low-temperature ore deposits, associated in veins with other lead sulphosalts: *stibnite, blende, galena, quartz* and *siderite.*

BOULDER

Rock fragment with diameter exceeding 250mm.

BOULDER CLAY

Sedimentary rock. Glacial deposit. Contains undecomposed fragments of parent rock owing to absence of chemical decomposition. Constituents vary widely depending on rock over which the ice has travelled. Usually a stiff *clay* with varying proportions of rock boulders which are angular, faceted or striated by attrition. Sometimes intercalated beds of sand and gravel.
Occurrence: England—North of imaginary line between London and Bristol. North Sea Basin. Regions bordering Arctic and Antarctic ice-caps.

BOURNONITE

$PbCuSbS_3$ Lead copper sulphantimonide. Orthorhombic. Large, intergrown crystals produce radiating effect; usually short prismatic, tabular; also massive. Black to greyish-black. Metallic adamantine lustre. Steel-grey streak. Subconchoidal to uneven fracture. Good cleavage in one direction. Brittle. Hardness 2·5–3·0. SG 5·8–5·9. F

1·0. Decomposes in nitric acid to give bluish-green solution which becomes cloudy from precipitation of sulphur and lead.
Occurrence: England—Cornwall (Liskeard). USA—Arizona, Arkansas, California, Colorado, Montana, Nevada, Utah. Canada—Ontario. Australia—New South Wales (Broken Hill), Victoria. Czechoslovakia. France. Germany. Hungary. Italy. Rumania. Spain. Bolivia. Chile. Mexico. Peru. With differing associated minerals in hydrothermal ore veins full of local cavities formed at medium temperatures.

BOWENITE

Variety of *serpentine*. Dense, felt-like aggregates of colourless *serpentine* fibres with occasional *magnesite* patches, *talc* flakes and *chromite* grains. Massive with very fine granular texture. Apple-green or translucent greenish-white. Hardness 5·5–6·0. SG 2·594–2·787. Used for carvings and jewellery.
Occurrence: USA—Pennsylvania (Delaware River), Rhode Island (Smithfield). New Zealand. Afghanistan—valley of Kabul River. China. As veins in foliated rock masses containing the same minerals as *bowenite*, but with *talc* as the dominant constituent.

BOWLINGITE

Synonym of *saponite*.

BOWRALITE

Igneous rock of the *syenite* clan. A coarse-grained holocrystalline hypabyssal rock. Contains automorphic *sanidine* tablets, subalkaline *amphibole*, usually *arfvedsonite* and some *aegirine*. *Quartz, perovskite, zircon* and *ilmenite* may be accessories.
Occurrence: Australia—New South Wales (Bowral).

BRACCIANITE

Variety of *cecilite* containing no *melilite*.
Occurrence: Italy—North West of Rome (Lake Bracciano.)

BRACHYPINACOID

Short type of pinacoid.

BRAGGITE

$8[(Pt, Pd, Ni)S]$ with $c4$ per cent of nickel and $c20$ per cent of palladium. Platinum palladium nickel sulphide. Tetragonal. Rounded grains and prisms. Steel-grey. Metallic lustre. SG 8·9–10·0.

Occurrence: South Africa—Transvaal (Potgietersrust, Rustenburg). In concentrations of platiniferous *norites* with *sperrylite* and *cooperite* (*of Wartenweiler*).

BRAMMALITE

Variety of *illite*. A sodium potassium aluminosilicate. Unlike *illite* it contains more sodium than potassium. Forms soft, fibrous incrustations of small, compact, white *tuffs*.

Occurrence: Wales—Llandebie. In *illite* crevices in *coal*-measure *shales*.

BRANDBERGITE

Igneous rock of the *granite* clan. A granophyric hypabyssal rock. Irregular outlines formed by phenocrysts of *feldspars* with *quartz* grains and *biotite* aggregates in groundmass of innumerable, small *biotite* flakes with traces of *arfvedsonite*, *zircon*, *magnesite* and *albite*.

Occurrence: South West Africa—Brandberg.

BRANDISITE

Variety of *clintonite*. Hexagonal prisms. Yellowish-green or leek-green to reddish-grey. Perfect basal cleavage. Hardness 5·0 (of base), 6·0–6·5 (of sides). SG 3·042–3·06.

Occurrence: Austria—Tyrol (Mount Monzoni in the Fassathal). In white *limestone*, either disseminated or in grouped crystals, in geodes among crystals of *fassaite* (*of Werner*) and black *spinel*, often intimately associated with *leuchtenbergite*.

BRANNERITE

UTi_2O_6 with rare earths thorium and yttrium replacing uranium, calcium replacing rare earths and ferrous iron replacing titanium. Uranium titanium calcium oxide of extremely variable composition with minor yttrium, thorium and ferrous iron. Triclinic. Rounded detrital pebbles and crystals. Black; externally brownish-yellow due to alteration. Dark greenish-brown streak. Conchoidal fracture. No cleavage. Hardness 4·5. SG 4·5–5·43. Radioactive. Decomposes in hot concentrated sulphuric acid.

Occurrence: USA—Idaho (Kelly Gulch in Western Custer County). Canada—Ontario (Elliot Lake district). South Africa—Transvaal (Witwatersrand district).

BRAUNITE (of Haidinger) or MARCELINE (of Beudant)

$8[Mn^{2+}Mn^{3+}_6SiO_{12}]$ Manganous silicate, often with appreciable amounts of ferric iron. Tetragonal. Octahedral crystals; also massive. Brownish-black. Submetallic lustre. Brownish-black streak. Uneven fracture. Brittle. Perfect cleavage in one direction. Hardness 6·0–6·5. SG 4·75–4·82. Gelatinises when boiled with hydrochloric acid. Gives reddish-violet borax bead in oxidising flame.

Occurrence: Australia—New South Wales. New Zealand—Wellington. Germany—Harz Mountains. Thuringia. Italy—Island of Elba, Piedmont. Norway—Upper Telemark. Sweden—Grythyttan, Jakobsberg, Långban, Örebro. India. Usually of secondary origin, but may be primary mineral occurring in veins transversing *porphyry*.

BRAVAISITE

Interstratification of *illite* and *montmorillonite*. Magnesium potassium aluminosilicate. Thin layers and schistose masses of fine, crystalline fibres, mostly parallel in position. Grey to greenish-grey. Hardness 1·0–2·0. SG 2·6. Unctuous to touch. Paste-like when wet. Fuses easily to white *glass*. Gives off water and becomes brown when heated in closed tube.

Occurrence: France—Allier Department (Noyant). In layers in *coal* and bituminous *schists*.

Name has been used as synonym of *illite*.

BRAVOITE

$4[(Ni, Fe)S_2]$ Iron nickel disulphide. Member of *pyrite* family. Cubic. Crusts of nodular masses with radially fibrous or columnar structure. Surfaces sometimes show individual cubic, octahedral or pyritohedral crystal faces. Steel-grey. Metallic lustre. Grey streak. Conchoidal to uneven fracture. Perfect cleavage in one direction. Hardness 5·5–6·0. SG 4·62. Brittle. Dissolves in hot nitric acid with evolution of hydrogen sulphide fumes.

Occurrence: England—Derbyshire (Mill Close Mine). USA—Alaska (Canyon Creek Valley, Copper River Valley, Spirit Mountain). Germany—Prussia (Mechernich), Siegen district (Victoria Mine near Müsen). Spain—Leon Province (near Villamanin).

BRAZILIAN EMERALD

Transparent green variety of *tourmaline*.

BRAZILIANITE (of Pough and Henderson)

$NaAl_3(PO_4)_2(OH)_4$ Basic sodium aluminium phosphate. Monoclinic. Short prismatic crystals with prism zone striated; also elongated or globular with radial, fibrous structure. Chartreuse-yellow to pale yellow. Vitreous lustre. Colourless streak. Conchoidal fracture. Good cleavage in one direction. Brittle. Hardness 5·5. SG 2·983.

Occurrence: USA—New Hampshire (Palermo Mine near North Groton in Grafton County, Smith Mine near Newport). Brazil—Minas Gerais (Conselheira Pena). In cavities in *pegmatite* with various minerals.

BRAZILIAN RUBY

Gemstone variety of *topaz*. An altered *topaz* formed by heating yellowish crystals to 450 degrees C. Crystals become colourless on heating, but develop salmon-pink to purple-red on cooling.

Occurrence: Brazil—Ouro Preto.

BRAZILIAN SAPPHIRE

A transparent, blue variety of *tourmaline*.

BRECCIA

Sedimentary rock of *rudaceous* type. Clastic rock made up of coarse, angular or sub angular fragments of varied or uniform composition. Fragments not water-worn. Edgewise breccias result from vertical packing of fragments.
Occurrence: Worldwide.

BRECCIA (Volcanic)

Pyroclastic rock. Built up of small fragments from broken blocks of lava or pyroclastic rock lying in a fine *tuff* matrix. A class of *rhyolite tuff*.
Occurrence: Worldwide. Associated with volcanic activity.

BREITHAUPTITE (of Chapman)

Synonym of *covelline*.

BREITHAUPTITE (of Frobel)

2[NiSb] Nickel antimonide. Hexagonal. Crystals rare, thin tabular or prismatic; usually aborescent and disseminated, massive. Light copper-red inclining strongly to violet on fresh fracture. Metallic, splendent lustre. Reddish-brown streak. Uneven to small conchoidal fracture. No cleavage. Hardness 5·5. SG 8·23. Brittle. F 1·5–2·0. Dissolves in nitric acid and aqua regia.
Occurrence: Canada—Ontario (Cobalt). Germany—Harz Mountains (Andreasberg). Italy—Sardinia (Mount Narbo near Sarrabus). In *calcite* veins, usually associated with *silver* minerals, *ullmanite*, *blende*, *galena* and *niccolite*.

BREUNNERITE

2[(Mg, Fe) CO$_3$] with ferrous carbonate from 5–30 per cent in series extending to *siderite*. Variety of *magnesite*. White, yellowish, brownish, rarely black and bituminous; often becoming brown on exposure. SG 3·0–3·5.
Occurrence: USA—Louisiana (Delacroix). Australia—New South Wales. Austria—Styria (Gleinalm, Ochsenkogel). Sweden—Northern Jämtland (Lejarklumpen, Säkok Ruopsok).

BREWSTERITE

(Sr, Ba)Al$_2$Si$_6$O$_{16}$.5H$_2$O Barium strontium aluminosilicate, usually with some calcium. Monoclinic. Prismatic, flattened crystals, with some faces vertically striated. White, inclining to yellow and grey. Vitreous lustre, pearly on one plane. Uneven fracture. Brittle. Perfect cleavage in one direction. Hardness 5·0. SG 2·45. Effloresces in unheated, dried air. F 3·0. Swells up in flame and fuses to white enamel.
Occurrence: Scotland—Argyllshire (Strontian). Northern Ireland—County Antrim (Giant's Causeway). France—Department of the Isère. Pyrenees (near Barèges), South West of Mont Blanc (Col du Bonhomme). Germany—Breisgau (near Freiburg).
Member of *heulandite* group of *zeolite* family.

BREWSTERLINITE

Colourless, transparent fluid which may be a lower paraffin. Forms liquid inclusions in crystals of *topaz*, *chrysoberyl*, *quartz* and *amethyst*. Mostly microscopic, but can be up to 12·5mm across. Generally layered. Thousands of inclusions in a single crystal. Volatilised in heat.

Occurrence: Scotland. Canada—Quebec. Australia. Brazil. USSR—Siberia.

BRILLIANT

Very bright.

BRITTLE

Easily shattered or broken.

BRITTLE MICAS

Minerals of the *mica* family conforming to the general *mica* formula X$_2$Y$_{4-6}$Z$_8$O$_{20}$(OH,F)$_4$, but with X=largely calcium. Group comprises *margarite*, *clintonite* and *xanthophyllite*.
Term also refers to group of minerals resembling the *micas* in form, cleavage and structure but yielding brittle laminae. Main member of the series is *chloritoid*.

BROCHANTITE

4[Cu$_4$SO$_4$(OH)$_6$] Basic copper sulphate. Monoclinic. Stout prismatic to acicular crystals, sometimes elongated or tabular in loosely coherent aggregates or groups and drusy crusts; massive, granular. Emerald-green to blackish-green; also light green. Vitreous lustre; pearly on cleavage face. Pale green streak. Uneven to conchoidal fracture. Perfect cleavage in one direction. Hardness 3·5–4·0. SG 3·97. When heated in closed tube at high temperatures yields water which has acidic properties.
Occurrence: England—Cornwall, Cumberland (Roughten Gill near Caldbeck). USA—Arizona, California, Colorado, Idaho, New Mexico, Utah. Australia—New South Wales (Broken Hill). South West Africa—Tsumeb. France. Germany. Italy. Rumania. Spain. Chile. USSR. Algeria. Congo. A secondary mineral found in oxidised zones of *copper* deposits, especially in arid regions.
Often confused with *antlerite*.

BROEGGERITE or BRÖGGERITE

4[(U,Th)O$_2$] Variety of *uraninite* containing up to 14 per cent thorium dioxide.

Occurrence: Norway—Ånnerôd Peninsula (Elvestad, Huggenaeskilen).

BRÖGGERITE

Synonym of *broeggerite*.

BROMARGYRITE

Synonym of *bromyrite*.

BROMLITE

Synonym of *alstonite*.

BROMYRITE or BROMARGYRITE

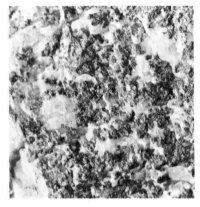

AgBr Generally mixture of silver chloride and silver bromide with proportion of bromine greater than chlorine. Silver bromide. Cubic. Cubic crystals in parallel groups; usually massive, in crusts or waxy or hornlike masses or coatings, columnar or stalactitic. Colourless when pure and fresh; usually grey or yellowish or greenish-brown. Resinous to adamantine lustre, hornlike. Tough, uneven to subconchoidal fracture. No cleavage Hardness 2·5. SG 6·474. Sectile and ductile. Very plastic. Chemical identification as for *cerargyrite*.
Occurrence: USA—Arizona (Tombstone). Australia—New South Wales (Broken Hill). France—Finistere (Huelgoet). Germany—Nassau (Dernbach). Chile. Mexico. USSR—Ukraine (Donetz Basin). Secondary mineral occurring in oxidised zone of *silver* deposits, especially in arid regions.
End member of *cerargyrite-bromyrite* series.

BRONZITE

An iron-bearing variety of *enstatite*.

BRONZITE (of Finch)

Synonym of *clintonite*.

BRONZITFELS or BRONZITITE

Igneous rock of the *ultrabasic* clan. A coarse-grained plutonic rock. Contains *bronzite*, usually altered to *serpentine*, with small amounts of *olivine*, *picotite*, *chromite*, *hornblende* and *diallage*.
Occurrence: New Caledonia. Rhodesia—Bushveld and Stillwater Complexes, Great Dyke.

BRONZITITE

Synonym of *bronzitfels*.

BROOKITE

8[TiO$_2$] Titanium dioxide. Orthorhombic. Crystals, usually tabular and elongated with striated prism faces. Yellowish-brown to iron-black. Metallic adamantine to submetallic lustre. Uncoloured to greyish and yellowish streak. Subconchoidal to uneven fracture. Brittle. Indistinct cleavage. Hardness 5·5–6·0. SG 4·14. Infusible. Chemical identification as for *rutile*.
Occurrence: England—Dartmoor, Yorkshire. Wales—near Snowdon, Tremadoc. USA—Arkansas (Magnet Cove in the Ozark Mountains). Europe—throughout Alps. Czechoslovakia—Bohemia, Moravia. Italy—Lombardy, Mount Etna, Piedmont. Brazil—Minas Gerais. USSR—Ural Mountains (Miask).
Polymorphous with *rutile* and *anatase*. Variety is *arkansite*.

BROWN COAL

Synonym of *lignite*.

BROWN SPAR

An iron-bearing variety of *dolomite* which turns brown on exposure.

BRUCITE (of Beudant)

Mg(OH)$_2$ Magnesium hydroxide. Hexagonal. Usually broad tabular crystals; commonly foliated massive or fibrous with separable and elastic fibres. White, inclining to pale green, grey or blue. Waxy to vitreous lustre, pearly on cleavages. White streak. No fracture. Perfect cleavage in one direction. Hardness 2·5. SG 2·39. Sectile. Foliae separable and flexible. Infusible. In flame glows brightly and ignited material reacts alkaline. In closed tube becomes opaque and friable, sometimes turning grey to brown.

Occurrence: Scotland—Shetland Islands (Swinna Ness). USA—California, Nevada, New Jersey, New York, Pennsylvania. Canada—Quebec (Asbestos). Austria—Styria, Tyrol. Italy—Mount Vesuvius, Sardinia (Teulada). Sweden. USSR—Ural Mountains (Slatoust region). In low-temperature hydrothermal veins in *serpentine* and chloritic or dolomitic *schists*. Varieties are *nemalite* and *ferrobrucite*.

BRUNCKITE

Pulverulent, massive variety of *blende*. White. Dull lustre. Clings to tongue.
Occurrence: Peru—West of Cerro de Pasco (Cercapuquio Mine).

BRUNSVIGITE

2[De^{2+},Mg,Al)$_6$(Si,Al)$_4$O$_{10}$(OH)$_8$] with silicon = 2·8–3·1. Variety of *chamosite* and member of *chlorite* family. Fine, scaly masses.
Occurrence: Northern Ireland—Mountains of Mourne. USA—Massachusetts (Westfield in Hampden County), Virginia (Goose Creek in Loudoun County). Germany—Harz Mountains (Radauthal).

BRUSHITE

·4[CaHPO$_4$·2H$_2$O] Hydrated calcium acid phosphate. Monoclinic. Minute, needle-like or prismatic to tabular crystals; also earthy to powdery or foliated. Colourless to pale yellow. Vitreous lustre, pearly on cleavage. No fracture. Perfect cleavage in one direction. Hardness 2·5. SG 2·328. In flame, exfoliates and fuses. Yields acid water when heated in closed tube. Readily soluble in hydrochloric acid.
Occurrence: Austria. Czechoslovakia. France—Hérault (Minerva Cave in Cevennes Mountains), near Limoges (Quercy). Germany. Caribbean and West Indies—Aves, Mona and Sombrero Islands. Algeria—near Oran. As efflorescences and cavity linings in insular and continental phosphate deposits.

BUCHITE

High grade contact metamorphic rock. A vitrified, hornfelsic rock produced from *phyllite*, *sandstones*, granitic rocks or other material by intense local heat due to contact with *basalt* magma or to the thermal effects of friction in *mylonite* crush-belts. Occurs as xenoliths, usually in *basalt* or *diabase*.

Occurrence: Worldwide. Associated with igneous intrusions.

BUCHOLZITE

Synonym of *fibrolite*.

BUCHONITE

Igneous rock of the *gabbro* clan. A blackish-grey, fine-grained porphyritic extrusive rock. A *hornblende*-bearing *labradorite-nepheline* rock with accessory *augite, analcite, apatite* and *biotite*.

Occurrence: Germany—near Fulda (Buchonia), near Peppenhausen (Calvarienberg).

BUNSENITE

4[NiO] Nickel oxide. Cubic. Octahedral crystals, sometimes with modifying dodecahedron or cube. Dark pistachio-green. Vitreous lustre. Brownish-black streak. Hardness 5·5. SG 6·898. Infusible. Dissolves with difficulty in hydrochloric acid.

Occurrence: Germany—Saxony (Johanngeorgenstadt). With native *bismuth* and nickel and cobalt arsenates in oxidised portion of nickel-uranium vein.

BUSTAMITE

6[(Mn,Ca)SiO₃] with ratio of calcium to magnesium greater than 1 : 2. Calciferous variety of *rhodonite* similar to *wollastonite*. Triclinic. Prismatic crystals; also fibrous. Greyish-red or pink, tarnishing brown on cleavage surface. Perfect cleavage in one direction. SG *c*3·4.

Occurrence: England—Cornwall (Altarnun). USA—New Jersey (Franklin Furnace). Sweden—Långban. With *rhodonite* and *tephroite*.

BYTOWNITE

8[(NaSi,CaAl)AlSi₂O₈] *Plagioclase feldspar* containing 10–30 per cent molar proportion of *albite*. Triclinic. Usually massive. Grey, dark grey, bluish; rare crystals of clear pale yellow. Perfect cleavage parallel to basal pinacoid. SG 2·72. Other properties resemble *labradorite*.

Occurrence: USA—Oregon. France—Corsica. Germany—Harz Mountains. Iceland. Poland—Silesia (Neurode). Mexico. As primary constituent of basic and *ultra-*

basic igneous rocks.

BYTOWNITFELS or BYTOWNITITE

Igneous rock of the *gabbro* clan. An *anorthosite*, a leucocratic, coarse-grained plutonic rock.

Occurrence: Norway.

BYTOWNITITE

Synonym of *bytownitfels*.

CABOCHON

A gem cut in convex form, highly polished but not faceted.

CACOXENITE

Fe³⁺₄(PO₄)₃(OH)₃ · 12H₂O Basic hydrated ferric iron phosphate. Hexagonal. Tiny, acicular crystals in tufted or radial aggregates or as fibrous coating. Yellow to brownish, reddish or golden-yellow; rarely green. Silky lustre. No fracture. No cleavage. Hardness 3·0–4·0. SG 2·2–2·4. Turns brown in flame and fuses to black magnetic bead. Easily soluble in acids.

Occurrence: Czechoslovakia—Bohemia. France—Morbihan (Rochefort-en-Terre). Germany—Rhine district (Eleonore Mine, Rothläufchen Mine). Sweden—Kirunava-

ara. A secondary mineral occurring with *dufrenite, strengite, beraunite, wavellite* and *limonite*.

CAHNITE

Ca₂AsO₄BO₂ · 2H₂O Basic calcium borate-arsenate. Tetragonal. Twinned crystals with well-developed faces individually interpenetrating symmetrically. Colourless to white. Vitreous lustre. No fracture. Perfect cleavage in one direction. Hardness 3·0. SG 3·56. Brittle. F 3·0. Easily soluble in dilute hydrochloric acid. When heated in closed tube, yields water and becomes opaque, but does not fuse.

Occurrence: USA—New Jersey (Franklin Furnace). In cavities in *axinite* veins or with *rhodonite* or on *garnet* crystals lining drusy cavities in *franklinite* ore.

CAIRNGORM STONE

Variety of *quartz*. Gemstone. Smoky, yellow or brown colour, often transparent, varying to brownish-black and nearly opaque in thick crystals. Colour probably result of radiation bombardment from surrounding rocks.

Occurrence: Scotland—Cairngorm Mountains.

CALAMINE

British synonym of *smithsonite*.

CALAVERITE

2[(AuTe)₂] Gold ditelluride; sometimes argentiferous with ratio of silver to gold = 6 : 1. Monoclinic. Bladed crystals, lathlike or short to slender prisms. Strongly striated faces; also massive, granular to indistinct crystalline. Brass-yellow to silver-white. Metallic lustre. Yellowish to greenish-grey streak. Subconchoidal to uneven fracture. No cleavage. Hardness 2·5–3·0. SG 9·24. Very brittle. F 1·0. Decomposes in hot concentrated sulphuric acid with separation of gold and formation of red solution.

Occurrence: USA—California (Calaveras County, El Dorado County), Colorado

(Boulder County, Hinsdale County, Teller County). Canada—Ontario (Boston Creek, Kirkland Lake). Western Australia—Kalgoorlie, Mulgabbie. Philippine Islands—Mountain Province (Antamok). In low-temperature veins associated with *pyrite* and other sulphides and often free *gold*.

CALCAREOUS

Rich in calcium carbonate.

CALCAREOUS SINTER, CALCAREOUS TUFA or TRAVERTINE

Variety of *calcite*. A cavernous and irregularly banded structure. Consists of cellular deposits of calcium carbonate with variable content of fossil leaves, twigs, moss and other organic matter, encrusted with a hard coating of *tufa*.
Occurrence: England—Derbyshire (Matlock), Yorkshire (Knaresborough). France—Auvergne. Italy—Tivoli, Tuscany.

CALCAREOUS TUFA

Synonym of *calcareous sinter*.

CALC-FLINTA

High grade metamorphic rock. Derived from calcareous *mudstone*. Fine-grained texture. Contains *feldspars* and rock-silicate minerals. Flinty, *calc-silicate horn-*

fels formed by action of certain heated gases on the rocks they penetrate along cracks and fissures.
Occurrence: England—Cornwall (Bodmin and Camelford areas).

CALCICLASITE, CALCICLASITITE, ANORTHITFELS or ANORTHITITE

Igneous rock of the *gabbro* clan. A leucocratic granular plutonic rock. Contains largely *anorthite*.
Occurrence: USA and Canada—Great Lakes area.

CALCICLASITITE

Synonym of *calciclasite*.

CALCICRETE

Synonym of *calcrete*.

CALCIKERSANTITE

Igneous rock of the *gabbro* clan. A hypabyssal rock, a variety of *kersantite*. Differs from *kersantite* in that *plagioclase* is *bytownite* or *labradorite*, usually containing considerable *augite*.
Occurrence: France—Kersanton, Markirch.

CALCIO-ANCYLITE

$(Sr,Ca)_3Ce_4(CO_3)_7(OH)_4 \cdot 3H_2O$ with pro-

portion of calcium greater than strontium. Variety of *ancylite* with calcium in excess of strontium. Same physical and chemical properties as *ancylite*.
Occurrence: Western USSR. In granitic boulders.

CALCIOCELSIAN

$4|(Ba,Ca)Al_2Si_2O_8|$ with $c4$ per cent calcium oxide. Variety of *celsian*. Scarcely observable cleavage.
Occurrence: Australia—New South Wales (Broken Hill). With *plagioclase*, but is virtually indistinguishable except for higher barium content than surrounding mineral. Member of *feldspar* family.

CALCITE, CALCSPAR, ICELAND SPAR or VATERITE

$CaCO_3$ Calcium carbonate. Hexagonal. Crystals extremely varied from tabular to needle-like; scalenohedrons and rhombohe-

impalpable. Colourless, white or pale tints. Vitreous lustre. White to greyish streak. Conchoidal fracture. Rhombohedral cleavage. Brittle. Hardness 3·0 on cleavage, but varying on different surfaces. SG 2·7. Dissolves with effervescence in cold dilute hydrochloric acid. Easily scratched.

Below left: dogtooth calcite
Below: phantom calcite

drons most common; microcrystalline to coarse: massive from coarse granular to

Colours brick-red in flame. Frequently fluorescent.
Occurrence: Worldwide in all types of rocks and environments.
Varieties are *agaric mineral*, *anthraconite*, *aphrite*, *satin spar*, *calcareous sinter*, *lublinite*, *rock meal*, *thinolite*, *verd antique* and *pisolite*.

CALCRETE or CALCICRETE

Conglomerate formed by cementation of superficial gravels with calcium carbonate.
Occurrence: Widespread in arid regions, including South West USA, Mexico and North East Africa.

CALC-SILICATE HORNFELS

Medium to high grade contact metamorphic rock. Texture variable, but generally fine-grained. Derived from *marls* and other calcareous sediments and therefore contains a great variety of minerals, mostly calc-silicates.
Occurrence: Worldwide.

CALCSPAR

Synonym of *calcite*.

CALDERITE (of Fermor)

$Mn_3Fe_2^{3+}Si_3O_{12}$. Cubic. Massive, dark yellowish to red. A manganese *garnet*.
Occurrence: South West Africa—Otjosondu.

CALDERITE (of Piddington)

Variety of *andradite*.
Occurrence: India—Central Province.

CALEDONITE

$2[Cu_2Pb_5(SO_4)_3CO_3(OH)_6]$. Basic lead copper carbonate-sulphate. Orthorhombic. Crystals usually small, striated, elongated, in divergent groups; occasionally as coatings; rarely massive. Deep verdigris-green or bluish-green. Resinous lustre. Greenish-white streak. Uneven fracture. Brittle. Perfect cleavage in one direction. Hardness 2·5–3·0. SG 5·76. Dissolves with effervescence in nitric acid leaving deposit of lead sulphate.
Occurrence: England—Cumberland (Red

Gill near Caldbeck) Scotland—Dumfriesshire, Lanarkshire. USA—Arizona, California, Idaho, Montana, New Mexico, Utah. Canada—British Columbia (Beaver Mountain). South West Africa—Tsumeb. France—Department of Rhône. Italy—Sardinia. USSR—Ural Mountains (Beresovsk). Secondary mineral found widely in small amounts in the oxidised zone of *copper*-lead deposits.

CALICHE

Sedimentary rock. Consists of *alluvium* cemented together with sodium nitrate, chloride and other soluble salts. Owing to recrystallisation, high-grade saline layers nearly free from debris are sometimes associated with normal type of deposit. Term can be used to describe deposits of *calcrete*.
Occurrence: Chile. Peru.

CALIFORNITE

Variety of *idocrase*. Gemstone with *jade*-like characteristics. Massive. Green. SG 3·25–3·35.
Occurrence: USA—California (Butte County, Fresno County, Siskiyou County).

CALOMEL

$2[Hg_2Cl_2]$ Mercuric chloride. Tetragonal. Crystal form variable, often tabular and prismatic; also massive, earthy. Colourless, white, greyish and yellowish-white, yellowish to ash-grey or brown. Adamantine lustre. Pale yellowish-white streak. Conchoidal fracture. Good cleavage in one direction. Hardness 1·5. SG 7·15. Plastic. Sectile. Blackens when treated with alkalis. When heated in closed tube, volatilises without fusion and condenses as white sublimate.
Occurrence: USA—Arkansas (Pike County), California (San Mateo County), Oregon (Denio district), Texas (Brewster County). France—Herault (Montpellier). Spain—Castile (Ciudad Real Province). Yugoslavia—Servia (Avala). Mexico—Queretaro (near Zimapan).

Secondary mineral formed from alteration of mercury-containing minerals.

CALTONITE

Igneous rock of the *gabbro* clan. Bluish-black, compact or ophitic extrusive rock. Contains *olivine*, *analcite*, *pyroxene* (usually *augite*), optional accessory *magnetite*, *apatite* and *palagonite* and abundant *plagioclase*, either *labradorite* or *bytow*-
Occurrence: England—Derbyshire (Calton Hill). USSR—Siberia (Tsheskaja Bay).

CAMBRIAN

See *geological time*

CAMPANITE

Igneous rock of the *syenite* clan. A microlitic extrusive rock. A type of *leucite-tephrite*. Contains large *leucite* crystals.
Occurrence: France—Somme.

CAMPTONITE

Igneous rock of the *diorite* clan. A dark grey to black, holocrystalline-porphyritic to saccharoidal hypabyssal rock. Contains phenocrysts of *biotite* and *hornblende* in a groundmass of *feldspars*, *amphibole* and *pyroxene*.
Occurrence: USA—New Hampshire (Campton Falls). Norway—Kristiania (Kjose station).

CAMPTOSPESSARTITE

Igneous rock of the *diorite* clan. A melanocratic, medium-grained equigranular hypabyssal rock. Contains *titanaugite*.
Occurrence: Germany—Saxony (Golenz near Bautzen).

CAMPYLITE

$2[Pb_5((As,P)O_4)_3Cl]$ Lead chloroarsenatephosphate. Variety of *mimetite*. Small, barrel-shaped crystals of *pyromorphite-mimetite*. Brown or yellowish.

Occurrence: England—Cumberland (Dry Gill near Caldbeck). France—le Rozier, Villevielle.

CANADITE

Igneous rock of the *syenite* clan. A plutonic rock, a variety of *litchfieldtite*. Contains pure *albite*, soda-*microcline*, *nepheline*, *amphibole* and *biotite* with accessory *cancrinite* (0–17 per cent), *zircon* and *apatite*.
Occurrence: Sweden—Almunge district (Byske).

CANCRINITE

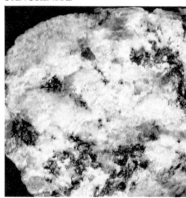

$c4[NaAlSiO_4].CaCO_3.H_2O]$ Hydrated sodium aluminium calcium silicate with carbon dioxide. Hexagonal. Massive. Commonly yellow, but also white and red. Subvitreous or slightly pearly or greasy lustre. Colourless streak. Uneven fracture. Brittle. Perfect prismatic cleavage. Hardness 5·0–6·0. SG 2·4–2·5. F 2·0. Effervesces with hydrochloric acid and forms a jelly on heating. Yields water when heated in closed tube.
Occurrence: USA—Maine (Litchfield, West Gardiner). Norway—Langesund Fiord. Rumania—Transylvania. USSR—Ural Mountains (Miask). As constituent of igneous rocks of the *nepheline-syenite* type. Member of *feldspathoid* family.
Variety is *microsommite*.

CANNEL COAL

Sedimentary rock. A dull, lustreless variety of *coal*. Consists of much-altered plant material and a large proportion of volatile constituents. Massive, unlaminated forma-tion with glassy and conchoidal fracture. Water transported and deposited as organic sediments or accumulations in stagnant ponds. Burns with bright, smoky flame.
Occurrence: Worldwide.

CANTALITE

Igneous rock of the *granite* clan. An extrusive rock. Transparent variety of *pitchstone* or *rhyolite-pitchstone*. Sometimes referred to as yellow *quartz*.
Occurrence: France—Cantal.

CAPE RUBY

Synonym of *pyrope*.

CAPILLARY

See *crystal*.

CAPILLITITE

$2[Mn,Zn,Fe)CO_3]$ Manganese zinc ferrous iron carbonate. Variety of *rhodochrosite*. Identical in form to *rhodochrosite*, but with yellowish to grey colour.
Occurrence: Argentina — Catamarca (Capillitas). Associated with *rhodochrosite*.

CAPPELENITE

Yttrium barium borosilicate with small amounts of calcium, potassium, sodium and thorium. Formula uncertain. Hexagonal. Thick, prismatic crystals. Greenish-brown. Vitreous to greasy lustre. Brownish streak. Conchoidal fracture. Brittle. No cleavage. Hardness 6·0–6·5. SG 4·407. Easily soluble in dilute hydrochloric acid. Swells up in flame and fuses with some difficulty to white enamel.
Occurrence: Norway—Langesund Fiord (Lille Arö). In small vein in *augite-syenite*.

CARBONACEOUS

Having a high carbon content.

CARBONACEOUS CHONDRITE

Type of *meteorite*.

CARBONADO

Variety of *diamond*. Black or greyish-black *bort*. A massive, sometimes granular to compact, structure.

CARBONATE

—CO_3^{2-} Radical from carbonic acid.

CARBONATE-APATITE or DAHLLITE

Member of *apatite* series. Conforms to general formula for *apatite* with basic carbonate substituting for the anion. Formula approaches $Ca_{10}(PO_4)_6(CO_3).H_2O$ in extreme cases. Hexagonal. SG 2·9–3·1. Dissolves with effervescence in dilute hydrochloric acid.
Occurrence: USA—Arkansas (Magnet Cove in the Ozark Mountains), Maine, New Hampshire (Palermo County), Wyoming (Beartooth Mountains). France—Mouillac. Greenland—Kangerdluarsuk. Norway—Oedegaarden. USSR—Kursk, Podolia (Ushitsa River).

CARBONATE SCAPOLITE

Member of *scapolite* series with carbonate ion as sole anion; contains considerable sodium. Tetragonal. Massive. Clear, colourless, yellowish or greenish. Prismatic cleavage.
Occurrence: Austria—Steiermark (Koralpe, Schwanberg). Finland—Pargas. At contact of *pegmatite* and *marble*.

CARBONATE VISHNEVITE

Member of the solid solution series *cancrinite*. Contains 20–50 per cent *cancrinite*. SG c2·4.
Occurrence: Scotland—Loch Assynt. Finland—Kuusama (Iivaara). USSR—Ilmen Mountains. In *nepheline-syenites* and ultra-alkaline plutonic rocks.

CARBONIFEROUS

See *geological time*.

CARBONISATION

Decomposition of animal or (more often) plant remains to pure carbon as a form of fossil.

CARBUNCLE

Term describing *almandine* gemstone *garnet* cut in hollow cabochon form. Old name for all fiery-red precious stones.

CARINTHINE, or BARROISITE or KARINTHINE

$2[Ca, Na)_{2.5}(Mg, Fe^{2+}, Al)_5(Si, Al)_8O_{22}(OH)_2]$ with silicon c7, magnesium c3·5, ferrous c0·5 and ratio of calcium to sodium c2:1. Dark green variety of *hornblende* with properties intermediate between *hornblende* and *glaucophane*.
Occurrence: USA—New Jersey (Franklin). Austria—Sau Alpe (Carinthia). Norway—Arendal. Sweden—Nordmark.

CARMELOÏTE

Igneous rock of the *gabbro* clan. A porphyritic extrusive rock. A type of *olivine-basalt* with *olivine* phenocrysts altered to *iddingsite*.

CARNALLITE

12|KMgCl$_3$.6H$_2$O| Hydrated potassium magnesium chloride. Orthorhombic. Forms pseudohexagonal, pyramidal crystals; usually massive, granular. Colourless to milk-white, often reddish with metallic schiller. Greasy, dull to shining lustre. Conchoidal fracture. No cleavage. Hardness 2·5. SG 1·602. Deliquescent in moist atmosphere. Bitter taste. Fuses easily. Soluble in water.
Occurrence: USA—New Mexico, Texas, Utah (Grand County). Germany—Stassfurt, Upper Silesia. Spain—Barcelona, Galicia, Lerida. USSR—Saratov. Ethiopia—Danakil Plain. Tunisia. Associated with *sylvite*, *halite*, *polyhalite* and other salts in upper layers of saline deposits of marine type.

CARNELIAN or CORNELIAN

Variety of *chalcedony*. Transparent. Yellowish-red, flesh-red or deep clear red

colour resulting from *iron* impurities in lattice. Ornamental use.
Occurrence: Worldwide, especially Scotland, USA, Germany, South America and deserts of Middle East.

CARNOTITE (of Friedel and Cumenge)

2|K$_2$ (UO$_2$)$_2$ (VO$_4$)$_2$.3H$_2$O| Hydrated potassium uranium vanadate; water is zeolitic. Orthorhombic. Forms powdery or minute crystal plates. Canary-yellow. Dull or earthy lustre; pearly or silky when coarsely crystalline. Powdery and crumbling fracture. Distinct cleavage parallel to basal pinacoid. Soft. Sectile. SG 4·1. Infusible. Easily acid-soluble and residue from evaporation of acid fluorescent. Cold borax bead fluoresces green.
Occurrence: USA—Arizona, Colorado, Nevada, New Mexico, Pennsylvania, Utah. Canada—Ontario (Elliot Lake). South Australia—Radium Hill. Mexico. Congo—Katanga. As impregnation or lenses in Jurassic *sandstone* or *pitchblende* deposits. Sometimes deposits considered as carnotite are really series of uranium minerals which can only be distinguished from pure carnotite by X-ray studies. Calcium equivalent is *tyuyamunite*.

CARPHOLITE

MnAl$_2$Si$_2$O$_6$(OH)$_4$ Hydrated manganese aluminium silicate with traces of fluorine. Monoclinic. Radiated and stellated tufts and groups of acicular crystals. Pure straw-yellow to wax-yellow. Silky, glistening lustre. Hardness 5·0–5·5. Very brittle. SG 2·935. F 3·5. In flame swells up and fuses to brown *glass*. Gives acidic water when heated in closed tube that attacks *glass* (free fluorine).
Occurrence: France—Ardennes (near Meuville), Beaujolais. Germany—Harz Mountains (Wippra). In tufts on *granite* with *quartz* and *fluorite*.

CARPHOSIDERITE

((H$_2$O)Fe$_3$(SO$_4$)$_2$((OH)$_5$H$_2$O)) Basic hydrated ferric iron sulphate. Hexagonal. Earthy or minutely scaly crusts and aggre-

Above: carpholite
Right: cassiterite

gates; reniform. Golden-yellow or dark straw-yellow. Dull to resinous or glistening lustre. No fracture. Scarcely observable cleavage. Hardness 4·0–4·5. SG 2·496–2·905. Insoluble in water. Dissolves in hydrochloric acid.
Occurrence: USA—Utah (Eureka Hill Mine in Tintic district). Cyprus—near Kynussa. Finland—Otravaara region. Greenland—Upernivik district. As incrustations in localities where *pyrite* is weathered mixed with *quartz*, *limonite* and *jarosites*.
Member of *alunite group*.

CARROLLITE

8|CuCo$_2$S$_4$| Member of *linnaeite* family with *copper* substituting for cobalt in the lattice. SG 4·83. Other properties as *linnaeite*.
Occurrence: USA—Maryland (Carroll County). Germany—Westphalia (Siegen district). Sweden—Kalmar.

CARYOCERITE

Calcium thorium rare earths silicate-borate, tantalate and fluoride. Formula uncertain but near that of *melanocerite* with more thorium. Orthorhombic. Tabular, rhombohedral crystals with brilliant but striated faces. Nut-brown. Vitreous to greasy lustre. Brown streak. Conchoidal fracture. Brittle. No cleavage. Hardness 5·0–6·0. SG 4·295.
Occurrence: Norway—Langesund Fiord (Store-Arö).

CARYOPILITE

Mn$_4$Si$_3$O$_{10}$.3H$_2$O Hydrated manganese

silicate containing 5 per cent magnesium oxide. Massive. Stalactitic and reniform shapes, compact within; outer forms show concentric radiate-fibrous structure. Brown fracture. Hardness 3·0–3·5. SG 2·83–2·91. Easily soluble in strong acids.
Occurrence: Sweden—Wermland (near Pajsberg at Harstig Mine).
Possibly isostructural with *bementite* (*of Lévy*).

CASCADITE

Igneous rock of the *syenite* clan. A melanocratic, porphyritic hypabyssal rock. Contains phenocrysts of *biotite* in a nearly aphanitic groundmass of *pyroxene*, soda-*orthoclase* and accessory *magnetite*, *apatite* and *olivine*.

Occurrence: USA—Montana (Cascade Creek).

CASSITERITE or STANNITE (of Breithaupt)

$2|SnO_2|$ Tin dioxide. Tetragonal. Short or long prismatic crystals; also massive, in radially fibrous crusts or concretionary masses; also granular, coarse to fine. Yellowish or reddish-brown to brownish-black; occasionally red, yellow, grey or white. Adamantine to metallic adamantine lustre; usually splendent. Subconchoidal to

uneven fracture. Imperfect cleavage in one direction; indistinct in others. Hardness 6·0–7·0. SG 6·99. Brittle. Infusible. Fragments placed in dilute hydrochloric acid with metallic zinc become coated with dull grey deposit of metallic tin.
Occurrence: Worldwide. Most important tin ore occurring in high-temperature hydrothermal veins or metasomatic deposits. Genetically closely associated with highly siliceous igneous rocks.
Varieties are *toad's eye tin* and *wood tin*.

CASTOR or CASTORITE

Variety of *petalite* occurring as distinct, attached, transparent crystals. SG 2·386.
Occurrence: Italy—Island of Elba.

CASTORITE

Synonym of *castor*.

CATACLASITE

Low to high grade dislocation metamorphic rock. Formed by ruptural deformation of brittle parent rocks without recrystallisation or chemical reaction. Extreme cataclasis with accompanying decrease in grain size and development of banded structure leads to gradation into *mylonites*.
Occurrence: Worldwide.

CATACLASTIC

High grade dynamic metamorphic rock. Brittle minerals within rock have been broken and flattened at right angles to the pressure stress, giving rock a characteristic granular appearance.
Occurrence: Worldwide.

CATAPLEIITE

$2[Na_2,Ca)ZrSi_3O_9.2H_2O]$ Hydrated zirconium calcium sodium silicate with calcium replacing sodium in the lattice. Monolinic, passing to hexagonal at higher temperatures. Thin, tabular hexagonal prisms. Light yellow to yellowish-brown, greyish-blue or violet. Dull lustre, weakly vitreous on fracture surface. Pale yellow

streak. Conchoidal fracture. Brittle. Perfect cleavage in one direction. Hardness 6·0. SG 2·8. F 3·0. Yields water when heated in closed tube. Easily soluble in dilute hydrochloric acid.
Occurrence: Denmark—Arö Islands. Greenland (low-temperature orthorhombic form). Norway—Langesund Fiord (Låven Island).

CATAWBERITE

High grade contact metamorphic rock. Contains *talc* and *magnetite*.
Occurrence: USA—South Carolina (along Catawba River and Lake Catawba).

CAT GOLD

Synonym of *mica*.

CATLINITE

Sedimentary rock. Red-coloured, indurated variety of siliceous clay. Used by Dakota Indians to make sacred pipes.
Occurrence: USA—Minnesota.

CATOPHORITE

$Na_2CaFe_4^{2+}$ $(Fe^{3+},Al)Si_7AlO_{22}(OH.F)_2$ Basic aluminium calcium sodium and iron silicate with magnesium as possible accessory. A soda-*iron amphibole* between *arfvedsonite* and *barkevikite* in optical properties. Monoclinic. Rose-red, dark red-

dish-brown, yellow or bluish-black in *iron*-rich varieties. Perfect cleavage in one direction. Hardness 5·0. SG 3·2–3·5.
Occurrence: USA—Idaho (Saw-Tooth Mountains), Montana (Shields River Basin), Texas (near Fort Davis). Norway—Oslo region (Sande Cauldron). Sweden—Island of Alnö. USSR—Ukraine (Mariupol). Kenya—around Lake Naivasha. Malawi—around Lake Malawi. Sierra Leone—Songo. In *ultrabasic* igneous rocks such as *theralite* and *shonkinite*.

CAT'S EYE

A crystal penetrated by a series of microscopic tubes or fibres running in a direction parallel to the vertical axis of the crystal. Cabochon stones cut with canals or needles running parallel to the base show a single ray of light crossing the gem. Term generally restricted to *chrysoberyl* cat's eye, but *quartz, amphibole, beryl, tourmaline, scapolite, apatite, diopside, kornerupine, kyanite, prehnite* and *orthoclase* exhibit chatoyancy.

CAT SILVER

Synonym of *mica*.

CATTIERITE

$4[CoS_2]$ Cobalt disulphide. Cubic. Granular. Pinkish. Cubic cleavage.

Occurrence: Congo—Shinkolobwe. In drill cores with *dolomite* and *pyrite*. Member of *pyrite* family.

CAVALORITE

Igneous rock of the *diorite* clan. A granular plutonic rock. Contains *orthoclase* in excess of *oligoclase*, abundant *hornblende* and accessory *quartz, apatite, magnetite* and *ilmenite*.
Occurrence: Italy—near Bologna (Monte Cavaloro in the Reno Valley).

CAVE PEARLS

Concretions of calcium carbonate which may be *calcite* or *aragonite*. Pearly lustre. Reasonably soft with strong cleavage. Golden-yellow variety has been faceted as collector's stone.
Occurrence: USA—Idaho (especially near Eagle Rock). Czechoslovakia—Bohemia (Karlsbad). Formed by agency of water in *limestone* caves.

CEBOLLITE

$Ca_5Al_2Si_3O_{12}(OH)_4$ Hydrated calcium aluminium silicate. Possibly orthorhombic. Compact, fibrous aggregates. White to greenish or brownish. Dull lustre. Hardness 5·0. SG 2·96. F. 5·0. Fuses with extreme difficulty to clear glass. Dissolves with gelatinisation in hydrochloric acid.
Occurrence: USA—Colorado (Beaver Creek and Cebolla Creek in Gunnison County).
An alteration product of *melilite*.

CECILITE

Igneous rock of the *gabbro* clan. Dark grey, dense texture extrusive rock. Phenocrysts rate. Contains nearly 50 per cent *leucite* with *augite, melilite, nepheline, olivine, anorthite, magnetite* and *apatite*.
Occurrence: Italy—common in regions associated with volcanic activity.
Variety of *braccianite*.

CELADONITE

$cKMg_3Fe_3Si_9O_{25}(OH)_2.9H_2O$ Hydrated iron potassium magnesium silicate. Monoclinic. Earthy or in minute scales. Deep olive-green, celandine-green or apple-green. Very soft. SG 2·8–2·9. Greasy to touch.
Occurrence: Scotland—Sgurr Mhor, Tay Bridge, Tayport. Northern Ireland—County Antrim (Giant's Causeway). USA—Lake Superior region. Austria—Tyrol (Fassathal). Italy—near Verona (Monte Baldo).
A decomposition product of *hornblende* and *augite*.
A *mica* near *glauconite*.

CELESTINE or CELESTITE

$4[SrSO_4]$ Strontium sulphate with barium

Above and below: celestine

Above: celestine

replacing strontium in series extending to *barytes*. Orthorhombic. Usually elongated, tabular crystals, commonly striated; also fibrous masses or granular massive; also rounded lenticular crystals or aggregates. Colourless to pale blue; sometimes white, reddish, greenish or brownish. Blue colour often unequally distributed in growth zones which distinguishes mineral from *barytes*. Vitreous lustre, inclining to pearly on cleavages. White streak. Uneven fracture. Perfect cleavage in one direction. Hardness 3·3–5·0. SG 3·97. F 3·0. Frequently decrepitates in flame and fuses to white pearl. Slowly soluble in alkali carbonate solutions.

Occurrence: Widespread throughout Bri-

tain, North America (especially USA) and Europe (especially Mediterranean region) Chiefly in sedimentary rocks or as primary mineral in hydrothermal veins.

CELESTITE

Synonym of *celestine*.

CELSIAN

4[BaAl$_2$Si$_2$O$_8$] Barium aluminium silicate. Monoclinic. Stout, short-prismatic or slender acicular crystals; usually massive. Usually colourless; sometimes reddish-brown or grey from included impurities. Somewhat greasy lustre on cleavage surface; streak and fracture as for *orthoclase*. Perfect cleavage in one direction. Hardness 6·0–6·5. SG 3.37. Chemical identification as for *orthoclase* except gives yellowish-green colour in flame.

Occurrence: Wales—Caernarvonshire (Benallt Mine). Australia—New South Wales (Broken Hill). South West Africa—Otjosondu. Italy—Piedmont (Candoglia). Sweden — Jakobsberg. Japan — Kaso Mine.

Uncommon member of the *feldspar* family. Varieties are *calciocelsian* and *kasoite*.

CEMENTATION

Process of binding together of sedimentary particles by some secondary material, such as that released from suspension or solution in percolating water. Important component of diagenesis.

CEMENTSTONES

Sedimentary rock. A marly deposit with lime, silica and alumina mixed in such proportions that it can be directly used for manufacture of hydraulic or Portland cement. Very hard. Occurs as nodular masses distributed irregularly throughout a thicker deposit of less calcareous *clay* or *shale* where it has crystallised without causing any increase in original volume of sediment.

Occurrence: England—between Dorset and Yorkshire (throughout Lower Lias).

CERARGYRITE, HORN SILVER or KERARGYRITE

Ag(Cl.Br) with proportion of chlorine greater than bromine. Silver chloride. Cubic Massive. Colourless when pure and fresh, developing greenish-grey tint with increasing substitution of bromine. Becomes violet-brown or purple on exposure to light. White or grey and shiny streak. Other properties as *bromyrite*. Fuses without decomposition in closed tube. When placed on strip of zinc and moistened with a drop of water, swells up, turns black and is entirely reduced to metallic *silver*.

Occurrence: England. USA—Arizona, California, Colorado, Idaho, Nevada, New Mexico. Australia—New South Wales (Broken Hill). Czechoslovakia. France. Germany. Italy. Spain. Argentina. Bolivia. Chile. Mexico. Peru. USSR—Altai Mountains. Secondary mineral found in oxidised zone of *silver* deposits, especially in arid regions.

End member of *cerargyrite-bromyrite* series.

CERITE

$c(Ce.Ca)_2Si(O.OH)_5$ Silicate of cerium group metals with calcium; aluminium or

iron may be present, but are not essential. Orthorhombic. Crystals rare, highly modified, short prismatic; usually granular massive. Between clove-brown and cherry-red, passing into grey. Dull adamantine or resinous lustre. Greyish-white streak. Splintery fracture. Brittle. No cleavage. Hardness 5·5. SG 4·86. Borax bead forms yellow globule in oxidising flame and becomes almost colourless on cooling. Gelatinises in hydrochloric acid.

Occurrence: Sweden—Westmanland (near Riddarhyttan).

CERUSSITE or BLACK LEAD

4[$PbCO_3$] Lead carbonate. Orthorhombic. Modified prismatic or twinned crystals, often in cruciform or radiate arrangements; granular massive, compact; sometimes stalactitic. Colourless to white and grey or smoky; sometimes tinged dark grey, black or bluish-green from included organic and inorganic impurities. Adamantine lustre, inclining to vitreous, resinous or pearly. Colourless to white streak. Conchoidal fracture. Very brittle. Distinct cleavage in two directions. Hardness 3·0–3·5. SG 6·55. Soluble with effervescence in dilute nitric acid. Gives brilliant yellow encrustation when heated on charcoal with sulphur and potassium iodide.
Occurrence: Britain. USA. Canada—British Columbia. Australia—New South Wales (Broken Hill). New Caledonia. Europe. South West Africa—Tsumeb. Rhodesia—Broken Hill. USSR—Transbaikalia, Ural Mountains. Congo—Mindouli. Secondary lead mineral found in upper oxidised portion of ore deposits with mainly *anglesite* and *limonite*.

CERVANTITE or STIBICONITE

$Sb^{3+}Sb^{4+}_2(O,OH,H_2O)_7$ Antimony oxide. Orthorhombic. Acicular crystals or powdery or fibrous crusts. White or canary-yellow. Greasy or pearly lustre; also white or earthy. Light yellow to white streak. Hardness 4·0–5·0. SG 6·64.
Occurrence: England—Cornwall. USA—California, Idaho, New Mexico, Utah. Australia—Tasmania, Victoria. New Zealand—Waikare. France—Isère, Puy-de-Dôme. Germany—Bavaria, Prussia, Westphalia. Italy—Tuscany. Spain—Galicia. Bolivia. Borneo. Mexico. Peru. Algeria.
Formed from oxidation of primary *antimony* ores.

CESAROLITE

$PbMn_3O_7 . H_2O$ Hydrous lead manganese oxide. Friable masses resembling coke and botryoidal crusts. Steel-grey. Dull to submetallic lustre. Hardness 4·5. SG 5·29. Infusible. Soluble in hydrochloric acid with evolution of chlorine.

Occurrence: Tunisia—Tunis (Sidi-amor-ben-Salem). In cavities in *galena* deposits.

CEYLONITE or PLEONASTE

8[$(Mg,Fe)Al_2O_4$] with ratio of magnesium to iron c3:1. Iron rich variety of *spinel*. Very dark green, almost black. Rarely used as gemstone as too dark.
Occurrence: Sri Lahka (gem gravels).

CHABAZITE

$(Ca,Na,K)_7 Al_{12} (Al,Si)_2 Si_{26}O_{80} . 40H_2O$ Hydrous calcium sodium potassium and aluminium silicate. Hexagonal. Rhombohedral, cube-like, lenticular crystals; also massive, compact. Colourless, white, pink, yellowish, brownish. Vitreous lustre. White streak. Uneven fracture. Brittle. Rhombohedral cleavage, not conspicuous. Hardness 4·0–5·0. SG 2·1–2·2. Decomposes in hydrochloric acid with separation of shiny silica. Often fluorescent blue after heating with charcoal, especially area in contact with charcoal.
Occurrence: Scotland—Isle of Skye, Renfrewshire. Northern Ireland—County Antrim (Giant's Causeway). USA—Connecticut, Massachusetts, New Jersey, New York. Canada—Nova Scotia (Bay of Fundy). Australia—Victoria (Richmond). Czechoslovakia—Bohemia (Aussig). France—Haute Saône, Plombiers. Germany—Annarode. Greenland. Iceland. India—near Bombay (Poona). In basic igneous rocks with *analcite*, *stilbite*, *harmotome*, *laumontite* and *heulandite* and occasionally in *gneiss* or *schists*.
Member of *zeolite* family. Varieties are *phacolite*, *seebachite* and *haydenite*.

CHALCANTHITE

2[$CuSO_4 . 5H_2O$] Hydrous copper sulphate. Triclinic. Short prismatic crystals or cross-fibre grainlets; also stalactitic or reniform; also granular massive. Sky-blue; sometimes a little greenish. Vitreous lustre. Colourless streak. Conchoidal fracture. Imperfect cleavage in one direction. Hardness 2·5. SG 2·286. Loses water on heating but does not fuse.
Occurrence: England—Cornwall. Ireland—County Wicklow. USA—Arizona, California, Colorado, Montana, Nevada, North Carolina, Pennsylvania, Tennessee. Cyprus. Czechoslovakia—Bohemia. Germany—Harz Mountains. Italy—Mount Vesuvius. Spain—Rio Tinto. Chile—especially Chuquicamata. Secondary mineral found in arid climates or in rapidly oxidising sulphide deposits.

CHALCEDONY, MYRICKITE or QUARTZINE

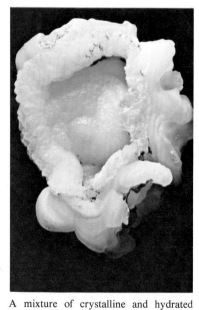

A mixture of crystalline and hydrated silicas. Crypto-crystalline structure. Colour very variable. Rather waxy lustre.
Occurrence: Worldwide. As nodules in *limestone* and other sedimentary rocks or in cavities in amygdaloidal rocks. Varieties are *agate, banded agate, carnelian, chryso-*

prase, dendritic agate, flint, moss agate, onyx, sapphrinine, sard and *sardonyx.*

Above: rainbow chalcedony

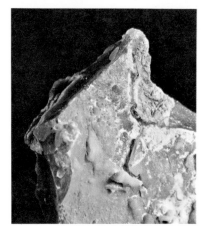

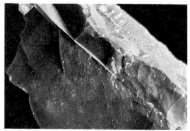

CHALCHIHUITL

Green *jadeite* of the ancient Aztecs. Also synonym of *chalchuite.*

CHALCHUITE or CHALCHIHUITL

Variety of *turquoise*. A bluish-green gemstone.
Occurrence: Argentina—Los Cerillos, Sante Fé.

CHALCOCITE

96|Cu₂S| Cuprous sulphide with small amounts of *silver* or *iron*. Orthorhombic. Short prismatic to thick tabular crystals; massive compact, impalpable. Blackish lead-grey. Metallic lustre. Blackish lead-grey streak. Conchoidal fracture. Indistinct cleavage in one direction. Hardness 2·5–3·0. SG 5·5–5·8. Fairly brittle. Imperfectly sectile. F 2·0–2·5. After heating on charcoal, globule of *copper* in reducing flame.
Occurrence: England—Cornwall (Cambourne, Redruth, St. Ives, St. Just). USA—Alaska, Arizona, Connecticut, Montana, Nevada, New Mexico, Tennes-

see, Utah. South West Africa—Tsumeb. Czechoslovakia. Germany. Italy—Tuscany. Rumania. Chile. Mexico. Peru. USSR—Ural Mountains. Congo—Mindouli. Principally as supergene mineral in enriched zone of sulphide deposits.

CHALCOMENITE

4|CuSeO₃ . 2H₂O| Copper selenite dihydrate. Orthorhombic. Minute acicular crystals with prism faces often striated or rounded. Bright blue. Transparent. Vitreous lustre. No fracture. No cleavage. Hardness 2·0–2·5. SG 3·35. Easily fusible. In closed tube yields water and sublimate of selenium dioxide.
Occurrence: Argentina—La Rioja, Mendoza. Bolivia—near Colquechaca (Hiaco Mine). With *azurite* and *cerussite* as oxidation product of primary selenides of *copper* in veins.

CHALCOPHANITE

ZnMn₃O₇ . 3H₂O Hydrated zinc manganese oxide. Probably hexagonal. Minute tabular crystals with octahedral habit; commonly drusy botryoidal or stalactitic crusts; also massive, dense, granular, platy or fibrous. Bluish to iron-black. Metallic lustre. Chocolate-brown, dull streak. No fracture. Perfect cleavage in one direction. Hardness 2·5. SG 4·0. Flexible when as thin plates. Soluble in hydrochloric acid with the evolution of chlorine. Yields oxygen and water when heated in closed tube, exfoliates slowly and changes to golden-brown colour.
Occurrence: USA—Colorada (Wolftone Mine at Leadville), New Jersey (Sterling Hill). Australia—Tasmania (Dundas). Poland—Olkusch. Israel. Secondary mineral found associated with hydrous oxides of manganese and *iron*.

CHALCOPHYLLITE

|Cu₁₈Al₂(AsO₄)₃(SO₄)₃(OH)₂₇ . 33H₂O| Basic hydrated copper aluminium sulphate-arsenate. Hexagonal. Tabular, six-sided crystals of rhombohedral aspect; also foliated massive, in druses or as rosette-like

aggregates. Emerald-green or grass-green to bluish-green. Transparent to translucent. Vitreous to subadamantine lustre, becoming pearly on partial dehydration. Streak paler green than colour. Perfect cleavage in one direction. Hardness 2·0. SG 2·67 in fully hydrated material, decreasing with loss of water. When heated in closed tube, yields much water and leaves residue of olive-green isotropic flakes.
Occurrence: England—Cornwall (Liskeard, Redruth, St Day). USA—Arizona (Bisbee), Nevada (Majuba Hill in Mineral County), Utah (Tintic district). Austria—Tyrol. France—Var. Germany—Saxony. Hungary—Herrengrund, Moldava. USSR—Ural Mountains. Secondary mineral found in oxidised zone of *copper* deposits.

CHALCOPYRITE or COPPER PYRITES

4|CuFeS₂| Copper iron sulphide. Tetra-

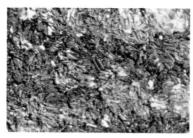

gonal. Crystals common as bisphenoids resembling tetrahedra; also massive, compact. Brass-yellow or golden-yellow, frequently with iridescent tarnish. Metallic lustre. Greenish-black streak. Uneven fracture. Brittle. Indistinct cleavage. Hardness 3·5–4·0. SG 4·1–4·3. Soluble in nitric acid with separation of sulphur. When heated in closed tube, decrepitates and gives sublimate of *sulphur*.
Occurrence: Worldwide. Usually in mesothermal or hypothermal deposits.
Most widely occurring *copper* mineral.

CHALCOSIDERITE

$Cu(Fe,Al)_6(PO_4)_4(OH)_8 \cdot 4H_2O$ Basic hydrated copper ferric iron and aluminium phosphate with proportion of ferric iron greater than aluminium. Member of *turquoise* group. Triclinic. Crusts or sheaf-like groups of distinct short-prismatic crystals. Light siskin-green. Transparent. Vitreous lustre. White or greenish to pale green streak. Perfect cleavage in one direction. Hardness 4·5. SG 3·22. Chemical identification as for *turquoise*.
Occurrence: England—Cornwall (Wheal Phoenix). USA—Arizona (Bisbee). Ger-

many—Saxony (Schneckenstein), Westphalia (Siegen). With *dufrenite*, *andrewsite* and *goethite* as secondary mineral in gossan.

CHALCOSTIBNITE

$4|CuSbS_2|$ Copper antimony sulphide. Orthorhombic. Blade-shaped crystals flattened in one direction, sometimes elongated and striated; also massive. Colour between lead and iron-grey, sometimes with blue or green alteration coating. Metallic lustre. Grey streak. Subconchoidal fracture. Brittle. Perfect cleavage in one direction. Hardness 3·0–4·0. SG 4·95. F 1·0. When heated in closed tube, decrepitates then fuses. Decomposes in nitric acid with separation of sulphur and antimony trioxide.
Occurrence: Germany—Harz Mountains (Wolfsberg). Spain—Granada. Bolivia. Morocco—East of Casablanca. Lining cavities or veins, usually with *pyrite* or other associated minerals.

CHALCOTRICHITE

Variety of *cuprite*. Forms delicate, straight, interlacing, fibrous, cochineal-red crystals. Carmine-red streak.
Occurrence: USA—Arizona (Bisbee, Jerome, Morenci).

CHALK

Sedimentary rock. A type of *limestone*. Contains *c*90 per cent calcium carbonate, almost entirely as *calcite*. Derived from floating micro-organisms and some bottom-dwelling forms in a matrix of finely crystalline *calcite*, with detrital *quartz* and *glauconite* as main accessories. Fine-grained and somewhat friable to gritty texture. Deposited from solution under quiet conditions with no wave action. White to

light grey. White streak. Earthy fracture. No cleavage. Brittle. Hardness 1·0–2·5. SG 1·7–2·7.
Occurrence: Worldwide, forming thick, extensive beds.

CHALYBITE or SIDERITE

$Fe\,CO_3$ Iron carbonate with nearly 50 per cent iron and a little manganese, magnesium and calcium often present. Hexagonal-trigonal. Rhombohedral crystals; also

Above: chalybite

massive and granular. Pale yellowish or buff-brownish, and brownish-black or brownish-red. Pearly or vitreous lustre. White streak. Uneven fracture. Brittle. Perfect rhombohedral cleavage. Hardness 3·5–4·5. SG 3·7–3·9. When heated in blow pipe, blackens and becomes magnetic. Effervesces in hot hydrochloric acid.
Occurrence: British coalfields. USA—Eastern coalfields. In *coal* measures.
Varieties are *oligon spar*, *pistomesite* and *sideroplesite*.

CHAMOSITE

$2[(Fe^{2+},Fe^{3+},Mg,Al)_6(Si,Al)_4O_{10}(O,OH)_8]$ with silicon = 2·8–3·1 and ferric iron and

hydroxide = 8. Hydrous magnesium iron aluminium silicate. Monoclinic. Compact

or oölitic. Greenish-grey to black. Streak lighter than colour. Hardness *c*3·0. SG 3·0–3·4. Feebly attracted by magnet. Easily fuses. Gelatinises in dilute hydrochloric acid.
Occurrence: Czechoslovakia—Bohemia (Chrustenic). France—Vosges (Banwald). Germany—Thüringer Wald (Schmiederfeld). Switzerland—Valais (Chamoson near St Maurice). Important constituent of many oölitic *iron* ores.
Member of *chlorite group.* Variety is *berthierine* and *brunsvigite*.

CHAPMANITE

$Fe_5SB_2Si_5O_{20} . 2H_2O$ Hydrous ferrous iron silicate antimonate. Probably orthorhombic. Pulverulent. Yellowish-green to olive-

green. SG 3·58. Easily dissolves in hydrofluoric acid but insoluble in all other acids.
Occurrence: Canada—Ontario (Keeley Mine at South Lorrain). With native *silver* in oxidised part of *silver*-bearing vein.

CHARNOCKITE

High grade contact metamorphic rock. Mineral assemblage approximates *granite* clan of igneous rocks. A dark, granular plutonic rock. Contains mainly *hypersthene*, with *microcline-microperthite*, *quartz* and *iron* ores.
Occurrence: India—Madras (rock from St Thomas's Mount used for tomb of Job Charnock, founder of Calcutta), plateau of Jeypore Zemindari. Gabon—Como Basin. Western Ghana.
Member of *charnockite series*.

CHARNOCKITE Series

High grade contact metamorphic rocks. Coarse-grained to granular, granoblastic to foliated rocks. Contain *hypersthene* with varying proportions of *quartz*, potash-*feldspars*, *plagioclase* and *garnet*.
Occurrence: USA—Adirondack Mountains. Australia—Musgrave Range. South Africa—Natal. Finland and Sweden—Fennoscandian Shield. Brazil. USSR—Crystalline Shield. Central Sahara. India. Sri Lanka. Congo. Madagascar. Uganda.

CHATOYANCY

Ability to shine like a cat's eyes; having a changeable lustre or colour with an undulating narrow band of white light.

CHELEUTITE or KERSTENITE

$Co(As,Bi)_2$ with proportion of bismuth smaller than five per cent. Cobalt arsenide bismuthinide. Variety of *smaltite*. Twisted crystal masses. Cubical cleavage.
Occurrence: Germany—Saxony (Freiberg, Schneeberg).

CHERALITE

$2[(CaTh,Ce_2)(PO_4)_2]$ with some silica replacing phosphorous and calcium replacing cerium group of rare earths. Calcium thorium and cerium group of rare earths phosphate. Monoclinic. Brittle masses up to 50mm across. Dark to pale green. Resinous to vitreous lustre. White streak. Uneven fracture. Hardness 5·0. SG 5·3. Highly radioactive.
Occurrence: Southern India—Travancore. Distributed throughout kaolinised *pegmatite* dike outcropping or in gravel of adjacent streams with *tourmaline*, *chrysoberyl*, *zircon* and considerable *smoky quartz*. Isomorphous with *huttonite* and *monazite*.

CHERT

Sedimentary rock. Grey to black, homogeneous, with flat fracture. Consists of dense aggregates of microcrystalline *quartz* and chalcedonic silica mixed in any proportion with or without remains of siliceous and other organisms such as *radiolarites*, *diatomites* and sponge spicules.
Occurrence: Worldwide. Notable example is carboniferous *limestone* of North Wales. As nodules and beds in *limestone* formations.
Variety is *novaculite*.

CHERT IRONSTONE

Sedimentary rock. Ferruginous deposits of *chert*. Consists of alternate layers of *chert* and *siderite* or *chert* and *hematite*; sometimes ellipsoidal granules of *greenalite* embedded in *chert* matrix with some carbonate minerals. Markedly banded structure. *Iron*-rich layers deposited from colloidal solutions in rivers.
Occurrence: USA—Lake Superior region. South Africa. Scandinavia. Brazil. India.

CHESSYLITE

Synonym of *azurite*.

CHIASTOLITE or MACLE

Variety of *andalusite*. Stout crystals have angles and axis of different colour to rest

owing to rectangular arrangement of carbonaceous impurities through interior. A coloured cross is exhibited in the transverse section. Hardness 3·0–7·5, varying with degree of impurity. Has ornamental use, particularly as amulet.
Occurrence: England—Cornwall, Cumberland. USA—California (Fresno County, Mariposa County). USSR—Ural Mountains. In metamorphic rocks, often as rounded and knotty projections with *kyanite*, *sillimanite*, *garnet* and *tourmaline*.

CHIASTOLITE-SLATE

High grade contact metamorphic rock. A carbonaceous *shale* formation. Contains conspicuous crystals of *chiastolite* in generally cryptocrystalline groundmass.
Occurrence: England—Cornwall. Cumberland, Devonshire (Dartmoor).

CHIBINITE

Igneous rock of the *syenite* clan. A coarse-grained, hypautomorphic-granular plutonic rock. Contains mainly *microcline-microperthite*, *sodaclase*, *nepheline*, *pyroxene* and *eudialyte*.
Occurrence: USSR—Umptek (Chibinä).

CHIKLITE

$2[(Na, Ca, Mn)_3 (Mn, Fe^{2+}, Fe^{3+}, Al)_5 (Si,$

Al)$_8$O$_{22}$ (OH)$_2$| Hydrated sodium calcium manganese iron and aluminium silicate. A clino-*amphibole* high in manganese. Deep violet. Pearly lustre on cleavage faces. Pale violet streak. Prismatic cleavage.

Occurrence: India—Bhandara district (Chikla). In *pegmatite* cutting manganese ore-band with manganese-bearing minerals, *quartz* and *feldspars*.

CHILDRENITE

8[(Fe,Mn)AlPO$_4$(OH)$_2$. H$_2$O] with proportion of ferrous greater than manganese. Basic hydrated aluminium phosphate with divalent iron and manganese mutually substituting. Orthorhombic. Equant or pyramidal to short prismatic and thick tabular crystals; also platy. Brown and yellowish-brown. Vitreous to somewhat resinous lustre. White streak. Subconchoidal to uneven fracture. Poor cleavage in one direction. Hardness 5·0. SG 3·25. Swells up in flame and fuses to black magnetic bead. Yields water when heated in closed tube.

Occurrence: England—Cornwall (Crinnis Mine at St Austell), Devonshire (Tavistock). Germany—Saxony (Greifenstein near Ehrenfriedersdorf).

Isomorphous with *eosphorite*.

CHINA CLAY

Synonym of *kaolinite*.

CHINA STONE

Sedimentary rock. A semi-decomposed, granitic rock. Usually kaolinised. Firm, does not crumble readily. Used in manufacture of china. Sometimes contains white *mica* and *fluorite*.

Occurrence: Widespread, especially England (Cornwall), USA, France, Malaya, China.

CHINGLUSUITE

Na$_4$Mn$_5$Ti$_3$Si$_{14}$O$_{41}$.9H$_2$O Hydrated titanosilicate of manganese, sodium and accessory, non-essential ferric oxide, zirconium dioxide and rare earths. Amorphous grains. Black. Hardness 2·0–3·0. SG 2·151. Easily fuses to dark *glass*. Soluble in cold hydrochloric acid, hot nitric acid and with difficulty in hot sulphuric acid.

Occurrence: USSR—Kola Peninsular (Lovozero tundra along Chinglusuai River). In *pegmatite* veins intruding into *sodalite-syenite*.

CHKALOVITE

Na$_2$Be(SiO$_3$)$_2$ Sodium beryllium silicate. Orthorhombic. Large grains. White, semi-transparent. Vitreous lustre. Cleavage fair in one direction. Hardness 6·0. SG 2·662. Readily fuses to clear bead. Soluble in hydrochloric and nitric acids.

Occurrence: USSR—Kola Peninsular (Punkarnayv Mountain on Lovozero tundra). Occurs sparingly with mainly *ussingite*, *sodalite*, *eudialyte* and *natrolite* in *foyaite-pegmatite*.

CHLADNITE

A very pure non-ferriferous variety of *enstatite* occurring in most achondritic *meteorites*.

CHLOANTHITE

8[NiAs$_2$] Nickel arsenide. Cubic. Cubic crystals; also massive. Tin white. Metallic lustre. Greyish black streak. Poor cleavage. Hardness 5·5–6. SG 6·4–6·7.

Occurrence: USA—New Jersey (Trotter Mine in Franklin), New Mexico (Rose Mine in Grant County). Czechoslovakia—Bohemia (Andreasberg). Germany—Hessen (Bieber, Riechelsdorf). Hungary—Dobschau. Switzerland—Anniversthal.

Alteration product is *annabergite*.

CHLOR-APATITE

Ca$_5$(PO$_4$)$_3$(F,Cl,OH) with proportion of chlorine greater than that of fluorine or of hydroxide. Member of *apatite* series. SG 3·1–3·2.

Occurrence: Norway—between Lillesard and Bamle (especially Oedegaarden). Rare mineral found chiefly in veins in gabbroic rocks and in some *meteorites*.

CHLORITE

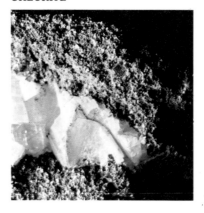

c(Mg,Fe)$_5$Al(AlSi$_3$)O$_{10}$(OH)$_8$ Hydrous aluminium iron magnesium silicate. General description of all allied minerals since properties very similar. Monoclinic. Tabular crystals; usually granular masses, disseminated scales and foliae, frequently encrustations. Various shades of green. Rather pearly lustre, subtranslucent to opaque. Pale green streak. Scaly, earthy fracture. Perfect cleavage parallel to basal pinacoid. Flexible, but not elastic flakes. Hardness 1·5–2·5. Scratchable with fingernail. SG 2·65–2·94. Feels greasy when granular or in scales. Gives water when heated in closed tube. Whitens in flame, but fuses only with great difficulty.

Occurrence: Detailed under localities for each group member.

Commonly secondary mineral of igneous rocks as alteration product of *mica* or *amphibole*. Abundant mineral of metamorphic origin in rocks of igneous or sedimentary parentage.

CHLORITE Group

Hydrous ferrous iron magnesium aluminosilicate; may have ferric iron, chromium or manganese. Monoclinic or pseudohexagonal. Green. Very soft with hardness *c* 2·0. Silicate sheet structure as the *micas*. Basal cleavage in crystallised forms. Tough and inelastic laminae. Series comprises *clinochlore, penninite, ripidolite, corundophilite, amesite, daphnite, cronstedtite, thuringite, stilpnomelane, diabantite, aphrosiderite, delessite* and *vermiculite*.

Often occurs as secondary mineral from alteration of mainly *pyroxene, amphibole, biotite* and *garnet*.

CHLORITOID or STRÜVERITE

4[(Mg,Fe^{2+})$_2$Al$_4$Si$_2$O$_{10}$(OH)$_4$] Hydrated magnesium aluminium ferrous iron silicate. Monoclinic. Crystals rare, tabular and hexagonal in outline; usually coarsely foliated massive, thin scales or small plates. Dark grey, greenish-grey, greenish-black, greyish-black or grass-green. Pearly lustre on cleavage face. Colourless, greyish or very slightly greenish streak. Scaly fracture. Basal cleavage less perfect than *micas*. Hardness 6·5. SG 3·52–3·57. Brittle laminae. Completely decomposes in sulphuric acid. In flame, nearly infusible, darkens and becomes magnetic.

Occurrence: Scotland—Shetland Islands

(Vanlup). USA—Massachusetts (Chester), Rhode Island, Virgina (Patrick County). Canada-Quebec (Brome County, Megantic County). France—Ardennes, Ile de Croix, Morbihan. Italy—Piedmont (St Marcel), Turkey—Gumuch-Dagh. USSR—Ural Mountains (Kosoibrod). A *brittle mica*. Variety is *ottrelite*.

CHLOROCALCITE

$KCaCl_3$ Potassium calcium chloride. Probably orthorhombic. Cube-like crystals with octahedral or dodecahedral-like truncations; also prismatic. White, sometimes stained violet. Transparent. Cube-like cleavage, with one direction better than other two. Hardness 2·5–3·0. Deliquescent. Bitter taste.
Occurrence: Germany—Prussia (Desdemona Mine in Leinetal). Italy—Mount Vesuvius.

CHLOROMELANITE

Variety of *jadeite* with considerable amounts of iron oxides. Dark green to black.
Occurrence: France—Dordogne, Morbihan. Switzerland. Mexico.

CHLOROPAL

Synonym of *nontronite*.

CHLOROPHAEITE

$c(Ca, Mg, Fe^{2+})_2 Fe^{3+}_2 Si_4 O_{13} . lOH_2O$ Hydrated calcium magnesium ferric and ferrous iron silicate. Amorphous. Granular massive. Dark green, olive-green changing rapidly to brown or black on exposure. Subresinous, rather dull lustre. Brownish streak. Conchoidal fracture. Cleavage in two directions. Hardness 1·5–2·0. SG 2·02.
Occurrence: Scotland—Argyllshire, Fifeshire, Isle of Rhum (Scuir Mohr), Midlothian (Kaimes Hills near Edinburgh), Northern Ireland—County Antrim (Giant's Causeway), Londonderry (Downhill). Republic of Ireland—Galway. Embedded or as coatings in geodes, fissures or amygdaloidal cavities. Possibly member of *chlorite group*. Composition near *delessite*.

CHLOROPHANE

Variety of *fluorite*. Emits bright green fluorescent light on heating.
Occurrence: With *fluorite*.

CHLOROPITE

Synonym of *delessite*.

CHLOROSPINEL

$8[Mg(Al, Fe)_2 O_4]$ with ratio of aluminium to ferric iron from 3 : 1. Iron-rich variety of *spinel*. Gemstone.
Occurrence: USSR—Ural Mountains (Slatoust).

CHONDRODITE

$2[Mg_5 Si_2 O_8 (F, OH)_2]$ with ratio of fluoride to hydroxide c 1 : 1. Basic magnesium fluorsilicate; may have some ferrous iron. Monoclinic. Flattened, often vicinal crystals; massive, compact; as embedded grains. Light to dark yellow, honey-yellow, deep garnet-red, brownish-red or hyacinthred. Vitreous lustre. Subconchoidal fracture. Cleavage sometimes distinct in one direction. Hardness 6·0–6·5. SG 3·1–3·2. Brittle. Gelatinises with acids. Infusible.
Occurrence: USA—New York (Brewster). Finland—Pargas. Italy—Monte Somma. Poland—Silesia (Strehlen). Sweden—Nya-Kopparberg (Kafveltorp). In *limestone* masses, sometimes embedded in *garnet*.

CHRISTOPHITE

Synonym of *marmatite*.

CHROME-AUGITE

Synonym of *lavrovite*.

CHROME-DIOPSIDE

$4[MgCaSi_2 O_6]$ with small amounts of chromate. Variety of *diopside*. A *clinopyroxene*. Bright green.
Occurrence: South Africa—Cape Province (Kimberley), Orange Free State (Jagersfontein), Czechoslovakia—Křemže. Germany—Kaiserstuhl, Nassau (Schwartz Stein). Burma. Arctic Ocean—Jan Mayen Island.

CHROME-MICA

Synonym of *fuchsite*.

CHROME-PHENGITE

Synonym of *mariposite*.

CHROME-SPINEL

Synonym of *picotite*. Sometimes synonym of *magnesiochromite*.

CHROME-TREMOLITE

$2[Ca_2 Mg_5 Si_8 O_{22} (OH)_2]$ with up to 2 per cent chromate. Variety of *tremolite*. A clino-*amphibole*. SG 3·0.
Occurrence: Finland—Outokumpu. Sierra Leone—Dilma. In *quartz* or *pegmatite* associated with such chrome-rich minerals as *chrome-diopside*, *fuchsite* and *uvarovite*.

CHROMITE

$8[FeCr_2 O_4]$ Iron chromate, usually with some aluminium and magnesium and thus

grading to *magnesiochromite*. Cubic. Crystals rare, octahedral; commonly massive, fine granular to compact. Black. Metallic lustre. Brown streak. Uneven fracture. No cleavage. Hardness 5·5. SG 4·5. Brittle. Insoluble in acids. When heated with sodium carbonate on charcoal is reduced to magnetic oxide.

Occurrence: USA—California, Maryland, North Carolina, Oregon, Pennsylvania, Texas, Wyoming. Canada—Quebec. Australia—New South Wales, New Caledonia. Rhodesia—Great Dyke. Bulgaria—Rhodope Foothills. France—Var. Norway—Ramberget (near Island of Hestmandö). Turkey—Antioch, Izmir. Yugoslavia—Uskub. Cuba. India. Philippine Islands. Principal ore of chromium occurring as accessory mineral in most *peridotites* and related rocks.

Member of *spinel* family. Varieties are *beresofite* (*of Simpson*), *alumoberesovite*, *alumochromite* and *magnesiochromite*.

CHROMITITE

Igneous rock of the *ultrabasic* clan. A dark-coloured, medium-grained holocrystalline automorphic hypabyssal rock. Essentially *chromite* with up to 5 per cent *biotite* or *pyribole*.

Occurrence: USA—Maryland. South Africa—Transvaal. Turkey.

CHROMOHERCYNITE

$8[Fe(Al,Cr)_2O_4]$ with ratio of aluminium to chromium at least 3 : 1. Variety of *hercynite* with chromium replacing aluminium. An isomorphous mixture of *chromite* and *hercynite*. Granular masses. Black. Shining, vitreous lustre. SG 4·415.

Occurrence: Madagascar—between Farafagana and Vangaindrano. With blocks of concretionary *limonite* and *magnetite* in red earth of gneissic region.

CHRYSOBERYL (of Werner), CYMOPHANE or HELIODOR

$4[BeAl_2O_4]$ Beryllium aluminate. Orthorhombic. Prismatic or tabular crystals; sometimes six-rayed, spokelike aggregates.

Green, greenish-yellow or brown. Transparent. Vitreous lustre. Colourless streak. Conchoidal fracture. Weak cleavage in three directions. Hardness 8·5. SG 3·71–3·72. Brittle. Not attacked by acids. Decomposed by fusion with potassium hydroxide. Gemstone.

Occurrence: Northern Ireland—Mourne Mountains. USA—Colorado, Connecticut, Maine, New Hampshire, New York. Canada—Quebec. Western Australia. Rhodesia—South of Gwelo. Finland—Helsinki. Italy—Lake Como, Veltlin. Norway—Saetersdalen. Switzerland—near St Gotthard. Brazil. USSR—Ural Mountains. Burma. Japan. Congo—Katanga. Ghana.

Variety is *alexandrite*.

CHRYSOCOLLA

$cCuSiO_3 . 2H_2O$ Hydrous copper silicate. Amorphous. Botryoidal and massive, often opal-like or enamel-like in texture; sometimes earthy. Green, bluish-green to turquoise or sky-blue; brown to black when impure. Vitreous, shining, earthy lustre. White streak when pure. Conchoidal fracture. Hardness 2·0–4·0. SG 2·0–2·238. Sectile. Translucent variety brittle. Blackens and yields water when heated in closed tube. Some varieties cut and polished for jewellery.

Occurrence: England—Cornwall (Lizard). USA—Arizona, Connecticut, Michigan, New Jersey, Pennsylvania, Utah, Wisconsin. Canada—Nova Scotia. Southern Australia. Austria—Tyrol. Germany. Hungary—Libethen. Italy—Mount Etna. USSR—Siberia. Chile. Encrusting or filling seams associated with other *copper* ores.

Name used in ancient times for mixture of *borax*, *chrysocolla* and *malachite* used for soldering *gold*.

Variety is *beaumontite* (*of Jackson*).

CHRYSOLITE (of Wallerius) or PERIDOT

$4[(Mg,Fe)_2SiO_4]$ An intermediate member of the *forsterite-fayalite* series containing 10–30 per cent *fayalite*. Orthorhombic. Striated, vertically flattened prismatic crystals. Oily bottle-green, often with brown tinge; rarely brown. Distinct cleavage in one direction. Hardness 6·5. Gemstone.

Occurrence: USA—Arizona, New Mexico. Australia—Queensland. South Africa—diamond fields of Cape Province and Orange Free State. Norway—Söndmöre. Brazil—Minas Gerais. Red Sea—Island of St John. Upper Burma. Congo.

A member of the *olivine* family. Variety is *hyalosiderite*.

CHRYSOPRASE

Gemstone variety of *chalcedony*. Apple-green colour is due to presence of nickel oxide.

Occurrence: USA—California (Tulare County), Oregon (near Nickel Mountain). Australia—Queensland (Rockhampton). Poland—Silesia (Frankenstein). Brazil—Goias. USSR—Ural Mountains.

CHRYSOTILE, α-CHRYSOTILE or CHRYSOTILE ASBESTOS

$2[Mg_3Si_2O_5(OH)_4]$ Hydrated magnesium

Above and below: chrysoprase
Right: chrysotile

silicate. Monoclinic. Delicately fibrous; fibres usually flexible and easily separating.

Greenish-white, green, olive-green, yellow or brownish. Silky or silky metallic lustre.

White streak. Fibrous fracture. Fibrous cleavage. Hardness 2·55. SG 2·219. Chemical identification as for *serpentine*.
Occurrence: USA—Arizona, New Jersey (Morris County), Pennsylvania (Chester County, Lancaster County). Canada—Ontario (Lanark County), Quebec (Gaspé Peninsula, Thetford area). South Africa—Transvaal. Coats seams in *serpentine*. Member of *serpentine* group. Varieties are *marmolite, retinalite, ishkyldite*.

δ-CHRYSOTILE

Synonym of *ishkyldite*.

CHRYSOTILE ASBESTOS

Synonym of *chrysotile*.

CIMINITE

Igneous rock of the *syenite* clan. A medium grey, compact and somewhat porphyritic, extrusive rock. Contains phenocrysts of *feldspar, olivine* and *augite* in an aphanitic groundmass of *orthoclase, labradorite, augite* and *olivine* and *magnetite* grains.
Occurrence: Italy—Viterbo region (Cimino volcano, Fontana Fiescoli).

CINNABAR

3[HgS] Mercuric sulphide. Hexagonal. Rhombohedral, thick tabular or stout to slender prismatic crystals; also crystalline incrustations or earthy coatings; granular, massive. Cochineal-red, often inclining to brownish-red and lead-grey. Adamantine lustre, inclining from metallic when dark-coloured to dull in friable varieties. Scarlet

streak. Subconchoidal, uneven fracture. Perfect cleavage in one direction. Hardness 2·0–2·5. SG 8·09. Volatile. In closed tube gives black sublimate of mercuric sulphide; with sodium carbonate in closed tube gives sublimate of mercury.
Occurrence: USA—Arkansas, California, Nevada, Oregon, Texas, Utah. Germany—Bavaria (Moschellandsberg). Italy—Gorizia, Tuscany. Spain—Ciudad Real (Almaden). Yugoslavia—near Belgrade. Mexico. Peru. Surinam. USSR—Ferghana, Ukraine. China. Mongolia. In veins and impregnations near recent volcanic rocks and hot-spring deposits.

CINNAMON STONE

Synonym of *hessonite*.

CIPOLINO

Metamorphic rock. A *marble* rich in silicate materials with layers rich in *micaceous* minerals. White with pale greenish shadings from green *talc*. Easily weathered.
Occurrence: Greece—Euboea Island. Italy.

CITRINE or FALSE TOPAZ

Variety of *quartz* with traces of ferric iron. Light golden-yellow to reddish-yellow. No cleavage. Gemstone, often artificially produced by heat treatment of *amethyst*.
Occurrence: Brazil.

CLARAIN

Sedimentary rock. A component of *bituminous coal*. Occurs in *bituminous coal* as bands of varying thickness consisting of very thin alternating bright and dull laminae. Satin-like lustre. Smooth, shining surface when broken at right angles to bedding plane. Consists of great variety of disintegrated plant substances in colloidal matrix formed from decomposition products of wood and soft tissues.

CLARKEITE

(Na₂,Ca)U₂O₇,nH₂O usually with some

potassium and/or lead. Hydrous uranium oxide with alkalis, lead and alkali earths. Water non-essential. Crystal system uncertain. Massive, dense. Dark reddish-brown to dark brown. Slightly waxy lustre. Yellowish-brown streak. Conchoidal fracture. Hardness 4·0–4·5. SG 6·39.
Occurrence: USA—North Carolina (Spruce Pine in Mitchell County). Alteration product of *uraninite* usually found between central core of *uraninite* and surface *gummite*. Ill-defined system needing further study.

CLASTIC

Built up from the products of weathering and/or erosion.

CLASTIC ROCKS

Sedimentary rocks composed of fragmental material derived from pre-existing rocks or from dispersed consolidation products of magmas. Detritus transported mechanically into place of deposition. Sub-divided according to grain size. Rock types include *breccia, conglomerate, sandstone* and *shale*.

CLAUDETITE

4[As₂O₃] Arsenic trioxide. Monoclinic. Crystals form elongated thin plates resembling *gypsum*. Colourless to white. Transparent. Vitreous to pearly lustre on cleavage surfaces. Fibrous fracture in one direction. Very flexible cleavage in one direction. Hardness 2·5. SG 4·15. Sublimes when heated in closed tube, condensing above in small octahedra of *arsenolite*. Soluble in hot solutions of dilute alkali.
Occurrence: USA—Arizona (Jerome County), California (Trinity County), Montana (Butte). France—Aveyron. Hungary—Schmölnitz. Portugal—San Domingo Mines. Spain—Andalusia. Secondary mineral from oxidation of other *arsenic* minerals found associated with *arsenic* minerals or native *sulphur*.

CLAUSTHALITE

4[PbSe] Lead selenide. Cubic. Massive, commonly fine granular; sometimes foliated. Lead-grey, somewhat bluish. Metallic lustre. Darker streak than colour. Granular fracture. Good cleavage in one direction. Hardness 2·5–3·0. SG 7·8. F 2·0. Soluble in nitric acid. Decrepitates when heated in closed tube.
Occurrence: Germany—Harz Mountains, Saxony. Spain—Huelva Province (Rio Tinto Mines). Sweden—Fahlun. Argentina—on Mendoza River (Cerro de Cacheuta). China—Kweichow.

Right: citrine

CLAY

Sedimentary rock. An *argillaceous rock* composed of fine-grained clastic sediments. Consists of minute, flaky crystals of minerals produced by chemical weathering of crystalline rocks. Cohesive and plastic properties derived from thin films of water enveloping flaky particles. When heated and water driven off, becomes hard and stone-like. Has adsorption and ion-exchange properties. Comprises *clay minerals*, with varying amounts of *quartz*, *feldspars*, other silicates, carbonates and ferruginous or organic matter.
Occurrence: Worldwide.

clay ironstone

CLAY IRONSTONE

Sedimentary rock. Sheet-like deposits or

concretionary masses of *argillaceous siderite* and of *limonite* associated with carbonaceous strata. Contains 20–30 per cent *iron*. Heavy, with compact or fine-grained texture. Deposited from solution in marine or fresh-water environments. **Occurrence:** Britain—Lias, Wealden. USA—Ohio, Pennsylvania. In *coal* measures. In nodules and uneven beds among carboniferous and other rocks.
Variety is *black-band ironstone*.

CLAY MINERALS

Finely crystalline, hydrous magnesium or aluminium silicates forming the major constituents of *clay*. Crystal structure of the two-layer or three-layer type. *Kandite* group (*kaolinite*, *dickite*, *chamosite*, *greenalite* and related *halloysite*) have two-layer structure and composition $Al_2O_3 . 2SiO_2 . 2H_2O$ with ferrous iron and magnesium substituting for aluminium. Three-layer types have two layers of silica tetrahedral interleaved with alumina di- and tri-octahedra. *Smectite* group (*nontronite* and *montmorillonite*) have lattice expandable by water adsorption. *Illite* group non-expandable minerals with complex structure resembling *muscovite*. Group includes *illite*, *glauconite* and related ferruginous *chlorites*. *Smectite* and *illite* groups have three-layer structure. Many clay minerals are of mixed-layer types of randomly or regularly interstratified growths of two or more clay minerals.

CLAY ROCK

Term for indurated *mudstones*.

CLAY-SLATE

Low grade regional metamorphic rock. A fine-grained *slate*. Derived from *argillaceous* sediments. Nearer to *clay* than *phyllite* in structure. Lustre of cleavage planes of more crystalline *slates* absent.
Occurrence: Worldwide.

CLAY TALLOW

Sedimentary rock. A *clay* with varying amounts of zinc silicates, *hemimorphite* or other zinc minerals or with zinc as essential constituent of a *clay mineral* such as *sauconite*. Yellowish, ash-grey or brown colour after drying. Fine-grained texture. Greasy to touch. Plastic. Shrinks and crumbles into small fragments on drying.
Occurrence: USA—Missouri.
Forms layers from 300mm to 600–900mm in thickness and lumps from 23 to 227kg.

CLEAVAGE

The plane of mechanical fracture in a rock or mineral. In rocks, cleavage planes are normally sufficiently closely spaced to break the rock into parallel-sided slices. Many minerals, when broken, display a plane parallel to a crystal face. Cleavage planes are developed along planes of symmetry in the atomic lattice. The perfection or otherwise of cleavage depends on the strength of the bonds in this plane.

CLEAVELANDITE

$NaAlSi_3O_8$ Sodium aluminium silicate. Variety of *albite*. Thin, elongated plates or blades. White.
Occurrence: USA—Connecticut, Maine, Massachusetts (Chesterfield), Virginia (Amelia).

CLEIOPHANE

Variety of *blende*. Colourless to pure white or light green. Very pure compound.
Occurrence: USA—New Jersey (Franklin Furnace). In *limestone* cavities.

CLEVEITE

Variety of *uraninite*. Up to 10 per cent cerium or yttrium group rare earths substituting, with yttrium group predominating.
Occurrence: USA—Texas (Baringer Hill in Llano County). Norway—near Arendal (Garta).

CLINKER

Pyroclastic rock fragments fused together.

CLINOCHLORE

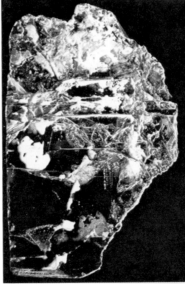

$2[(Mg,Fe^{2+},Al)_6(Si,Al)_4O_{10}(OH)_8]$ with silicon $= 2·8–3·1$. Hydrated magnesium aluminium ferrous iron silicate. Monoclinic. Tabular or modified prismatic crystals; rosette, fan-shaped or vermicular groups; massive, coarse scaly to fine granular and earthy. Shades of green, yellowish, white or rose-red. Pearly lustre on cleavage face. Greenish-white to uncoloured streak. Scaly, earthy fracture. Highly perfect cleavage in one direction. Flexible laminae, tough but slightly elastic. Hardness 2·0–2·25. SG 2·65–2·78. Tough to brittle. Wholly decomposed by sulphuric acid. Borax bead gives clear *glass* coloured by iron or chromium.
Occurrence: USA—New York (Brewster), Pennsylvania (Unionville, West Chester). Austria—Ziller Tal. Czechoslovakia—Moravia. Germany—Bavaria, Saxony. Italy—Piedmont (Ala). Sweden—Pajsberg. Switzerland—Zermatt. USSR—Ural Mountains (Achmatovsk). With chloritic or talcose rocks or *schists* and *serpentine*.
Member of *chlorite group*. Varieties are *kotschubeite* and *leuchtenbergite*.

CLINO-CHRYSOTILE

Intergrowth of monoclinic form of *chrysotile* with up to 50 per cent of orthorhombic form. Nearly same unit cell dimensions as monoclinic form.
Occurrence: Australia. Rhodesia. India. Swaziland.

CLINOCLASE

$4[Cu_3AsO_4(OH)_3]$ Basic copper arsenate. Monoclinic. Tabular or rhombohedral single crystals or rosettes; also crusts or coatings. Dark greenish-black to greenish-blue. Vitreous to pearly lustre on cleavage. Bluish-green streak. Uneven fracture. Brittle. Perfect cleavage in one direction. Hardness 2·5–3·0. SG 4·33–4·38. Chemical identification as for *olivenite*.
Occurrence: England—Cornwall (Ting Tang, Wheal Gorland, Wheal Unity), Devonshire (near Tavistock). USA—Nevada (Majuba Hill), Utah (Tintic district). France—Vosges. Germany—Saxony (Er-

gebirge). Japan—Kitabira. With *olivenite* and other secondary *copper* minerals.

CLINOENSTATITE

8[MgSiO$_3$] Polymorph of *enstatite*. Member of *clinopyroxene* family. Monoclinic. Similar to *diopside* but contains no calcium. SG 3·19.
Occurrence: Worldwide as constituent of meteoric *stones*.

CLINOFERROSILITE

8[FeSiO$_3$] Ferrous metasilicate. Monoclinic. Minute crystals. Colourless or faintly yellow.
Occurrence: USA—California, New York (Adirondack Mountains), Wyoming. Iceland. Kenya—Lake Naivasha. Mainly in lithophysae of *obsidians*. Member of *clinopyroxene* family.

CLINOHEDRITE

CaZnSiO$_3$(OH)$_2$ Basic calcium zinc silicate. Monoclinic. Variety of *tetrahedrite*.
Occurrence: USA—New Jersey (Franklin Furnace).

CLINOHUMITE

2[Mg$_9$Si$_4$O$_{16}$(F,OH)$_2$] with ratio of fluorine to hydroxide *c* 1. Monoclinic. Polysynthetic twinning lamellae; also massive. White, yellowish or greyish-white, shades of brown, yellow and red. Vitreous lustre. Subconchoidal fracture. Brittle. Sometimes distinct cleavage in one direction. Hardness 6·0–6·5. SG 3·1–3·3. Gives

fluorine reaction when heated with potassium bisulphate in closed tube. Gelatinises with acids.
Occurrence: Scotland—Kilbride (South Uist), Skye. Italy—Monte Somma. Spain—Andalusia. USSR—Siberia (Lake Baikal), Ural Mountains.

CLINOHYPERSTHENE

8[(Mg,Fe)SiO$_3$] with molar proportion of ferrous silicate from 20–50 per cent. Magnesium iron silicate. Dimorphous with *hypersthene*. Monoclinic.
Occurrence: Worldwide as common constituent of meteoric *stones*.

CLINOPTILOLITE

c(Na,Ca)$_{4-6}$Al$_6$(Al,Si)Si$_{19}$O$_{72}$. 24H$_2$O Silicon-rich variety of *heulandite*. Member of *zeolite* family. Monoclinic.
Occurrence: USA—Arizona (Dome), Wyoming.
Dimorphous with *mordenite*.

CLINOPYROXENE

Group name for monoclinic *pyroxenes*, including monoclinic forms of normally orthorhombic *pyroxenes*: enstatite, ferrosilite and hypersthene.

CLINOZOISITE

2[Ca$_2$Al$_3$Si$_3$O$_{12}$OH] Basic calcium alumin-

ium silicate. Monoclinic. Microscopic crystals elongated in one direction. Colourless, pale yellow or dark green. Perfect cleavage in one direction. Hardness 6·5. SG 3·21–3·38. Insoluble in hydrochloric acid.
Occurrence: Republic of Ireland—County Wicklow. Finland—Kalvia. India—Madras. Associated with low to medium-grade regionally metamorphosed igneous and sedimentary rocks.

CLINTONITE, BRONZITE (of Finch), HOLMITE or SEYBERTITE

H$_6$(Mg,Ca)$_{10}$Al$_{10}$Si$_4$O$_{36}$ Hydrated calcium magnesium aluminium silicate. Monoclinic. Thick, approximately hexagonal plates with rough lateral faces; also foliated masses. Yellowish, reddish-brown or copper-red. Submetallic lustre on basal pinacoid, resinous on cleavage surfaces; otherwise vitreous. Colourless or slightly yellowish or greyish streak. Very brittle fracture. Perfect basal cleavage. Hardness 4·0–5·0. SG 3·1. Infusible, but turns white and opaque in flame. Completely decomposes in concentrated hydrochloric acid.
Occurrence: USA—New York (Orange County). In *serpentine* with *amphibole*, *pyroxene*, *spinel*, *chondrodite* and *graphite*. Group name for *brittle micas*.
Varieties are *brandisite* and *xanthophyllite*.

CLUTHALITE

Variety of *analcite*. Small, opaque or subtranslucent crystals. Flesh-red. Fragile. Vitreous lustre. Hardness 3·5. SG 2·166.
Occurrence: Scotland—Dunbartonshire and Stirlingshire (Kilpatrick Hills). In amygdaloids.

COAL

Sedimentary rock. Formed by accumulation and decomposition of plant material. A black, firm and compact rock with brittle fracture and earthy texture. Dull or shining lustre. Contains less than 40 per cent inorganic matter (dry basis). Subdivides into *humic coals* (*lignite, bituminous coal* and *anthracite*), *sapropelic coals* (*cannel coal, boghead coal*) and coal-like substances such as *jet*.
Occurrence: Worldwide. Forms beds in rocks of all ages.
Generally deposited from fresh water environment.

COBALTITE

4[CoAsS] Cobalt sulpharsenide. Cubic. Striated, cubic crystals, pyritohedra or combinations; sometimes granular massive to compact. Silver-white inclining to red; steel-grey with violet tinge or greyish-black when much iron. Metallic lustre. Greyish-black streak. Uneven fracture. Perfect cleavage in one direction. Hardness

5·5. SG 6·33. Brittle. F 2·0–3·0. Fuses to weakly magnetic globule. Decomposes in nitric acid with separation of sulphur and arsenic oxide.
Occurrence: England—Cornwall (Botallack Mine). Canada—Ontario (Coleman Township). Western Australia—Ravensthorpe. Germany—Westphalia. Norway—Buskerud. Poland—Silesia. USSR—Transcaucasia. India—Rajputana.
In high-temperature deposits as disseminations in metamorphic rocks or in vein deposits.

COBBLE

A fragment of rock with a particle size between 65mm and 250mm.

COCCOLITE

A granular, white or green, variety of *diopside.*

COCCOLITH OOZE

A type of *ooze.*

COCITE

Igneous rock of the *ultrabasic* clan. A yellowish-green, fine-grained hypabyssal rock. Contains phenocrysts of *olivine* and *diopside* and smaller amounts of *olivine, pyroxene, magnetite, biotite* and *leucite* in groundmass of *augite, orthoclase* and *biotite.*

Occurrence: North Vietnam—Upper Tonkin (Coc Pia).

COFFINITE

cUSiO₄ Uranium silicate with some silicate replaced by hydroxyl. Tetragonal. Botryoidal masses and crystals. Black. SG c3·3.
Occurrence: USA—Colorado (Colorado Plateau), New Mexico, Utah. With *pitchblende* and/or carbonaceous matter in hydrothermal veins. Often partly or totally pseudomorphed into *uraninite.*

COKEITE

Natural coke formed by action of magmas on *coal* or by natural combustion of *coal* in mines.

Above: colemanite

COLEMANITE

2[Ca₂B₆O₁₁ . 5H₂O] Hydrated calcium borate. Monoclinic. Prismatic, highly modified crystals; also massive, cleavable to granular and compact; also crudely spherulitic aggregates. Colourless, milky-white, yellowish-white, grey or muddy. Brilliant, vitreous to adamantine lustre. White streak. Uneven to subconchoidal fracture. Perfect pinacoidal cleavage. Hardness 4·5. SG 2·4. When heated before blowpipe exfoliates, sinters and fuses imperfectly.
Occurrence: USA—Arizona, California, Nevada (Clark County). Argentina—Jujuy Province (Salinas Grandes).

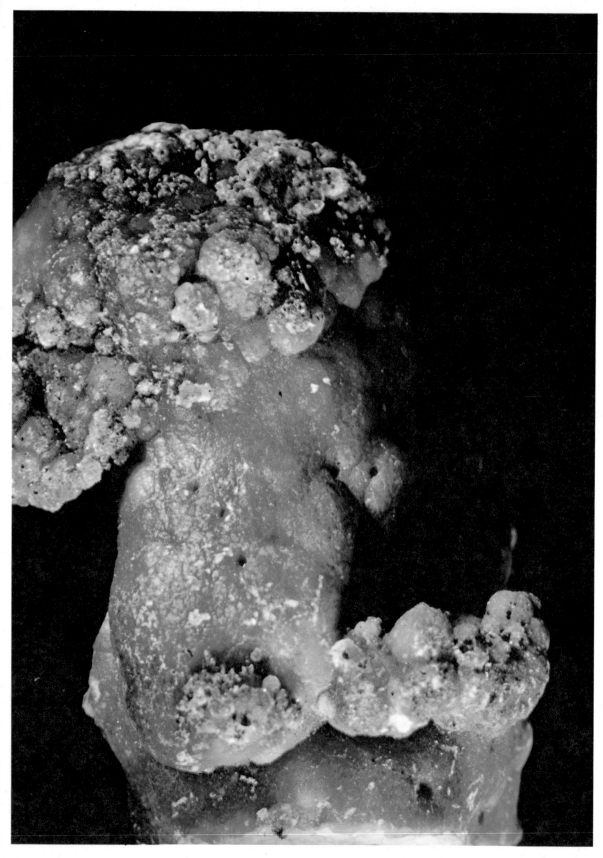

USSR—Kazakhstan (Inder region). Principally in geodal masses in bedded tertiary sediments associated with *celestine, gypsum* and *quartz*.
Varieties are *pandermite* and *priceite*.

COLLINSITE

$Ca_2(Mg,Fe)(PO_4)_2 \cdot 2H_2O$ Hydrated calcium magnesium phosphate. Triclinic. Bladed crystals. Pale brown. Silky lustre. Indistinct cleavage. Hardness 3·5. SG *c*3·0. Easily soluble in hydrochloric acid.
Occurrence: Canada—British Columbia (Francois Lake).

COLLOID, -AL

In a state between suspension and solution in which the particles are so small that there is no tendency to settle.

COLLOPHANE

Variety of *apatite*. Carbon-enriched intermediate member of *hydroxyl-apatite-fluorapatite* series. Amorphous. Massive, opaline or hornlike with dense, layered or colloform structure; sometimes concretionary, nodular, spherulitic or pulverulent. Greyish-white, yellowish or brown to black. Weakly vitreous to subresinous or dull lustre. White streak. Conchoidal, uneven fracture. No cleavage. Hardness 3·0–4·0. SG 2·5–2·9.
Occurrence: Worldwide as independent beds, nodules or concretions or constituting bulk of phosphate rock.

COLOPHONITE

Variety of *andradite*. A calcium-iron *garnet*. Coarse, granular crystals. Dark reddish or brownish colour with iridescent hues. Resinous lustre.
Occurrence: USA—New York (Lewis, Roger's Rock, Willsboro).

COLORADOITE

4[HgTe] Mercury telluride. Cubic. Massive.

Left: collophane.

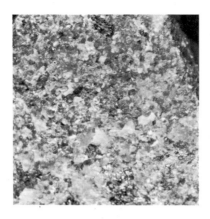

Above: colophonite. *Below:* coloradoite

sive, granular. Iron-black, inclining to grey. Metallic lustre. Uneven to subconchoidal fracture. Brittle and friable. No cleavage. Hardness 2·5. SG 8·04. In flame decrepitates slightly, fuses and yields metallic *mercury* as sublimate.
Occurrence: USA—Colorado (Boulder County). Western Australia—Kalgoorlie.

COLUMBITE or ILMENITE

4[(Fe,Mn)(Nb,Ta)$_2$O$_6$] with proportion of niobium (columbium) greater than tantalum. Iron manganese niobium (columbium) and tantalum oxide grading to *tantalite*. Orthorhombic. Short prismatic or equant crystals; massive compact or as disseminated granules. Iron-black or brownish-black. Submetallic to greasy or dull lustre. Reddish-brown to blackish-brown streak. Conchoidal, uneven fracture. Brittle. Pinacoidal cleavage, not con-

spicuous. Hardness 6·0–6·5. SG 5·4–6·4. Frequently tarnishes iridescent. Unaltered in flame. Partially decomposes by evaporation in hot concentrated sulphuric acid.
Occurrence: Worldwide. Mainly with *beryl, tourmaline, spodumene* and *cryolite* in *granite pegmatites* or washed from these rocks as alluvial deposits.

COLUMBRETITE

Igneous rock of the *syenite* clan. An ash-grey, compact extrusive rock. Contains *sanidine*, and altered *hornblende* in dense groundmass of *oligoclase, analcite*, small *augite* prisms and *magnetite* grains, with interstitial filling of *sanidine, leucite* and *plagioclase* laths.
Occurrence: Spain—Columbrete Islands (Bauzá).

COLUMN

Strictly a length of material with a constant, usually circular, cross-section. Normally used to describe situations such as the structure of the Giant's Causeway in Northern Ireland, the rocks there being columnar; more accurately, however, these shapes should be called prisms.

COLUMNAR

See Column.

COLUSITE

8[Cu$_3$(As,Sn,V)S$_4$] Copper sulphovanadate. Cubic. Massive. Bronze. Metallic lustre. Black streak. Uneven to hackly fracture. No cleavage. Hardness 3·0–4·0. SG 4·5.
Occurrence: USA—Colorado (Red Mountain in San Juan County), Montana (Butte).

COMENDITE

Igneous rock of the *granite* clan. A white porphyritic extrusive rock. Contains phenocrysts of *quartz*, alkali-*feldspar* and *pyroxene* in microgranitic groundmass of *quartz* and alkali-*feldspar*. Leucocratic variety of *pantellerite*.

Occurrence: Italy—South of Sicily (Comende on San Pietro Island).

COMMON GARNET

A brownish-red translucent or opaque variety of *almandine*.

COMPACT

Closely packed, with little space or water between particles.

COMPTONITE

Variety of *thomsonite* with deficiency of sodium. SG 2·37.
Occurrence: Italy—Monte Somma. Switzerland—Seeberg.

CONCHOIDAL

A term describing a fracture resulting in a curved, ribbed, 'shell-like' surface. Subconchoidal fracture is a less obvious form.

CONCRETION, -ARY

The process or result of concentration of certain components in a sediment.

CONGLOMERATE

Sedimentary deposit. A cemented *clastic rock* containing rounded fragments corresponding in grain size to gravels or pebbles. Monogenetic, consisting of fragments of one kind of rock or polygenetic, built up from fragments of different rocks. Cementing material siliceous or *argillaceous*.
Occurrence: Worldwide.

CONGLOMERATE (Volcanic)

Pyroclastic igneous rock. A type of *rhyolite-tuff* from accumulation of rounded materials ejected from volcanic vents or accumulation of eroded, waterlaid volcanic boulders and pebbles in paste of same material.
Occurrence: Worldwide. Associated with volcanic activity.

Above: connellite

CONGRESSITE

Igneous rock of the *syenite* clan. A pale pink, coarse-grained, usually foliated, plutonic rock. Contains much *nepheline* with mainly *orthoclase*, *albite*, *sodalite*, *muscovite*, *biotite* and *magnetite*. Related to *craigmontite*.
Occurrence: Canada—Ontario (Congress Bluff at Craigmont Hill).

CONICHALCITE

$4[CuCaAsO_4OH]$ often with phosphorus pentoxide. Basic calcium copper arsenate. Orthorhombic. Equant to short-prismatic crystals; usually botryoidal to reniform crusts and masses with radial fibrous structure. Grass-green to emerald-green. Vitreous to greasy lustre. Green streak. Uneven fracture. Brittle. No cleavage. Hardness 4·5. SG 4·33 (or 4·1 if fibrous). F 3·0. When heated in closed tube, decrepitates, turns black and gives off water.
Occurrence: USA—Arizona, Nevada, Texas, Utah. South West Africa—near Otavi. Germany—Saxony. Poland—near Kielce. Spain—Andalusia (Cordoba). USSR—Western Siberia. Chile—Tarapacá Province. Secondary mineral found in oxidised zone of *copper* deposits mainly with other *copper* minerals.

CONNARITE

$cNi_2Si_3O_6(OH)_4$ with c5 per cent of aluminium oxide. Hydrated nickel silicate. Possibly hexagonal. Small fragile grains and crystals. Yellowish and shades of green. Siskin-green streak. Perfect clinodiagonal cleavage. Hardness 2·5–3·0. SG 2·459–2·619.
Occurrence: Germany—between Freiberg

and Karl-Marx-Stadt (Hans Georg Mine at Röttis). Possible variety of *garnierite*.

CONNELLITE

$2[Cu_{19}Cl_4SO_4(OH)_{32} \cdot nH_2O]$ Hydrated basic copper chloride and sulphate. Hexagonal. Acicular crystals. Azure blue. Vitreous lustre. Pale greenish blue streak. Hardness 3·0. SG 3·36. Soluble in hydrochloric acid.
Occurrence: England—Cornwall (Wheal Unity and Wheal Damsel). USA—Arizona (Bisbee in Cochise County).

CONSOLIDATION

Change over a period of time of a fresh loose sediment into a compact hard rock.

CONTACT METAMORPHISM

Heat process usually associated with igneous intrusions by which rocks in the crust change in nature.

COOKEITE (of Brush)

$4[LiAl_4(Si,Al)_4O_{10}(OH)_8]$ with silicon $c3$. Hydrous lithium aluminium silicate. Minute scales or slender six-sided prisms. White to yellowish-green. Pearly lustre on cleavage. Perfect basal cleavage. Hardness 2·5. SG 2·7. Flexible. Inelastic. Exfoliates in flame, colouring it intense carmine-red.
Occurrence: USA—Maine (Hebron, Paris). Italy—Island of Elba. USSR—Eastern Transbaikal region. India and Pakistan—Kashmir. Alteration product of *tourmaline* occurring with *tourmaline, lepidolite, sapphire* or *spodumene*, often as coatings.
Member of *mica* family.

COOKEITE (of D'Achiardi)

Synonym of *foresite*.

COOMBE ROCK

Sedimentary rock. A rubbly mass of unrolled *flints* and fragments of *chalk* with some earthy interstitial matter. Result of movement by solifluction during glacial period.
Occurrence: England—bottoms of dry valleys in the South Downs and coastal plain at their foot.

COOPERITE (of Adam)

Synonym of *marmolite*.

COOPERITE (of Wartenweiler)

$2[PtS]$ Platinum sulphide. Tetragonal. Irregular grains and infrequently distorted crystal fragments. Steel-grey. Metallic lustre. Conchoidal fracture. Distinct cleavage in one direction. Hardness 4·0–5·0. SG 9·5. Gives fumes of hydrogen sulphide and deposit of platinum chloride in hydrochloric acid.
Occurrence: South Africa—Transvaal (Potgietersrust and Rustenberg districts). In concentrates from platiniferous *norites* with *sperrylite, braggite* and native *platinum*.

COPIAPITE

$R^{2+}Fe^{3+}_4(SO_4)_6(OH)_2 \cdot nH_2O$ where R^{2+} includes ferrous iron, magnesium, aluminium, copper or sodium. Basic hydrated ferric iron sulphate with varying proportions of the above-named elements. Triclinic. Tabular, sometimes highly modified crystals; massive, granular or incrusting; loose aggregates of minute scales. Shades of yellow; massive form greenish-yellow to olive-green. Pearly lustre on cleavage. Yellowish streak. Earthy, scaly fracture. Brittle. Perfect cleavage in one direction. Hardness 2·5–3·0. SG 2·08–2·17. Metallic taste. Dissolves in water to yellowish solution with acid reaction.
Occurrence: USA—Arizona (Bisbee, Jerome), California, Missouri, Nevada (Comstock Lode). Czechoslovakia—Vashegy. France—coal mines. Germany—Harz Mountains. Italy—Island of Elba. Chile—North Eastern Atacama Desert.
Secondary mineral associated with other sulphate minerals formed from oxidation of sulphides.

COPPAELITE

Igneous rock of the *ultrabasic* clan. A melanocratic porphyritic extrusive rock. Contains phenocrysts of *diopside* in holocrystalline groundmass of *melilite, diopside* and *phlogopite* with accessory *perovskite, apatite* and *magnetite*.
Occurrence: Italy—Umbria (Coppaeli di Sotto near Rieti).

COPPER

$4[Cu]$ Native copper. Cubic. Cubic, octahedral or tetrahedral crystals; massive scales, plates, lumps and arborescent aggregates. Copper-red, tarnishing readily to red, blue, green or black. Metallic, copper-red, shiny lustre. Metallic, shining streak. Hackly fracture. No cleavage. Hardness 2·5–3·0. SG 8·5–9·0. Ductile. Malleable. Excellent conductor of heat and electricity. F 3·0.
Occurrence: England—Cornwall. Scotland—Stirling. USA—especially Lake Superior region. Canada—Nova Scotia (Cap d'Or). Australia—New South Wales (Broken Hill). Germany—Hesse-Nassau, Rhineland. Italy—Tuscany. Bolivia. Chile. Mexico. USSR. Usually associated with basic igneous rocks.
Variety is *whitneyite*.

COPPERAS

Synonym of *melanterite*.

COPPER PYRITES

Synonym of *chalcopyrite*.

COPROLITE

Sedimentary deposits. Fossilised excrement of fishes, reptiles and mammals consisting of 50–60 per cent phosphate, six–eight per cent calcium carbonate, four–five per cent organic matter and moisture, and sand and remains of marine life. Irregular nodules, sometimes distinctly concretionary, varying from pea-size to weights in excess of one tonne. Light grey to brown, rarely jet-black. Hardness 2·0–4·0. SG 2·2–2·5.

Occurrence: England—Kent and Sussex Greensands. North Wales. USA—Alabama, Florida, North Carolina, South Carolina. Canada—Hudson Bay area. Belgium. France. USSR. On ocean floors, river beds and in fossil sediments or fossiliferous rocks of different ages.

COQUINA

Sedimentary rock. A loosely fragmented, shelly *limestone*. Coarse-grained texture. Porous. Friable. Composed of fragments of shells of living or recently extinct species of molluscs and coral.
Occurrence: USA—Florida.

CORAL

Hard substance formed from skeletons of marine polyps.

CORAL MUD and SAND

Sedimentary deposits. Contain abundant fragments of corals. Coral sand is coarse and found near coral reefs. Coral mud is finer-grained and found further out.
Occurrence: Worldwide. Around coral islands and coasts bordered by coral-reefs.

CORDIERITE, DICHROÏTE or IOLITE

$4[(Mg,Fe^{2+})_2Al_4Si_5O_{18}]$ Magnesium aluminium ferrous iron silicate. Orthorhombic. Short-prismatic, pseudohexagonal crystals; massive compact, disseminated or granular. Light blue, violet, smoky or greenish-blue. Vitreous to dull lustre. White streak. Uneven fracture. Brittle. Pinacoidal cleavage, sometimes conspicuous. Hardness 7·0–7·5. SG 2·6–2·7. Decrepitates on fusion with alkali carbonates. Colour varies markedly from violet-blue to greyish in different crystal directions. Has been used as gemstone.
Occurrence: USA—Connecticut, Massachusetts. Canada—Northwest Territories.

Finland—Orijärvi. Germany—Bavaria. Greenland—Ujordlersoak. Hungary. Italy—Tuscany. Norway—Kragerö. Spain—Cabo de Gata. Sweden—Tunaberg. Brazil (Gemstones). Japan. Sri Lanka (gemstones). Mineral of metamorphic rocks occurring with *quartz, feldspar, sillimanite, hornblende* and *andalusite*.

CORDYLITE

$c9[Ba_2(Ce,La)_3(CO_3)_5F_3]$ Barium and cerium group of rare earths fluorcarbonate. Hexagonal. Tiny, short prismatic and sceptre-like crystals with hexagonal dipyramidal terminations. Colourless to waxy-yellow when fresh, often ochre-yellow and dull from surface alterations. Greasy to adamantine lustre. Conchoidal fracture. Rather brittle. Distinct cleavage in one direction; possibly parting due to alteration. Hardness 4·5. SG 4·31. Easily soluble in acids. Decrepitates strongly in flame but does not fuse.
Occurrence: Greenland—Julianehaab district (Narsarsuk). In *pegmatite* veins in *nepheline-syenite* with *aegirine, ancylite* and *neptunite*.

CORKITE

$PbFe_3(PO_4)(SO_4)(OH)_6$ Basic lead ferric iron sulphate-phosphate. Hexagonal. Rhombohedral, generally pseudocubic crystals. Dark green, yellowish-green to pale yellow. Vitreous to resinous lustre.

Perfect cleavage in one direction. Hardness 3·5–4·5. SG 4·295. Easily soluble in warm hydrochloric acid. Fuses with difficulty in flame.
Occurrence: Republic of Ireland—County Cork. USA—Utah (Beaver County). Germany—Hesse-Nassau (Dernbach). Italy—Sardinia. Yugoslavia—Bosnia (near Ljubija). USSR—Kazakhstan.

CORNELIAN

Synonym of *carnelian*.

CORNETITE

$8[Cu_3PO_4(OH)_3]$ Basic copper phosphate, sometimes with small amounts of cobalt. Orthorhombic. Short prismatic to equant crystals, usually rounded; also minute crystal crusts. Deep blue, peacock-blue to greenish-blue. Vitreous lustre. No cleavage. Hardness c4·5. SG 4·10. Soluble in cold water.
Occurrence: USA—Nevada (Blue Jay and Empire-Nevada Mines). Congo—Katanga (near Lubumbashi). Zambia—Bwana Mkubwa.

CORNSTONE

Sedimentary deposit. A variety of *micrite* and dolomitic *micrite* formed under semi-arid conditions. A concretionary or other *limestone*-mass occurring as irregular nodules or as lenticular beds of c1·8m thickness in matrix of red *sandstone*. Often fluted, porous structure. Cross-bedding well-developed in some of the pebble and pellet layers.
Occurrence: England—Cheshire, Nottinghamshire. Scotland. Wales—Borderland. Republic of Ireland. Associated with Old Red Sandstone and new red sandstone (permo-trias) formations.

CORNUBIANITE

High grade contact metamorphic rock. Type of *hornfels*. A non-fissile, fine-grained rock. Contains *quartz, feldspar* and *mica*.
Occurrence: England—Cornwall.

CORNWALLITE

$2[Cu_5(AsO_4)_2(OH)_4]$ Basic hydrated divalent copper arsenate, sometimes with phosphate. Small, botryoidal crusts with radial, fibrous structure, somewhat resembling *malachite*. Verdigris-green to blackish-green. Conchoidal fracture. Not very brittle. No cleavage. Hardness 4·5. SG 4·166. F 2·0–2·5 to black *glass*.
Occurrence: England—Cornwall. With *olivenite* and *tenorite*.

CORONADITE

$cPbMn^{2+}Mn^{4+}_7O_{16} . H_2O$ Hydrated lead manganese oxide. Tetragonal. Massive as botryoidal crusts with fibrous structure. Dark grey to black. Dull to submetallic lustre. Brownish-black streak. No fracture. No cleavage. Hardness 4·5–5·0. SG 5·44. Borax bead turns reddish-violet in oxidising flame, colourless in reducing flame.
Occurrence: USA—Arizona (Clifton-Morenci district). Morocco—Imini (Bou Tazoult). In manganese deposits.

CORROSION

The effect of chemical action from components in the environment (such as oxygen in the atmosphere and dissolved substances in sea water) on the exposed layers of a material. The chemical nature, and often the physical behaviour, of these layers will be changed as the material is corroded.

CORSITE or NAPOLEONITE

Igneous rock of the *gabbro* clan. An ophitic, orbicular plutonic rock. Contains oval zonal masses of 25–75mm diameter showing alternate rings of *pyribole* and *feldspar* in groundmass of *labradorite*, *bytownite* and *hornblende* with small amounts of *hypersthene*, *quartz*, *apatite*, *pyrite* and ore.
Occurrence: Italy—Corsica (Santa Lucia di Tallano).

CORTLANDTITE or HUDSONITE

Igneous rock of the *ultrabasic* clan. A dark green, fine-grained hypabyssal rock with glistening bronze-coloured cleavage surfaces. Contains *hornblende* and *olivine* with small amounts of *mica*.
Occurrence: USA—New York. Finland—Suomusjärvi (Northern shore of Lake Pyhälampi).

CORUNDOLITE

Synonym of *emery*.

CORUNDOPHILITE

$2[(Mg,Fe^{2+},Al)_6(Si,Al)_4O_{10}(OH)_8]$ with silicon = 2·0–2·5 and proportion of magnesium greater than ferrous iron. Hydrated magnesium aluminium ferrous iron silicate. Monoclinic. 6-sided or 12-sided tablets or low prisms. Olive-green, leek-green or greyish-green. Pearly lustre on cleavage surface. Basal, eminent cleavage. Hardness 2·5. SG 2·9. Flexible, somewhat elastic laminae.
Occurrence: USA—Massachusetts (Chester), North Carolina (Asheville). With *corundum* or *emery*.
Member of *chlorite group*.

CORUNDUM

$2[Al_2O_3]$ Aluminium oxide. Hexagonal. Steep pyramidal, barrel-shaped, sometimes flat tabular or rhombohedral crystals; granular massive, in rounded grains. Shades of blue to colourless, yellow, golden, purple, green, pink to deep red, brown or black.

Adamantine to vitreous lustre. White streak. Uneven to conchoidal fracture. Brittle, but very tough when compact. No cleavage, but well-developed partings on rhombohedral planes. Hardness 9·0. SG 4·022. Infusible. Insoluble in acids. Crystals sometimes show colour zoning.
Occurrence: Worldwide. Characteristic mineral of rocks lacking in silicon oxide. Associated with *spinel, garnet, kyanite* and high-calcium *feldspars* in plutonic, *pegmatite* and metamorphic rocks.
Varieties are *oriental amethyst, oriental emerald, oriental topaz, sapphire, padparadschah* and *ruby* (gemstones) and *emery* (massive).

COSALITE

$2[CuPb_7Bi_8S_{22}]$ Lead copper bismuth sulphide. Orthorhombic. Prismatic crystals, frequently elongated to needle-like and capillary forms; massive, in radiating prisms, fibrous or feathery aggregates; also dense with indistinct crystalline structure. Lead-grey to steel-grey. Metallic lustre. Black streak. Uneven fracture. Hardness 2·5–3·0. SG 6·76. F 1·0. Soluble in nitric acid with separation of lead sulphate.
Occurrence: USA—Colorado, Washington. Canada—British Columbia, Ontario. Australia—New South Wales (New England range). Hungary—Vaskö. Rumania —Rezbanya. Sweden—Nordmark. Switzerland — Forno Glacier. Mexico — Chihuahua, Sinaloa. Madagascar—Amparindravato. In hydrothermal deposits formed at moderate temperatures in contact metamorphic deposits and *pegmatites*.

COSSYRITE

Variety of *aenigmatite* with little or no titanium dioxide. Forms minute, embedded crystals in trachyte lavas. Black. SG 3·74–3·75.

Occurrence: Italy—Island of Pantellaria.

COTUNNITE

4[PbCl$_2$] Lead chloride. Orthorhombic. Crystals often flattened and elongated; massive, granular. Colourless to white, yellowish or greenish. Adamantine, inclining to silky or pearly lustre. Colourless. Subconchoidal fracture. Perfect cleavage in one direction. Hardness 2·5. SG 5·8–5·81. Slightly sectile. Easily fusible. Soluble in water, dilute hydrochloric acid or nitric acid.

Occurrence: England—Cornwall. USA—Arizona (Bentley district in Mohave County). Italy—Mount Vesuvius. Chile—Antofagasta and Tarapacá Provinces. Peru—Pallasca Province. Under arid, saline conditions as sublimation product or associated with *cerussite, anglesite* and other secondary minerals as alteration product of *galena*.

COUNTRY ROCK

Rock bodies which enclose an intrusive mass of igneous rock or series of mineral veins.

COVELLINE, COVELLITE or BREITHAUPTITE (of Chapman)

6[CuS] Copper sulphide. Hexagonal. Crystals rare, horizontally striated, hexagonal plates; commonly massive or spheroidal with sometimes crystalline surface. Indigo-blue or darker. Submetallic to resinous lustre, slightly pearly on cleavage, subresinous or dull when massive. Lead-grey to black, shining streak. Uneven fracture. Highly perfect cleavage in one direction. Flexible in thin leaves. Hardness 1·5–2·0. SG 4·6–4·76. F 2·5. Burns with blue flame and reduces to metallic globule when heated on charcoal.

Occurrence: USA—Alaska, Colorado, Montana, Utah, Wyoming. New Zealand—Kawau Island. Austria—Salzburg. Germany—Baden, Hesse-Nassau,

Saxony. Italy—Sardinia. Yugoslavia—Serbia. Argentina. Philippine Islands. In veins in zones of secondary enrichment associated with other *copper* minerals.

COVELLITE

Synonym of *covelline*.

COVITE

Igneous rock of the *syenite* clan. A mottled black and white, fine-grained plutonic rock. Contains *orthoclase, nepheline, diopside* with *aegirine* borders, *hornblende, titanite* and accessory *apatite* and *magnetite*.

Occurrence: USA—Arkansas (Magnet Cove in the Ozark Mountains).

CRAIGMONTITE

Igneous rock of the *syenite* clan. A mesotype, coarse-grained plutonic rock. Contains *nepheline, oligoclase, muscovite* and *corundum*. Differs from *congressite* in containing more *nepheline* and less *plagioclase* and *corundum*.

Occurrence: Canada—Ontario (Craigmont Mountain).

CRANDALLITE

[CaAl$_3$(PO$_4$)$_2$(OH)$_5$.H$_2$O] Basic hydrated calcium aluminium phosphate with strontium, barium, iron or rare earths and hydroxyl replacing phosphate. Hexagonal. Crystals rare, minute trigonal prisms or fibrous rosettes; usually massive as nodular aggregates grading in dense agate-like forms; also banded spherules with radial, fibrous structure. Yellow to yellowish-white, white or grey. Lustre of crystals vitreous, massive vitreous to chalky and dull. Perfect cleavage in one direction. Hardness 5·0. SG 2·78. F 2·5. Fuses to white enamel.

Occurrence: USA—Nevada (Esmeralda County), South Dakota (near Harney City), Utah (Juab County, Utah County). Germany—Bavaria (Amberg), Hesse-Nassau (Ahlbach, Dehrn). Bolivia—Llallagua.

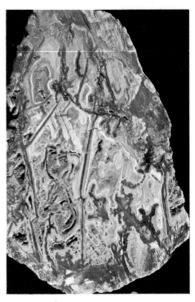

Above: crandallite

CREDNERITE

CuMn$_2$O$_4$ Copper manganese oxide. Possibly monoclinic. Thin, six-sided plates in radiating, hemispherical or spherulitic groupings; also as earthy coatings. Iron-black. Bright metallic lustre. Sooty black streak with brownish tint. Perfect cleavage in plane of plates. Hardness 4·0. SG 5·01. Soluble in hydrochloric acid with evolution of chlorine. Insoluble in nitric acid.

Occurrence: England—Somerset (Mendip Hills). USA—California County (Calistoga in Vapa County). Germany—Thuringia (Friedrichroda).

CRETACEOUS

See *geological time*.

CRICHTONITE

9[Fe$^{2+}_{16}$Fe$^{3+}_{14}$Ti$_{66}$O$_{169}$] Ferric ferrous iron titanate. Mineral near *ilmenite*, but distinct species. More titanium than *ilmenite*. Acute rhombohedra. Bright lustre. Basal cleavage. SG 4·689–4·79.

Occurrence: France—St Christophe-d'Oisans. Germany.

CRINANITE

Igneous rock of the *gabbro* clan. A melanocratic, fine-grained hypabyssal rock. Contains *olivine, augite*, basic *plagioclase* and considerable *analcite* and *zeolites*.

Occurrence: Scotland—Argyllshire (near Loch Crinan), Ayrshire coast (Greenan), Bute (Arran Island), Hebrides. Closely related to *teschenite*.

CRISTOBALITE or LUSSATINE

4[SiO$_2$] High-temperature form of *quartz*.

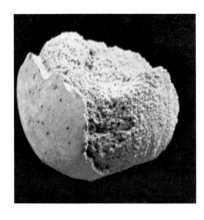

Cubic. Octahedral crystals. White. Dull lustre. No cleavage. Hardness 6·0–7·0. SG 2·27–2·34. Crystals show abnormal double refraction. Infusible.
Occurrence: Mexico—near Pachuca (San Cristobal Mountains). With *tridymite* in cavities in *andesite*.

CROCIDOLITE or BLUE ASBESTOS

Fibrous variety of *riebeckite* and member of *asbestos* group. Monoclinic. Long, delicate, easily separable fibres; also massive, earthy. Lavender-blue or leek-green. Silky, dull lustre. Lavender-blue or leek-green streak. Elastic fibres. Prismatic cleavage. Hardness 4·0. SG 3·2–3·3. Easily fuses with intumescence to black magnetic *glass*, colouring flame yellow. Unaffected by acids.
Occurrence: USA—Rhode Island (near Cumberland). Canada—Ontario (Ottawa County). South Africa—Transvaal, West Griqualand. Czechoslovakia—Moravia. France—Vosges (near Framont). Greenland.

CROCOITE or BERESOFITE (of Shepard)

$4[PbCrO_4]$ Lead chromate. Monoclinic. Prismatic, acicular crystals; massive columnar, granular, as crusts. Hyacinth-red or aurora-red. Adamantine, greasy lustre. Orange-yellow streak. Conchoidal,

uneven fracture. Sectile. Cleavage visible in two directions. Hardness 2·5. SG 5·9–6·1. F 1·5. When heated in closed tube decrepitates and blackens, but recovers original colour on cooling.
Occurrence: USA—Arizona (Maricopa County, Pinal County), California (Inyo County, Riverside County). Australia—Tasmania. Rhodesia—Umtali district. Rumania—Rézbánya. Brazil—Minas Gerais (near Congonhas do Campo). USSR—Ural Mountains. Philippine Islands—Luzon (North Camarines Province). Alteration product of *galena* occurring with *galena*, *quartz*, *pyrite* and *vanadinite*.

CROMALTITE

Igneous rock of the *ultrabasic* clan. A black to dark green, coarse to medium-grained hypabyssal rock. Contains *aegirine-augite*, *melanite*, *biotite*, *ilmenite*, *apatite* and *pyrite*. Very tough.
Occurrence: Scotland—Assynt (Bad na h'Achlaise).
A *garnet-bebedourite*.

CRONSTEDTITE

$c6[Fe_2^{2+}Fe_2^{3+}SiO_5(OH)_4]$ with small amounts of magnesium. Hydrated ferric ferrous iron silicate. Hexagonal. Forms vertically striated pyramidal crystals, tapering to one extremity; also fibrous, diverging groups, cylindroidal and reniform, or amorphous. Coal-black to brownish-black. Brilliantly vitreous lustre. Dark olive-green streak. Thin laminae elastic. Not brittle. Perfect basal cleavage. Hardness 3·5. SG 3·34–3·35. Gelatinises in concentrated hydrochloric acid. Borax bead gives reactions for iron.
Occurrence: England—Cornwall (Wheal Maudlin). Czechoslovakia—Bohemia

(Kuttenberg, Přibram). Brazil—Minas Gerais (Congonhas do Campo). Doubtful member of *chlorite group*. *Related to kaolinite.*

CROOKESITE (of Nordenskiöld)

$8[(Cu,Tl,Ag)_2Se]$ Copper thallium selenide with small amount of *silver*. Massive, compact. Lead-grey. Metallic lustre. Brittle. Hardness 2·5–3·0. SG 6·9. Easily fuses to greenish-black enamel, colouring flame green.
Occurrence: Sweden—Skrikerum.

CROSSITE

$2[Na_2(mg,Fe^{2+})_3(Fe^{3+},Al)_2Si_8O_{22}(OG)_2]$ with ratio of ferric iron to aluminium $c1$ and ratio of ferrous to magnesium $c1$. Hydrated sodium magnesium aluminium ferric and ferrous iron silicate. A clino-*amphibole* between *riebeckite* and *glaucophane*. Blue in parent rock, medium greyish-blue as powder. SG 3·184–3·206.
Occurrence: Wales—Anglesey (Monument Hill). USA—California (Berkeley). Yugoslavia—Serbia. In *schist*.

CRUST

The outer layers of the earth.

CRYOLITE

$2[Na_3AlF_6]$ Sodium aluminium fluoride with sodium deficiency and water replacing fluorine. Monoclinic. Modified cuboidal or short prismatic crystals with modified faces; massive, coarsely granular. Colourless to white, brownish, reddish or brick-red; rarely black. Vitreous to greasy lustre, somewhat pearly on basal plane. White streak. Uneven fracture. Brittle. No cleavage. Hardness 2·5. SG 2·97. Soluble in sulphuric acid with evolution of hydrogen fluoride. Fuses on charcoal to clear bead which becomes opaque on cooling.
Occurrence: USA—Colorado (Pikes Peak in El Paso County). Greenland—Frederikshaab district (Arksuk Fiord). Spain—Huesca Province (Pyrenees).

MONOCLINIC

a   Acicular: Corundum

b   Prismatic: Staurolite

c   Tabular Prismatic: Orthoclase

d   Tabular: Chloritoid

  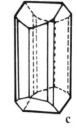 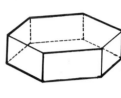

a   b   c   d

ORTHORHOMBIC

a   Acicular: Strontianite

b   Platy/Tabular: Lepidocrosite

c   Prismatic: Forsterite

d   Pyramidal: Humite

  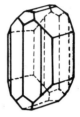 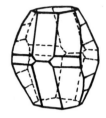

a   b   c   d

TETRAGONAL

a   Prismatic: Cassiterite

b   Pyramidal: Torbernite

c   Pseudo Cubic: Leucite

d   Tabular Prismatic: Calomel

e   Tetrahedral: Chalcopyrite

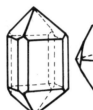

a   b   c   d   e

TRICLINIC

a   Bladed: Kyanite

b   Prismatic: Polyhalite

c   Tabular: Albite

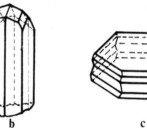 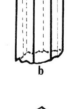 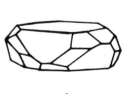

a   b   c

HEXAGONAL

a   Prismatic: Quartz

b   Pyramidal: Vanadinite

c   Rhombohedral: Siderite

d   Tabular: Ilmenite

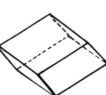

a   b   c   d

CUBIC

a   Cubic: Halite

b   Dodecahedral: Spessartite

c   Octahedral: Chromite

d   Trapezohedral: Sal Ammoniac

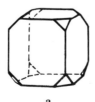 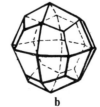 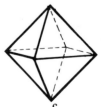 

a   b   c   d

USSR—Ilmen Mountains (near Miask). In *pegmatite*.

CRYPTOCRYSTALLINE

With a crystalline structure visible only high magnification.

CRYPTOMELANE

$K(Mn^{2+}(Fe,Cu)Mn^{3+}_7O_{16} . 2H_2)$ Hydrated potassium iron manganese manganate. Tetragonal. Massive, fine-grained, cleavable or fibrous. Resembles *psilomelane* in physical properties.
Occurrence: USA—Arizona (Tombstone). France—Romanèche. Germany—Nassau. India—Sitapar (monoclinic or pseudotetragonal form).
Isostructural with *coronadite* and *hollandite* and can form isomorphous mixtures with them.

CRYPTOPERTHITE

Type of *microperthite*. Consists of interlamination of *albite* and *orthoclase* as lamellae of submicroscopic thickness. Gives marked schiller to *feldspars*.
Occurrence: Norway—among *microperthites*. Worldwide in igneous rocks rich in alkali-*feldspars*.

CRYSTAL

Smooth faced solid shape assumed by a mineral when allowed to solidify in sufficient space. An external reflection of the arrangement of atoms within the mineral structure.
Minerals are divided into crystal systems by a consideration of symmetry:
Cubic—crystal form with three mutually perpendicular axes of equal length, e.g. *pyrite* and *galena*.

Tetragonal—crystal form with three mutually perpendicular axes, two of which are of equal length, e.g. *zircon* and *rutile*.
Hexagonal and **Trigonal**—crystal forms with three axes, each separated by 120 degrees, and a fourth axis perpendicular to the plane containing the other axes, e.g. *quartz* and *calcite*.
Orthorhombic—crystal form with three mutually perpendicular unequal axes, e.g. *barytes* and *natrolite*.
Monoclinic—crystal form with three unequal axes, two of which are perpendicular, the third at an oblique angle, e.g. *gypsum* and *muscovite*.
Triclinic—crystal form with three unequal axes, e.g. *axinite* and *turquoise*.
Crystals occur in a number of common habits:
Acicular—needle-like crystals (Hexagonal).
Bladed—flat elongated crystals (Monoclinic).
Capillary—very fine thin crystals (Hexagonal).
Columnar—column-like crystals (Hexagonal).
Cubic—six equal rectangular faces (Cubic).
Dodecahedral—six unequal rectangular faces (Cubic).
Octahedral—eight square faces (Cubic).
Platy—thin extensive plate-like crystals (Hexagonal).
Prismatic—crystals resemble joined prisms (Monoclinic, Orthorhombic or Hexagonal).
Pyramidal—crystals resemble joined pyramids (Tetragonal).
Rhombdodecahedral—twelve rhomb-shaped faces (Cubic).
Rhombohedral—six rhomb-shaped faces (Hexagonal).
Tabular—thick platy crystals (Monoclinic, Orthorhombic, Tetragonal or Hexagonal).
Tetragonal—type of prismatic or pyramidal crystal (Tetragonal).
Trapezohedral—twenty four trapezoid-shaped faces (Cubic).

CRYSTALLINE
Composed of crystals.

CRYSTALLINE LIMESTONE

General term for metamorphosed *limestones*. Mineral composition depends on character of original *limestones*, thermodynamic conditions under which metamorphism was affected and composition of material introduced from external sources (if any). Term includes *cipolino*, *marble*, *ophicalcite* and *predazzite*.

CRYSTAL TUFF

Type of volcanic *tuff*. Comprises detached crystals formed by explosive disruption of lava in an advanced state of crystallisation

at time of eruption. Contains crystals of *biotite, hornblende, quartz* and *feldspar* in glassy interstitial groundmass of fine *glass* dust, iron oxide and secondary *chalcedony* or hydrous silica.
Occurrence: Worldwide in regions associated with volcanic activity.

CUBANITE

$4[CuFe_2S_3]$ Copper iron sulphide. Orthorhombic. Striated, Elongated or thick tabular crystals; also massive. Brass to bronze-yellow. Metallic lustre. Greenish-black streak. Conchoidal fracture. No cleavage. Hardness 3·5. SG 4·03–4·18. Strongly magnetic. F 2·0. Chemical identification as for *chalcopyrite*.
Occurrence: USA—Alaska (Prince William Sound), New Mexico (Fierro). Canada—Ontario (Sudbury). Italy—Piedmont (Traversella). Norway—Sulitjelma. Sweden—Kaveltorp, Tunaberg. Brazil—Minas Gerais. Cuba. In deposits formed at relatively high temperatures associated with *chalcopyrite, blende, pyrite* and *pyrrhotite*.

CUBIC

See *crystal*.

CUMBERLANDITE

Undersaturated igneous rock of the *ultrabasic* clan. A greenish-black hypabyssal rock with a few white *plagioclase* crystals. Black, granular groundmass comprises *c*50 per cent *olivine*, 40 per cent *magnetite* and *ilmenite* and accessory *spinel*.
Occurrence: USA—Rhode Island (Cumberland).

CUMBRAITE

Igneous rock of the *granite* clan. A porphyritic extrusive rock. Contains phenocrysts of *bytownite-anorthite* in a groundmass of *labradorite, enstatite-augite* and much *glass* which is potentially *andesine, sanidine* and *quartz*.
Occurrence: Scotland—Firth of Clyde (Great Cumbrae).
Type of *dacite*.

CUMMINGTONITE (of Dewey)

$2[(Mg,Fe)_7Si_8O_{22}(OH)_2]$ with ratio of magnesium to iron greater than 1:1 Hydrated magnesium ferrous iron silicate. Monoclinic. Usually fibrous or fibrolamellar, often radiated. Grey to brown. Perfect prismatic cleavage. Hardness 5·0–6·0. SG 3·1–3·32. **Occurrence:** USA—Massachusetts (Cummington), Ohio (Baltimore). Greenland. Norway—Kongsberg. A clino-*amphibole*, identical with *anthophyllite* in composition and therefore an *amphibole-asbestos*.
Forms series with *grünerite*.

CUMMINGTONITE (of Rammelsberg)

Synonym of *rhodonite*.

CUPRITE

2[Cu_2O] Cuprous oxide. Cubic. Octahedral, dodecahedral or cubic crystals, sometimes highly modified; massive, granular or earthy. Various shades of red; sometimes almost black. Adamantine or submetallic to earthy lustre. Shining, several shades of brownish-red, streak. Conchoidal to uneven fracture. Brittle. Cleavage interrupted and rare in one direction. Hardness 3·5–4·0. SG 6·14. Soluble in hydrochloric acid. Concentrated solution yields heavy white precipitate of *copper* chloride when cooled.

Occurrence: England—Cornwall. USA—Arizona, California, Colorado, Idaho, Nevada, New Mexico, Pennsylvania, Tennessee. Australia—New South Wales, Tasmania. South West Africa—Otavi. France—Chessy. Germany—Westphalia. Rumania—Timisoara. Bolivia. Chile. Mexico. USSR—Ural Mountains. Japan. Congo—Katanga. In oxidised zone of *copper* deposits with mainly native *copper*, *malachite*, *azurite*, *chalcocite* and *chrysocolla*.

Varieties are *chalcotrichite, ruby copper* and *tile ore*.

CUPROTUNGSTITE

[$Cu_2(WO_4)(OH)_2$] Basic copper tungstate. Tetragonal. Microcrystalline masses and crusts, friable to compact. Pistachio-green to olive-green, leek-green or emerald-green. Vitreous lustre; also waxy to earthy. Greenish-grey to greenish-yellow streak. Easily soluble in acids. Fuses to black *glass* in flame.

Occurrence: USA—Arizona, California, New Mexico. Australia—New South Wales (Georgiana County). South Africa—Transvaal. Italy—Sardinia. Spain—Montoro (Sorpresa). Chile—near Santiago. Mexico—La Paz. Japan—Province Nagato.

Secondary mineral from alteration of *scheelite*.

CURITE

$Pb_2U_5O_{17}.4H_2O$ Hydrated lead uranium

oxide. Orthorhombic. Striated, prismatic crystals, usually massive; in aggregates of fine needles or saccharoidal; also compact earthy. Orange-red. Adamantine lustre. Orange streak. Cleavage observed in one direction. Hardness 4·0–5·0. SG 7·26. Easily soluble in acids. Liberates chlorine from strong solutions of hydrochloric acid. Blackens when heated in flame.
Occurrence: Canada—Great Bear Lake region, Quebec (Villeneuve). Congo—Katanga (Kasolo).
An alteration product of *uraninite*.

CUSELITE

Igneous rock of the *diorite* clan. A melanocratic, porphyritic to granular hypabyssal rock. Contains phenocrysts of *plagioclase* in granular groundmass of *plagioclase*, *orthoclase*, *biotite*, *augite*, *hornblende* and *diopside*.
Occurrence: Germany—between the Saar and Nahe Rivers (near Cusel, Remigiusberg).
Type of *lamprophyre*.

CUSPIDINE

$Ca_4Si_2O_7(F,OH)_2$ Basic calcium fluorsilicate. Monoclinic. Minute, spear-shaped crystals. Pale rose-red. Vitreous lustre. Pale rose-red streak. Uneven fracture. Brittle. Very distinct cleavage in one direction. Hardness 5·0–6·0. SG 2·853. Crystals usually altered on surface and covered with calcium carbonate shell. Readily soluble in nitric acid. Fuses with great difficulty.
Occurrence: Italy—Monte Somma. In cavities or embedded in granular, rock-like masses.

CUYAMITE

Igneous rock of the *gabbro* clan. A dark-coloured, holocrystalline xenomorphic-granular to ophitic hypabyssal rock. Contains *apatite*, *feldspar*, *magnetite*, *augite* and *analcite*.
Occurrence: USA—California (Cuyamas Valley in San Luis Obispo County).

CYANITE

Synonym of *kyanite*.

CYANOTRICHITE

$Cu_4Al_2SO_4(OH)_{12} \cdot 2H_2O$ Basic hydrated-copper aluminium sulphate. Orthorhombic. Acicular crystals; also massive. Sky blue. Silky lustre. Pale blue streak. SG 2·95. Soluble in hydrochloric acid.
Occurrence: Scotland—Lanarkshire (Leadhills). USA—Arizona (Grandview

Left: cuprite

Mine in Coconino County), Nevada (Majuba Hill).

CYMATOLITE

A mixture of *albite* and *muscovite*. Fibrous to wavy structure. White or slightly pinkish. Silky lustre. Hardness 1·5–2·0. SG 2·69–2·70.
Occurrence: USA—Massachusetts (Chesterfield, Goshen).

CYMOPHANE

Synonym of *chrysoberyl* (*of Werner*).

CYPRINE

Variety of *idocrase* containing traces of *copper*. Pale sky-blue or greenish-blue colour due to presence of *copper*.
Occurrence: Norway—Telemark.

CYRTOLITE

Radioactive variety of *zircon*. Contains uranium and yttrium. Crystals identical to *zircon*, but dull, convex, pyramidal faces. Reddish-brown. Used for recovery of rare earths.
Occurrence: USA—New York (Bedford). Forms rows of crystals with rounded faces in *pegmatite*.

DACHIARDITE

$c(Ca,K_2,Na_2)_3Al_4Si_{18}O_{45} \cdot 14H_2O$ Hydrated calcium potassium sodium and aluminium silicate. Zeolitic mineral with composition near *epistilbite* but higher alkali and silica content. Monoclinic. Octagonal prisms built-up of sectors of different optical orientation. Colourless. Platy, perfect cleavage. Hardness 4·0–4·5. SG $c2 \cdot 16$. Decomposes in hydrochloric acid. Decrepitates and fuses to white enamel in flame.
Occurrence: Italy—Island of Elba (Campo, San Piero). In granite *pegmatite*.

DACITE

Igneous rock of the *granite* clan. A fine-

grained extrusive rock. Usually contains phenocrysts of *quartz, plagioclase, pyroxene, hornblende* and minor *sanidine* and *biotite* in finely crystalline groundmass of alkali-*feldspar* and silica materials. Marked predominance of *plagioclase* over alkali-*feldspar*.
Variety is *cumbraite*.
Occurrence: Worldwide.

DAHAMITE

Igneous rock of the *granite* clan. A porphyritic hypabyssal rock. Variety of *paisanite* characterised by abundant *albite*.
Occurrence: Indian Ocean—Carlsberg Ridge (Dahamis).

DAHLLITE

Synonym of *carbonate-apatite*.

DALYITE

$K_2ZrSi_6O_{15}$ Potassium zircon silicate. Triclinic. Small, short prismatic crystals. Colourless and transparent. Vitreous lustre. Distinct cleavage in three directions. Hardness 7·5. SG 2·84.
Occurrence: Atlantic Ocean—Ascension Island (Green Mountain, Middleton Peak). In ejected blocks of alkali-*granites* with *microperthite*, *quartz*, *aegirine* and alkali-*amphibole*.

DAMKJERNITE

Igneous rock of the *ultrabasic* clan. A dark-coloured porphyritic hypabyssal rock. Contains phenocrysts of *biotite* and *pyroxene* in fine-grained groundmass of *pyroxene*, *magnetite* and *biotite* in base of *nepheline* (altered to *muscovite* and *chlorite*) and less *Microcline*.
Occurrence: Norway—West of Melteig (Fen region).

DAMOURITE or NACRITE

Variety of *muscovite*. A *hydromica*. Small scales, passing into fine scaly or fibrous and compact. Somewhat pearly or silky lustre.

SG 2·79–2·867. Folia less elastic than *mica*. Unctuous to touch.

Occurrence: USA—Maine (Hebron, Stoneham), Massachusetts (Sterling), North Carolina (Culsagee Mine), Pennsylvania (Unionville), South Carolina (Laurens County). Austria—Salzburg. France—Pontivy.

Derived from alteration of *kyanite, topaz* or *corundum*.

DANAITE

8[(Fe,Co)AsS] with up to 12 per cent cobalt in lattice. Cobalt-rich variety of *arsenopyrite*.

Occurrence: USA—New Hampshire (Franconia).

DANALITE

$(Fe,Zn,Mn)_8Be_6Si_6O_{24}S_2$ Zinc manganese beryllium and ferrous iron sulphide-silicate. Cubic. Octahedral or dodecahedral crystals with long, striated faces. Flesh-red to grey. Vitre-resinous lustre. Streak lighter than colour. Subconchoidal to uneven fracture. Brittle. No cleavage. Hardness 5·5–6·0. SG 3·427. Easily fuses to black enamel. Decomposes in hydrochloric acid, with evolution of hydrogen sulphide and separation of gelatinous silica.

Occurrence: USA—Colorado (El Paso County), Massachusetts (Cape Ann, Gloucester), New Hampshire (Bartlett). As constituent of *granite*.

Member of *helvine* family.

DANBURITE

4[$CaB_2Si_2O_8$] Calcium boron silicate. Orthorhombic. Small, well-developed, tetrahedral or cubic crystals; massive compact, nodular or fine fibrous. Yellow, greyish-yellow or brownish. Vitreous lustre. White streak. Conchoidal, uneven fracture. Brittle. No cleavage. Hardness 7·0. SG 2·9–3·0. F 3·5 to colourless *glass*. Imparts green colour to oxidising flame. Gelatinises with hydrochloric acid when previously ignited. Massive variety resembles

fine-grained *marble*. Crystals used as gemstones.

Occurrence: USA—Connecticut (Danbury), New York (Russell). Switzerland—South of Dissentis. Bolivia. Mexico—San Luis Potasí (Charcas). Burma. Japan—Miyazaki. Madagascar. With *gypsum, anhydrite, halite* and *carnallite*.

DANCALITE

Igneous rock intermediate in composition between the *syenite* and *diorite* clans. A porphyritic extrusive rock. Contains phenocrysts of sodic *oligoclase*, green *augite* and rare *amphibole* in trachytic groundmass of *plagioclase* laths and interstitial *analcite*.

Occurrence: Ethiopia—Dancala.

DANNEMORITE

2[(Fe,Mn,Mg)$_7Si_8O_{22}(OH)_2$] with iron $c5$, and magnesium and manganese $c1$. Manganese-rich member of *cummingtonite-grünerite* series with composition approaching *grünerite*. Monoclinic. Fibrous to asbestiform. Yellowish-brown to greenish-grey. Fuses to slag when heated in thin pieces.

Occurrence: Sweden—Dannemora, Nävekvarn (Uttersvik).

A clino-*amphibole*.

DAPHNITE

2[($Fe^{2+},Al)_6(Si,Al)_4O_{10}(OH)_8$] with silicon = 2·5–2·8. Hydrated aluminium ferrous iron silicate. Monoclinic. Small, spherical or botryoidal aggregates with concentric radiate-foliated structure. Dark green; basal plane olive-green and yellow normal to basal plane. Pearly lustre. Green streak. Somewhat flexible laminae. Basal, perfect cleavage. SG 3·172. Becomes black in flame but does not exfoliate and fuses easily to steel-grey bead.

Occurrence: England—Cornwall (Penzance). As incrustation on *quartz* and *arsenopyrite*. Member of *chlorite group*. Variety is *bavalite*.

DARAPSKITE

$Na_3NO_3SO_4.H_2O$ Hydrated sodium nitrate sulphate. Monoclinic. Tabular and pseudo-tetragonal crystals. Colourless. Perfect cleavage in one direction; cleavage or parting in second direction. Hardness 2·5. SG 2·2. Easily soluble in water.

Occurrence: USA—California (San Bernardino County). Chile—Antofagasta Province, Atacama Province, Chuquicamata. In nitrate deposits, especially those rich in sulphates.

DARWIN GLASS

Synonym of *queenstownite*.

DASHKESANITE

2[(Na, K)Ca$_2$(Fe^{2+}, Mg, Fe^{3+})$_5$(Si, Al)$_8O_{22}Cl_2$] with ratio of sodium to potassium $c1$, ratio of ferrous to magnesium to ferric iron $c3:1:1$ and ratio of silicon to aluminium $c3:$. Aluminium sodium potassium calcium magnesium ferric and ferrous iron chloride-silicate. Monoclinic. Prismatic crystals. Greenish-black. Perfect cleavage in one direction. Hardness 4·5–5·0. SG 3·59. Fuses to black *glass*. Decomposes in hydrochloric acid.

Occurrence: USSR—Transcaucasia (Dashkesan).

A clino-*amphibole*. Related to *hastingsite*.

DATOLITE

4[CaBSiO$_4$OH] Basic calcium boron silicate. Monoclinic. Highly modified tabular, prismatic or pyramidal crystals; massive compact, fibrous, granular, botryoidal. Pink, red, reddish-violet, pale green, olive-green, yellow, brown, colourless, greenish-white, grey. Vitreous to greasy, dull lustre. White streak. Conchoidal, uneven fracture. Brittle. No cleavage. Hardness 5·0–5·5. SG 2·9–3·0. F 2·0. Fuses with intumescence to clear *glass*, colouring flame bright green. Gelatinises with hydrochloric acid. Massive variety used as gemstone.
Occurrence: England—Cornwall. Scotland—Dunbartonshire and Stirlingshire (Kilpatrick Hills), Perthshire, Salisbury Craigs. USA—California, Connecticut, Lake Superior region, Massachusetts, Minnesota, New Jersey, New York. Australia—Tasmania. Austria—the Alps. Czechoslovakia—Bohemia (near Prague). France—Alsace. Germany—Bavaria, Harz Mountains, Thuringia. Italy—-the Alps. Norway—Arendal. Sweden—Utö. Secondary mineral occurring in veins and cavities in basic eruptive rocks.
Variety is *bakerite*.

DAUPHINITE

Synonym of *anatase*.

DAVAINITE

Igneous rock of the *ultrabasic* clan. A black, granular plutonic rock. Contains *pyroxene*, brown *hornblende* and a little *orthopyroxene*.
Occurrence: Scotland—North East of Loch Beinn (Garabal Hill).
Rock possibly metamorphic, in which case classed as *amphibolite*.

DAVIDITE

UFe$^{2+}_3$Fe$^{3+}_2$Ti$_8$O$_{22}$ with uranium oxide *c* 18 per cent and cerium group of rare earths. Uranium ferric ferrous iron titanate. Imperfectly studied. Forms rounded grains or big, rough, cuboidal crystals. SG *c* 4·48. Highly radioactive.
Occurrence: Southern Australia—Radium Hill. Mozambique—Tete district (chemically similar material). In *pegmatite* formations.

DAVYNE

(Na, K, Ca)$_{6-8}$Al$_6$Si$_6$O$_{24}$(SO$_4$, Co$_3$, Cl$_2$)$_{1-2}$ Member of *cancrinite* series containing considerable chlorine. Hexagonal. Crystals resemble *nepheline*. Colourless to white. Vitreous to pearly lustre on cleavage. Conchoidal fracture. Perfect basal and prismatic cleavage. Hardness 5·5. SG 2·4. Fuses in flame with intumescence to clear, slightly blebby *glass*, colouring flame yellow.
Occurrence: Italy—Monte Somma. With *nepheline*.

DAWSONITE

4[(NaAlCO$_3$(OH)$_2$)] Basic sodium aluminium carbonate. Orthorhombic. Thin incrustations or rosettes of bladed to acicular crystals or tufts of fine needles. Colourless to white. Vitreous lustre, silky in fine aggregates. Colourless streak. Perfect cleavage in one direction. Hardness 3·0. SG 2·44. Soluble with effervescence in acids. When heated in flame swells, but does not fuse.
Occurrence: Canada—Quebec (Montreal). Italy—Tuscany (Santa Fiora in Siena). Northern Albania—Drin Valley (Komana). Algeria—Alger (East of Tenès).

DECOMPOSITION

Process of chemical breakdown following the action of heat, chemical reagents or living matter.

DECREPITATION

Fracture of crystals on heating, often violent, due to forces of expansion of components such as water.

DEHYDRATION

Removal of water, usually when it is chemically combined in a substance.

DELAFOSSITE

CuFeO$_2$ Copper iron oxide. Hexagonal. Tabular to equant crystals; massive as botryoidal crusts. Black. Metallic lustre. Black streak. Brittle fracture. Imperfect cleavage in one direction. Hardness 5·5. SG 5·41. Weakly magnetic. Easily fusible. Becomes magnetic on heating. Readily soluble in hydrochloric and sulphuric acids; insoluble in nitric acid.
Occurrence: USA—Arizona (Bisbee), Idaho (near Salmon), Nevada (Eureka, Kimberley). Germany—Oberpfalz (Pfaffenreuth). Spain—Sevilla (Pedroso). Mexico—Sonora (Copreasa Mine in Sonoripa district). USSR—Siberia.

DELDORADITE

Igneous rock of the *syenite* clan. A leucocratic, coarse grained plutonic rock. Contains *microperthite* with *cancrinite* and accessory *biotite, aegirine, apatite, titanite* and *iron* ore.
Occurrence: USA—Colorado (Deldorado Creek).

DELESSITE or CHLOROPITE

2[(Mg,Fe^{2+},Fe^{3+},Al)$_6$(Si,Al)$_4$O$_{10}$(O,OH)$_8$]

with silicon greater than 3·1 and ferric iron and hydroxide = 8. Hydrated magnesium aluminium ferric and ferrous iron silicate. Monoclinic. Massive, with short fibrous or scaly, feathery texture, often radiated. Olive-green to blackish-green. Grey or green streak. Hardness 2·5. SG 2·89. Easily soluble in acids with deposition of silica.
Occurrence: Scotland. Canada—Nova Scotia. France—near Miélan (La Grève). Germany—Zwickau. Coats or fills cavities of amygdaloidal *porphyry*.
Member of *chlorite group*.

DELIQUESCENCE

Conversion to liquid on exposure to air, often by means of solution in atmospheric water.

DELLENITE

Igneous rock of the *granite* clan. A fine-grained extrusive rock. Contains basic *plagioclase* with *orthoclase*, occasional *quartz* and some *hornblende*. Type intermediate between *rhyolite* and *dacite*.
Occurrence: Sweden—Helsingland (Dellen).

DELTAIC

Type of sediment laid down in river delta. Structurally and texturally complex.

DEMANTOID

Variety of *andradite* with calcium replaced by traces of chromic acid. Cubic. Massive, as rolled pebbles. Grass-green to emerald-green. Brilliant lustre. Hardness 6·5. SG 3·82–3·85. Gemstone showing characteristic inclusion of *asbestos* fibres with radiating arrangement.
Occurrence: USSR—Ural Mountains. Congo.

DENDRITIC

'Tree-like', branching.

DENDRITIC AGATE

Variegated variety of *chalcedony*. Has brown or black dendritic markings due to visible impurities distributed throughout the mass.
Occurrence: Worldwide. Forms nodules.

DENSE

A material with a high density.

DENSITY

Mass (or less strictly, weight) per unit volume, measured in kilogrammes per cubic metre (kg/m^3) or in grammes per cubic centimetre (g/cm^3).
See specific gravity.

DENSITY, RELATIVE

See specific gravity.

DENUDATION

The end result of processes such as weathering and erosion which cause a lowering of the land surface.

DERMOLITH

Synonym of *pahoehoe*.

DESCLOIZITE

4[Pb(Zn,Cu)VO$_4$OH] with proportion of copper smaller than zinc. Basic lead copper zinc vanadate. Orthorhombic. Small, pyramidal, prismatic crystals, rarely tabular or short prismatic; drusy crusts, stalactitic or massive fibrous, granular, friable. Cherryred to yellowish-brown, brownish-red to blackish-brown, green or black. Greasy lustre. Orange to brownish-red or yellowish streak. Small conchoidal to uneven fracture. Brittle. No cleavage. Hardness 3·0–3·5 on fracture surfaces, harder on external faces. SG 6·2. Easily soluble in acids. Easily fusible.
Occurrence: England—Cheshire, Shropshire (near Shrewsbury). USA—Arizona (Pinal County), New Mexico. Rhodesia—Broken Hill. South West Africa—Otavi region. Austria—Carinthia (Obir). Argentina—Sierra Córdoba. Mexico—San Luis Potosi. Algeria. Congo—Katanga. Tunisia. Secondary mineral

found in oxidised zone of ore deposits chiefly associated with *vanadinite*.
Forms series to *mottramite*.

DESERT ROSES

Concretions of *barytes* or sometimes *gypsum* in *sandstone,* or clusters of platy crystals containing sand. *Barytes* acts as cement for sand. Forms reddish-brown or pink rosettes. Sandy texture.
Occurrence: USA—Oklahoma (Norman). *Gypsum* 'roses' occur in many areas where *gypsum* has been dissolved in percolating waters which are drawn to the surface by capillary action and evaporate.

DESTINEZITE

[Fe^{3+}$_2$PO$_4$SO$_4$OH.5H$_2$O] Basic hydrated iron sulphate-phosphate. Triclinic. Microcrystalline in reniform, nodular or earthy masses, or as microscopic six-sided plates of various habits. Yellow to yellowish-brown and brown, reddish-brown, greenish-yellow, pale greenish or yellowish-white. Dull lustre. Earthy, uneven to conchoidal fracture. Pulverulent to brittle. Hardness less than 3·0. SG 2·0–2·4. Easily soluble in acids. When heated in closed tube, affords much acid water and turns brownish-red.
Occurrence: USA—California (San Benito County). Austria—Styria (Leoben). Belgium—Argenteau (Visé), Védrin. Czechoslovakia—Bohemia. France—Finistère (Huelgoat), Isère (Peychagnard). Germany—Thuringia (Ansbach). Hungary (Eisenbach, Vashegy). In gossan of pyritic deposits, as recent deposit in mine workings or formed by action of sulphate solutions on secondary phosphate minerals.

DETRITUS

Particles of minerals or rocks which have been derived from pre-existing rock by processes of weathering and/or erosion.

DEVITRIFICATION

Loss of glassy nature as crystallisation occurs.

DEVONIAN

See *geological time*.

DEVONITE

Igneous rock of the *gabbro* clan. A porphyritic hypabyssal rock. Contains colourless phenocrysts of potash-rich *labradorite* in dark green groundmass of altered *plagioclase* and *augite*. Type of *diabase-porphyry*.

Right: desert rose

123

Occurrence: USA—Missouri (Mount Devon in Madison County).

DEWEYLITE or GYMNITE

Mixture of *stephanite*, *chrysotile* and *lizardite*. Amorphous, resembling gum arabic. Brittle and often much cracked. Whitish, yellowish, wine-yellow, greenish or reddish. Greasy lustre. Imperfect conchoidal fracture. Hardness 2·0–3·5. SG 2·0–2·2. Decomposes in hydrochloric acid. Gives off much water when heated in closed tube. **Occurrence:** USA—Massachusetts (Middlefield), Pennsylvania (Texas). Austria—Tyrol. Germany—Bavaria (Passau). Member of *serpentine* group.

DIABANTITE

$2[(Mg,Fe^{2+},Al)_6(Si,Al)_4O_{10}(OH)_8]$ Hydrated aluminium magnesium ferrous iron silicate. Possibly monoclinic. Massive compact, fibrous or with foliated, radiating and concentric structure. Dark green to greenish-black. Perfect basal cleavage. Hardness 2·0–2·5. SG 2·7–2·93. Dissolves in hydrochloric acid leaving silica skeleton. **Occurrence:** USA—Connecticut (Farmington Hills), Massachusetts (Turner's Falls). Germany—Frankenwald. In seams, clefts and amygdaloidal cavities in *diabase*. Member of *chlorite group*.

DIABASE

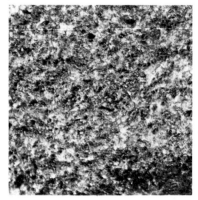

General term used in Britain and Europe to describe rocks of doleritic composition altered to such an extent that few of the original minerals have survived. In USA term applies to rocks described as *dolerites* by British petrologists.

DIABOLEITE

$Pb_2CuCl_2(OH)_4$ Lead copper hydroxide-chloride. Tetragonal. Tabular crystals with square outline, sometimes subparallel aggregates of thin plates. Deep blue. Vitreous to adamantine lustre. Pale blue streak. Conchoidal fracture. Perfect cleavage but not easy in one direction. Hardness 2·5. SG 5·42. When heated in closed tube breaks into cleavage fragments with evolution of water followed by vaporisation of lead chloride. Residue melts to form brown liquid which becomes bright green *glass* on cooling. **Occurrence:** England—Somerset (Mendip Hills). USA—Arizona (Mammoth Mine near Tiger).

DIADOCHITE

$cFe^{3+}_2PO_4SO_4OH.5H_2O$ Basic hydrated ferric iron sulphate phosphate. Amorphous. Same properties and composition as *destinezite*. Waxy, horn-like or subvitreous lustre. Hardness 3·0–4·0. **Occurrence:** With *destinezite*, particularly in Germany—Thuringia (Ansbach).

DIAGENESIS

Processes which affect sedimentary materials after deposition, excluding erosion and weathering as well as true metamorphism into which it merges. Consolidation and cementation main processes; many others of some degree of importance.

DIALLAGE

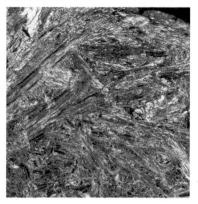

Variety of *augite* with composition near *diopside*, but with much alumina. Lamellar or thin-foliated; also fibrous. Greyish-green to bright grass-green and deep green; also brown. Lustre often pearly, sometimes metalloidal or exhibiting schiller from presence of microscopic inclusions of secondary origin. No cleavage but parting parallel to orthopinacoid. Hardness 4·0. SG 3·20–3·35. **Occurrence:** Scotland—Balta Islands. USA—Maine, Massachusetts. New Zealand—Dun Mountain. Germany—ehrenburg. Italy—Island of Elba. Norway—Kyrkjebo. Poland—Silesia (Buchberg). Sweden—Åkerö. In *gabbro* and other related rocks.

DIALLAGITE

Igneous rock of the *ultrabasic* clan. A fine-grained hypabyssal rock. Contains mostly *diallage* with mainly accessory *hornblende, garnet, pyroxene, olivine, feldspar* and *pyrite*. **Occurrence:** New Caledonia—Thio. Rhodesia—Bushveld Complex.

DIALOGITE

Synonym of *rhodochrosite*.

DIAMANTIFEROUS

Diamond-bearing.

DIAMOND

$8[C]$ Carbon. Cubic. Crystals predominantly octahedral, surfaces usually much curved and often striated; massive as rounded or irregular grains or pebbles, often with internal radial structure. White to bluish-white, shades of yellow or brown;

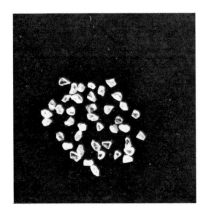

occasionally orange, pink, mauve, green, blue, red or black. Adamantine to greasy lustre. Ash-grey streak. Conchoidal fracture. Brittle. Perfect octahedral cleavage. Hardness 10·0. SG 3·5–3·53. Unattacked by acids or alkalis.
Occurrence: USA—Arkansas, California, Colorado, Georgia, North Carolina, Oregon, Virginia, Wisconsin. Australia—New South Wales, Queensland, Victoria. Rhodesia. South Africa—Cape Province (Kimberley, Namaqualand), Orange Free State (Jagersfontein), Transvaal (near Lichtenburg, Premier). South West Africa—Lüderitz Bay. Bolivia. Brazil—mainly Minas Gerais. Venezuela. China. India. Indonesia—Borneo. Africa—particularly Botswana, Lesotho. In pipes, alluvial fan, wave-concentrated deposits on shore-lines, in *conglomerates* and *meteorites*.
Non-gemstone varieties are *bort, hailstone bort, shot bort* and *stewartite*.

DIASPORE or KAYSERITE

4[AlÒ.OH] Hydrous aluminium oxide. Orthorhombic. Thin, platy and elongated crystals, sometimes acicular; massive, stalactic or disseminated. White, greyish-white, colourless, greenish-grey, hair-brown, yellowish, lilac or pink. Brilliant lustre, pearly on cleavage, vitreous elsewhere. White streak. Conchoidal fracture. Very brittle. Pinacoidal, conspicuous

cleavage. Hardness 6·5–7·0. SG 3·3–3·5. Not attacked by acids, but soluble in sulphuric acid after ignition.
Occurrence: England—Cornwall (Botallack, Kenidjack). USA. South Africa—West Griqualand. France—with *bauxite*. Greece—Islands of Naxos, Nikaria and Samos. Greenland. Hungary—near Schemnitz. Poland—Silesia. Southern Norway. Sweden—Wermland. Switzerland—Tessin. Turkey—Ephesus. USSR—Ural Mountains. China. Japan. Mainly with *corundum, emery, dolomite, margarite, chlorite* and *magnetite*. Dimorphous with *boehmite*.

DIATOMACEOUS EARTHS

Sedimentary rocks. Siliceous organic deposits formed by accumulation of frustules of diatoms. Composed of nearly pure silica. Friable and earthy.
Occurrence: Worldwide in oceans, freshwater lakes and swamps.

DIATOMITE

Sedimentary rock. A type of *diatomaceous earth*. Marine deposit. A fairly hard rock containing abundant remains of diatoms lithified by *calcite*, silica or calcium phosphate.
Occurrence: USA—California, Maryland, Virginia.

DIATOMS

Microscopic plants.

DICHROÏTE

Synonym of *cordierite*.

DICKÏTE

4[Al₂Si₂O₅(OH)₄] Basic aluminium silicate. Monoclinic. Well-crystallised; commonly piles of platelets. SG 2·6. Properties similar to *kaolinite*.
Occurrence: Wales—Anglesey (Amlwch). USA—Colorado (Red Mountain), Pennsylvania (Schuylkill County). Poland—

Silesia (Ruben Coal Mine in Neurode). Japan—Shôkôzan.
Member of *kaolinite* family.

DIDYMITE

Variety of *muscovite*. Contains calcium carbonate as impurity. Forms greenish or greyish-white fine scales.
Occurrence: Austria—Tyrol (Zillerthal).

DIGENITE

4[Cu₂S] Copper sulphide, usually with some copper deficiency. Cubic. Crystals rare, octahedral; usually massive. Blue to black. Conchoidal fracture. Brittle. No cleavage. Hardness 2·5–3·0. SG 5·546. Soluble in nitric acid. When heated on charcoal, melts to globule which boils with spurting.
Occurrence: USA—Alaska (Kennecott), Arizona (United Verde Mine at Jerome), Montana (Butte). South West Africa—

Ehlerts, Khan, Tsumeb. Sweden—Kiruna. Mexico—Sonora (Cananea). With *chalcocite*, *chalcopyrite* and other *copper* ores as replacement of *bornite*.

DILUTE

A dilute solution is one with a low concentration of the solute in question. In this context the word will normally be applied to testing reagents; it is rarely used to describe the concentration of a component in a rock.
See *solution*.

DIMORPHITE

As_4S_7 Arsenic sulphide. Orthorhombic. Dipyramidal crystals, usually as groups of minute, parallel individuals. Orange-yellow. Almost adamantine lustre. Brittle. No cleavage. Hardness 1·5. SG 2·58. Melts on heating, turns red, then brown, gives off abundant yellow fumes, ignites and burns without leaving a residue.
Occurrence: Italy—Phlegraean Fields (Solfatara). At a fumarole with *realgar, sal ammoniac*, sulphur and various sulphates.

DIMORPHOUS

See *polymorphous*.

DIOPSIDE, ALACOLITE or MALACOLITE

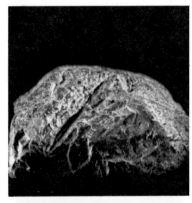

$4[MgCaSi_2O_6]$ with some ferrous oxide. Magnesium calcium silicate. Monoclinic. Prismatic, often slender crystals; also lamellar massive to granular and columnar. White, yellowish, greyish-white to pale

green, dark green to nearly black; sometimes transparent and colourless. Vitreous lustre. White or grey streak. Uneven fracture. Brittle. Perfect prismatic cleavage. Hardness 5·5. SG 3·2–3·38. F 3·75. Gives magnetic globule when fused on charcoal. Gemstone.
Occurrence: USA—California, New York. Canada—Quebec (Three Rivers). Austria—Tyrol. Finland. Italy—Piedmont (Ala Valley, Traversella), Rumania. Sweden—Nordmark. Switzerland—Zermatt. Brazil—Minas Gerais. USSR—Siberia (Lake Baikal), Ural Mountains. Sri Lanka.
A *clinopyroxene*. Varieties are *chrome-diopside, coccolite, enstatite-diopside, violane, sahlite, mansjöite, alalite*.

DIOPSIDE-JADEITE

Synonym of *tuxtlite*.

DIOPSIDITE

Igneous rock of the *ultrabasic* clan. A fine-grained hypabyssal rock. Contains essentially *diopside* with accessory *iron* ore, *ceylonite* and *garnet*.
Occurrence: France and Spain—the Pyrenees. USSR—Ural Mountains (Krasnouralsky).

DIOPTASE

$6[CuSiO_2(OH)_2]$ Hydrous copper silicate. Hexagonal. Crystals commonly small, prismatic; also indistinct crystalline aggregates or massive. Emerald-green to dark green. Vitreous lustre. Green streak. Conchoidal, uneven fracture. Brittle. Perfect cleavage in one direction. Hardness 5·0. SG 3·3. Decrepitates in flame, which is coloured emerald-green. Infusible. Gelatinises in hydrochloric acid. Has been used as gemstone.

Occurrence: USA—Arizona. Rumania—Rézbanya. Chile—Atacama Desert. USSR—Kirghiz Steppes, Siberia (Transbaikal). Congo—Katanga, basin of Niari River. In *limestone* with *quartz* and other *copper* minerals.

DIORITE

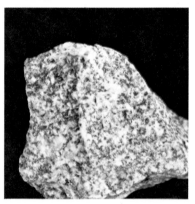

Clan of igneous rocks in which *plagioclase feldspar* is in excess of alkali-*feldspar* and *quartz* is absent. *Plagioclase* is within *oligoclase-andesine* range associated with mafic minerals (mainly *hornblende, biotite, augite* and *hypersthene*) and accessories (mainly *sphene, apatite* and *magnetite*). Type-rocks of three sub-divisions are *diorite* (coarse-grained plutonic), micro-*diorite* (medium-grained hypabyssal) and *andesite* (fine-grained extrusive).

DIPYRAMID, -AL

In form of two pyramids base to base.

DIPYRE or MIZZONITE

Calcium sodium aluminosilicate, sulphate, carbonate and chloride. Intermediate between *meionite* and *marialite* with ratio from 3:1 to 1:1. Tetragonal. Small, prismatic or elongated crystals with prism faces vertically striated. Colourless to white. Vitreous lustre. Distinct cleavage in two directions. Hardness 5·5–6·0. SG 2·633.

Occurrence: Canada—Ontario (Haliburton, Monmouth Township), Quebec (Huddersfield Township). Australia—Queensland. Czechoslovakia—Moravia (Ruda). France and Spain—the Pyrenees. Southern Norway. Ghana. In *limestone* and crystalline *schists*.
Member of *scapolite* group.

DISCOID

Disc-shaped.

DISLOCATION

Slippage along a crystal plane or on a larger scale a bedding plane in a rock formation.

DISLOCATION METAMORPHISM

Same as dynamic metamorphism.

DISSEMINATED

Spread out, dispersed, as applied particularly to particles of ore mineral in the substrate rock.

DISTHENE

Synonym of *kyanite*.

DITRÓITE

Igneous rock of the *syenite* clan. A coarse to fine-granular plutonic rock. Contains *nepheline*, *microline-microperthite*, *sodalite* and *cancrinite* with *biotite* or *aegirine-augite* rimmed with *arfvedsonite*.
Occurrence: Rumania — Transylvania (Ditró).

DIVALENT

Otherwise 'bivalent'; associated in pairs.

DIXENITE

$c(Mn, Fe, Cu)_{15}As_5^{3+}O_{13}(OH)_7((Si, P, As)O_4)_3$ Basic hydrated silicate. phosphate and arsenate of manganese iron copper arsenic oxide. Hexagonal. Rhombohedral scales occur as aggregates of thin foliae. Black. Resinous to metallic lustre. Hardness $3 \cdot 0 - 4 \cdot 0$. SG $4 \cdot 2$.
Occurrence: Sweden—Långban. In crevices in *dolomite* with native lead or in *hematite*.

DODECAHEDRAL

See *crystal*.

DOLERINE

Metamorphic rock. A variety of *talc-schist*. Contains *feldspar* and *chlorite* as chief varietal minerals.

Occurrence: Italy and Switzerland—the Pennine Alps.

DOLERITE

Igneous rock of the *gabbro* clan. A dark-coloured, generally fresh basaltic, medium-grained hypabyssal rock. Consists essentially of *labradorite* and *pyroxene* with optional *olivine*. Term sometimes expanded to include all rocks of appropriate composition regardless of age, texture or mode of occurrence. Latter synonymous with American usage of *diabase*.

DOLOMITE, PICRITE (of Brongniart) or TARASPITE

$CaMg(Co_3)_2$ Calcium magnesium carbonate. Hexagonal. Rhombohedral crystals with curved surfaces; massive coarsely crystalline, compact, granular, friable. Colourless and transparent when pure; grey, greenish, yellowish-brown or brown

with *iron* impurities; pink, rose or flesh-red with manganese impurities. Vitreous, pearly lustre. White or grey streak. Conchoidal to subconchoidal fracture. Brittle. Perfect rhombohedral cleavage. Hardness $3.5 - 4 \cdot 0$. SG $2 \cdot 9$. Infusible in flame, but glows brightly. Finely powdered mineral dissolves readily with effervescence in warm acids.
Occurrence: Of crystals only. England—Cornwall, Cumberland. USA. Canada—Nova Scotia, Ontario. Austria—Salzburg, Tyrol. France—Alsace, Savoie. Germany—Saxony. Italy—Piedmont, Tuscany. Rumania. Switzerland. Brazil. Colombia. Mexico. Algeria. Crystallised as pearly masses in old sedimentary rocks. Forms group with *ankerite* and *kutnahorite*. Varieties are *miemite, pearl spar, rhomb spar, bitter spar* and *brown spar*.

DOLOMITE ROCKS

Sedimentary rocks of secondary origin. A type of magnesian *limestone* near *dolomite* in composition. Formed from *limestone*, coral or *marble* by action of solutions containing magnesium. Mostly mixtures of *dolomite* and *calcite*. Siliceous types have *diopside* and other magnesium silicates.
Occurrence: Worldwide among rocks of all geological ages or as massive rocks forming extensive beds.

DOME

A type of rock fold—an anticlinal structure which plunges in all directions.

DOMEYKITE

16[Cu₃As] Copper arsenide with some nickel and cobalt. Cubic. Massive, reniform or botryoidal. Tin-white to steel-grey; tarnishes yellowish to pinchbeck-brown and iridescent. Metallic lustre but dull on exposure. Uneven fracture. Hardness 3·0–3·5. SG 7·2–7·9. F 2·0. Insoluble in hydrochloric acid but soluble in nitric acid.

Occurrence: England—Cornwall (near Helston). USA—Michigan (Houghton County). Canada—Ontario (Michipicoten Island on Lake Superior). Germany—Saxony (Zwickau). Sweden—Långban. Chile—Atacama Desert, Coquimbo, Rancagua. Mexico—Cerrode Paracatas. Intergrown with *algodonite*.

Isomorphous with *algodonite*.

DOMITE

Igneous rock of the *syenite* clan. A light-coloured, fine-grained extrusive rock. Porous and friable. Contains sodic *sanidine*, *oligoclase*, *biotite* and sometimes *hornblende* with *tridymite* in pores in the rock.

Occurrence: France—Auvergne (Puy de Dôme).

DOPPLERITE (of Haidinger)

Hydrocarbon compound. A calcium salt of a humic acid. Amorphous. Earthy; elastic when dry or as partly jelly-like masses or thin plates. Brownish-black when fresh. Greasy lustre; subvitreous to adamantine when dry. Dull brown streak. Hardness 0·5; hardness after drying 2·0–2·5. SG 1·089–1·466 (after drying). Tasteless. Insoluble in alcohol or ether.

Occurrence: Republic of Ireland—County Limerick. Austria—Styria (near Aussee). Switzerland—Appenzell (Bad Gonten), Obbürgen, Unterwalden (near Stansstad). In *peat* beds.

Formed from plant carbohydrates by bacterial action, or chemical reaction between carbohydrates and amino acid.

DORGALITE

Igneous rock of the *gabbro* clan. A holocrystalline extrusive rock. Contains phenocrysts of *olivine* and abundant *plagioclase* in groundmass of *magnetite, olivine, augite* and *labradorite*.

Occurrence: Italy—Sardinia (Monte Pirische near Dorgali).

DRAKONITE

Igneous rock of the *syenite* clan. A light, white, yellowish or grey porphyritic extrusive rock. Contains phenocrysts of *sanidine*, *plagioclase* (*oligoclase* to *labradorite*), *biotite* and/or *hornblende* in groundmass of alkali-*feldspar*, *diopside* and alkali-*amphibole*.

Occurrence: Germany—Siebengebirge (Drachenfels).

DRAVITE

*c*NaMg₃Al₆B₃Si₆O₂₇(OH)₄ Hydrous sodium magnesium aluminium and boron silicate. Magnesium-rich variety of *tourmaline*. Brown, greenish-black or brownish-black. White in calcium-rich variety. Gemstone.

Occurrence: England—Cornwall (Dinas Head). USA—New York (Gouverneur). Austria—Carinthia (Unterdrauburg in the Drave district). Czechoslovakia—Dobrowa.

DRUSE

A cavity in a rock or mineral often filled or partially filled with large crystals.

DUCTILE

Capable of being fashioned into a new form; capable of being drawn out into a thread.

DUFRENITE

Fe²⁺Fe³⁺₄(PO₄)₃(OH)₅·2H₂O Basic hydrous ferric ferrous iron phosphate. sometimes with appreciable aluminium. Monoclinic. Crystals rare, indistinct, rounded in subparallel or sheaf-like aggregates; commonly botryoidal masses or crusts with radial, fibrous structure. Dark green or olive-green to greenish-black in

fresh material and olive-brown to reddish-brown with increasing oxidation. Vitreous to silky or earthy lustre. No fracture. Two cleavages (one perfect) parallel to fibre direction; traces of third cleavage at right angles to other two. Hardness 3·5–4·5. SG 3·1–3·34. Soluble in dilute acids.

Occurrence: England—Cornwall (Wheal Phoenix). USA—Alabama (Rock Run in Cherokee County). Germany—Hesse, Saxony, Thuringia, Westphalia.

Secondary mineral found with *limonite* in gossan of veins and in *iron* ore deposits. Closely related to *frondelite*.

DUFTITE

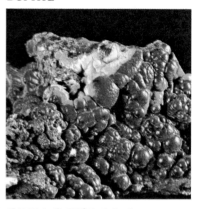

4[CuPbAsO₄OH] Basic lead copper arsenate. Orthorhombic. Aggregates. Olive green to greyish green. Vitreous to dull lustre. Pale green to white streak. Hardness 3·0. SG 6·98.

Occurrence: South West Africa—Tsumeb.

DUMALITE

Igneous rock intermediate in composition between the *syenite* and *diorite* clans. A fine-grained extrusive rock. A *trachyandesite*. Contains potash-*feldspar*, acid-*plagioclase*, *augite* and minor ore in glassy interstitial material, potentially *nepheline*.

Occurrence: USSR—Caucasus (Dumala).

DUMONTITE

Pb₂(UO₂)₃(OH)₄(PO₄)₂.3H₂O Basic hydrated lead uranium phosphate. Orthorhombic. Small flattened and elongated crystals. Ochre-yellow. Ochre-yellow streak. Soluble in acids. When heated in closed tube, loses water and turns orange.

Occurrence: Congo—Katanga (Chinkolobwe). With *torbernite*.

DUMORTIERITE

4[(Al,Fe)₇BSi₃O₁₈] Boron aluminium ferric iron silicate. Orthorhombic. Small, prismatic or pseudohexagonal crystals; massive fibrous, columnar or radial aggregates. Pink, reddish-violet, lavender, greenish or darkish blue. Vitreous to silky lustre. White to bluish-white streak. Conchoidal

fracture. Brittle. Distinct pinacoidal cleavage. Hardness 7·0. SG 3·3. Cobalt solution turns blue. Whitens when heated on charcoal, but colour returns on cooling. Ornamental use.
Occurrence: USA—Arizona (Yuma County), California (Los Angeles County), Nevada (Oreana), New York Island (near Harlem), Washington. Canada. France—near Lyons. Norway—Tvedestrand. Poland—Silesia (Wolfshau). China. Sri Lanka. Madagascar—near Soavina. In *pegmatite* veins and lenticular masses with *quartz, muscovite, andalusite* and *kyanite*.

DUNGANNONITE

Undersaturated igneous rock of the *diorite* clan. A white or grey, coarse-grained plutonic rock. Contains dominant *plagioclase*, with *biotite, muscovite, calcite, magnetite* and large *corundums, scapolite* and *nepheline*.
Occurrence: Canada—Ontario (Dungannon Township). As interfoliated masses or bands in *scapolite*-bearing *nepheline-*

syenite or in independent masses covering large areas.

DUNHAMITE

$PbTeO_3$ Lead tellurium oxide. Pale brown. Alteration product of *altaite*.
Occurrence: USA—New Mexico (Organ Mountains).

DUNITE or OLIVINE ROCK

Igneous rock of the *ultrabasic* clan. A plutonic, yellowish-green on fresh fractures, with greasy to vitreous lustre. Xenomorphic-granular rock. Contains *olivine* and *chromite* with accessory *magnetite, ilmenite, pyrope* and *spinel*.
Occurrence: New Zealand—Dun Mountain. Forms a mass 1 200m thick.

DUPARCITE

Synonym of *idocrase*.

DURAIN

Sedimentary rock. A constituent of *bituminous coal*. Consists of small, resistant débris of plants, such as fragments of cuticle, spore-cases, bits of resin or remains of algal colonies and *fusinite*. Close, firm, granular texture. Irregular and rough fracture with dull or matt surface.
Occurrence: In bands or lenticles of varying thickness in *bituminous coal*. Often flecked with hair-like streaks of bright *coal*.

DURANGITE

$4[NaAlAsO_4F]$ Sodium aluminium fluorarsenate with lithium substituting for sodium and ferric iron substituting for aluminium. Monoclinic. Oblique pyramidal crystals with dull and rough faces. Light and dark orange-red. Vitreous lustre. Cream yellow streak. Uneven fracture. Brittle. Distinct cleavage in one direction. Hardness 5·0. SG 3·94—4·07 (increasing with increasing iron content). F 2·0. Soluble in sulphuric acid.
Occurrence: Canada—Nova Scotia (near Lake Ramsay in Lunenberg County).

Mexico—Durango (Barranca Tin Mine). With *cassiterite, hematite, amblygonite* and *topaz*.

DURBACHITE

Igneous rock of the *syenite* clan. A porphyritic, hypabyssal rock. Contains phenocrysts of twinned *orthoclase* in groundmass of coarse flakes of *biotite* and *orthoclase* with accessory *plagioclase, hornblende, titanite, zircon* and occasionally a little *quartz*.
Occurrence: Germany—Baden (near Durbach between Ödsbach and Reichenbach).

DYKE

Sheet of igneous rock intruding through other strata.

DYNAMIC METAMORPHISM

Process by which strong but local forces (as in thrusting) change the nature of rocks in the crust.

DYSANALYTE

$8[(Ca,Ce,Na)(Ti,Nb,Ta)O_3]$ Calcium cerium sodium tantalate-niobate(columbate)-titanate. Proportion of niobium (columbium) is usually greater than tantalum and cerium or sodium may be absent. Variety of *perovskite* containing niobium (columbium).
Occurrence: USA—Arkansas (Magnet

Cove in the Ozark Mountains). Germany—Kaiserstuhl (Vogtsburg). Sri Lanka—Uva Province. In contact metamorphosed *limestone*.

DYSCRASITE

Ag_3Sb Silver antimonide. Orthorhombic. Pyramidal crystals, usually massive, foliated or granular. Silver-white, usually tarnishing lead-grey to yellowish or blackish. Metallic lustre. Silver-white streak. Uneven fracture. Distinct cleavage in two directions, imperfect in third. Hardness 3·5–4·0. SG 9·74. Sectile. F 1·5. Decomposes in nitric acid leaving deposit of antimony oxide.
Occurrence: USA—Nevada (Reese River district). Canada—Ontario (Cobalt). Australia—New South Wales (Broken Hill). France—Alsace (Ste Marie aux Mines). Germany—Baden (Wolfach), Harz Mountains (St Andreasberg). As vein mineral in *silver* deposits with antimonian *silver*, other *silver* minerals and *galena*, generally in a *calcite* gangue.

DYSLUITE

$8[(Zn,Mn)(Al,Fe)O_4]$ Zinc manganese aluminium and ferric iron oxide. A manganese-bearing variety of *gahnite* (*of Von Moll*).
Occurrence: USA—New Jersey (Frank-

lin, Sterling Hill). In crystalline *limestone* wall rock adjacent to ore body and occasionally in ore.

DYSSNITE

$c(Mn,Fe)_2Si_2O_7$ Trivalent manganese ferric iron silicate. Alteration product of *fowlerite*. An iron-black ore. SG 3·67.
Occurrence: USA—New Jersey (Franklin).

EARTH, RARE

See rare earth.

EARTHY

Earth-like.

EARTHY CALAMINE

Synonym of *hydrozincite*.

EASTONITE (of Hamilton)

Magnesium sodium potassium aluminosilicate. A silver-white *mica* similar in composition to a *vermiculite*. A *hydrobiotite*.
Occurrence: USA—Pennsylvania (Easton).

EASTONITE (of Winchell)

$cK_2Mg_5Al_4Si_5O_{20}(OH)_4$ Hydrated potassium magnesium aluminium silicate. Variety of *biotite*.
Occurrence: USA—Pennsylvania (Easton).

ECKERMANNITE

$2[Na_3(Mg,Li)_4 (Al, Fe)Si_8O_{22} (OH,F)_2]$ Basic sodium magnesium lithium aluminium and ferric iron fluorsilicate. Magnesium-rich clino-*amphibole* with included lithium. Ferrous iron replaces magnesium in series grading to *arfvedsonite*. Monoclinic. Dark bluish-green. Properties as for *arfvedsonite*. Hardness 5·0–6·0. SG 3·00.
Occurrence: USA—Colorado (Boulder County) (*asbestos* type). Southern Sweden—Norra Kärr. Burma—Tawmaw.

ECLOGITE

High grade regional metamorphic rock. Coarsegrained, occasionally fine, texture. Contains mostly *omphacite* and *pyrope*-rich *garnet* with main accessories of *rutile*, *quartz*, *amphibole*, *muscovite*, *clinozoisite* and *diamond*.
Occurrence: Worldwide. Notable examples: Scotland—Inverness-shire (Glenelg). USA—California. New Caledonia. Germany. Norway. Usually as lenses or layers in *schist* or *peridotite*. Also forms nodules, xenoliths or inclusions in *kimberlite, tuff* or *breccia*.

EDENITE

$NaCa_2Mg_5(Si_7AlO_{22})(OH,F)_2$ Basic sodium calcium magnesium and aluminium fluorsilicate. Iron-free variety of *hornblende*. Colourless, white, grey to pale green. SG 3·0–3·059. All other properties same as *hornblende*.
Occurrence: USA—New York (Edenville). Canada—Ontario (Eganville). Japan—Niigata Prefecture (Kotaki).
A clino-*amphibole*. Variety is *pargasite*.

EDINGTONITE

$2[BaAl_2Si_3O_{10}.4H_2O]$ Hydrated barium aluminium silicate, sometimes with appreciable calcium oxide. Tetragonal. Minute and inconspicuous crystals with slightly curved faces; also massive. White, greyish-white or pink. Vitreous lustre. White streak. Subconchoidal to uneven fracture. Brittle. Perfect cleavage in one direction. Hardness 4·0–4·5. SG 2·694. Gelatinises with hydrochloric acid. Fuses to colourless mass when heated at high temperature.
Occurrence: Scotland—Dunbartonshire and Stirlingshire (Kilpatrick Hills). On *thomsonite* associated with *harmotome*, *calcite*, *analcite* and *prehnite*.
Member of *zeolite* family.

EDOLITE

Variety of *hornfels*. Consists essentially of *feldspar* and *mica*. *Cordierite* and/or *andalusite* may be present.
Occurrence: Italy—the Alps (Edelo).

EFFERVESCENCE

Process whereby bubbles of gas escape from a liquid body.

EFFLORESCENCE

Loss at a crystal surface of water, leaving a powdery residue.

EFFUSIVE

Synonym of extrusive.

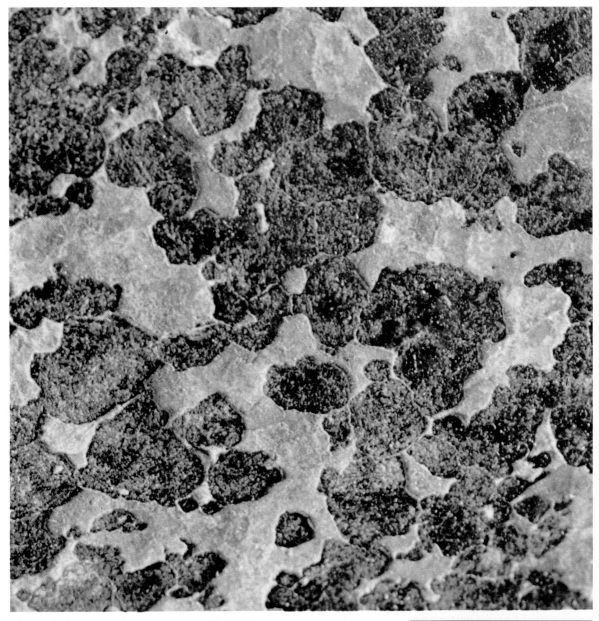

Above: eclogite
Far right: elaterite

EKANITE

$(Ca,Fe,Pb)(Th,U)Si_8O_{20}$ Thorium uranium calcium iron and lead silicate. Metamict, recrystallising tetragonal on heating. Green. Transparent to translucent. Some specimens exhibit four-rayed star. Hardness 6·0–6·5. SG 3·28. Markedly radioactive.
Occurrence: Sri Lanka—Raknapura district (Eheliyagoda). In gem gravels.

ELAEOLITE

Synonym of *nepheline.*

ELASTIC

Able to return to original form after deformation.

ELATERITE

Variety of *bitumen.* Amorphous. Dark brown. Brown streak. Soft and elastic with consistency like india-rubber when fresh; hard and brittle with conchoidal fracture on exposure. Melts without decrepitation in candle flame.
Occurrence: England—North Derbyshire. In carboniferous *limestone.*

ELBAITE (of Vernadsky)

Synonym of *rubellite.*

ELECTRIC CALAMINE

American synonym of *hemimorphite.*

ELECTRUM

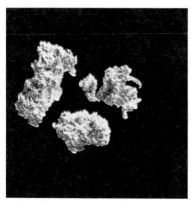

4[Au, Ag] Argentiferous gold. Cubic. Massive. Pale yellow to yellowish white. SG 12·5–15·5.
Occurrence: USA—Nevada.

ELEOLITE

A dark coloured variety of *nepheline* with a greasy lustre.

ELIE RUBY

Synonym of *pyrope*.

ELKHORNITE

Igneous rock of the *syenite* clan. A mottled, medium-grained hypabyssal rock. Contains predominantly *microcline* and soda-*orthoclase* with small amounts of *labradorite* and *augite*, with or without *biotite* and *titanite*.
Occurrence: USA—Montana (Elkhorn district).

ELLESTADITE

$2(Ca_5(S,Si)_3O_{12}(O,OH,Cl,F))$ Member of *apatite* group with silicate and sulphate substituting almost completely for phosphate. String-like. Pale rose. Resembles *wilkeite*.
Occurrence: USA—California (Crestmore). With *wollastonite*, *idocrase* and *diopside*.

ELLIPSOID,-AL

A surface of which all plane sections are ellipses or circles.

ELVAN or ELVANITE

Igneous rock of the *granite* clan. A porphyritic hypabyssal rock. Contains phenocrysts of *quartz* and *orthoclase*, often rounded or corroded, in microcrystalline groundmass of the same minerals. *Tourmaline* occurs as isolated crystals or radiating groups.
Occurrence: England—Cornwall, Devonshire (Dartmoor).

ELVANITE

Synonym of *elvan*.

EMERALD

$2[Be_3(Al,Cr)_2Si_6O_{18}]$ Beryllium chromium aluminium silicate. Gemstone variety of *beryl*. Bright emerald-green colour from chromate or trace of vanadium and *iron*.
Occurrence: USA—Connecticut, North Carolina. Australia—New South Wales, Western Australia. Rhodesia—Belingwe Reserve. South Africa—Transvaal. Austria—Salzburg. Norway—near Oslo. Brazil—Bahia, Minas Gerais, Rio Doce. Colombia—Cosquez, El Chivor, Muso. USSR—Takovaya River. India. Algeria. Mozambique. Zambia.

EMERALD NICKEL

Synonym of *zaratite*.

EMERY, EMERY-ROCK or CORUNDOLITE

Impure variety of *corundum*. Admixture of *corundum*, *magnetite*, *hematite* and *spinel*. Massive, fine to coarse granular. Black to greyish-black. Metallic lustre. Yellowish-brown, blackish-brown or black streak. Uneven fracture. Brittle to tough. Indistinct cleavage. Hardness 7·0–9·0. SG 3·7–4·3.
Occurrence: USA—Massachusetts (Chester), New York (Westchester County). France—Le Croustet. Germany—Saxony. Greece—Islands of Naxos and Samos. USSR—Ural Mountains. Asia Minor. In *limestones* as result of contact metamorphism of highly aluminous sediments.

EMERY-ROCK

Synonym of *emery*.

EMMONSITE (of Hillebrand)

$Fe_2(TeO_3)_3.2H_2O$ Hydrated ferric telluride with selenium substituting for *tellurium*. Possibly monoclinic. Fibrous crystals with small botryoidal surface; isolated globular masses; compact crystalline aggregates; radial or lichen-like groups of rough, acicular crystals or druses of thin scales. Yellowish-green. Perfect cleavage in one direction, two others at angles of 85 degrees and 95 degrees to first. Hardness *c* 5·0. SG 4·52. Melts easily to deep red liquid, loses water and leaves anhydrous residue. Easily soluble in concentrated acids.
Occurrence: USA—Arizona (near Tombstone), Colorado (Cripple Creek in Teller County), Nevada (Goldfield), New Mexico (near Silver City). Honduras—Tegucigalpa (Ojojoma). Secondary mineral from oxidation of tellurides and native *tellurium* occurring with *tellurite*, *gold*, *limonite*, *quartz* and *iron* tellurides.

EMPLECTITE

$4[CuBiS_2]$ Copper bismuth sulphide. Orthorhombic. Striated, prismatic or flattened crystals. Greyish to tin-white. Metallic lustre. Conchoidal to uneven fracture. Brittle. Perfect cleavage in one direction, distinct in second direction. Hardness 2·0. SG 6·38. F 1·0. Fuses with frothing and spurting. Decomposes in nitric acid with separation of sulphur.
Occurrence: Germany—Black Forest (near Freudenstadt), Saxony (Annaberg, Johanngeorgenstadt, near Schwarzenberg). Norway—Telemark. Chile—near Copiapó (Cerro Blanco).

ENARGITE

Cu_3AsS_4 Copper arsenic sulphide with up

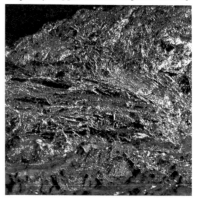

to 6 per cent antimony substituting for arsenic. Orthorhombic. Crystals common, prismatic, sometimes tabular with several grooved, almost curving prisms and flat, truncating base; also massive, granular. Greyish-black to iron-black. Metallic lustre. Greyish-black streak. Uneven fracture. Brittle. Perfect prismatic cleavage. Hardness 3·0. SG 4·4–4·5. Soluble in aqua regia. When heated in closed tube decrepitates and gives sublimate of sulphur.
Occurrence: USA—Alaska, Arkansas, California, Colorado, Louisiana, Missouri, Montana (Butte), Nevada, Utah. South West Africa–Tsumeb. Austria—Tyrol. Hungary. Italy—Sardinia. Yugoslavia. Argentina. Chile—especially Chuquicamata. Mexico—Chihuahua, Sonora. Peru. Formosa. Philippine Islands. In medium-temperature ore veins with other copper minerals and sulphides.

ENCRUSTATION

Same as incrustation.

ENDELLITE

$Al_2Si_2O_5(OH)_4 . 2H_2O$ Hydrated aluminium silicate. Hydrated form of *halloysite*. Translucent and porcelain-like. Conchoidal fracture. SG 2·11–2·17. Readily dehydrates to *halloysite*.
Occurrence: USA—Iowa (Anamosa), Maryland (Rockdale), Utah (Eureka). Libya—Djebal Deber.
Derived from basic igneous and metamorphic rocks.

ENDERBITE

Metamorphic rock. A type of *charnockite* with dioritic mineral composition. Coarse-grained texture. Contains *andesine* in excess of *microcline*, *pyroxene* (typically *hypersthene*), with *perthite* and *antiperthite*.
Occurrence: Antarctica—Enderby Land.

ENDIOPSIDE

Synonym of *enstatite-diopside*.

ENDLICHEITE

$2[Pb_5 (V,As)O_4)_3Cl]$ Variety of *vanadinite* with *arsenic* substituting for vanadium in the lattice up to 1 : 1. Various shades of yellow.
Occurrence: USA—New Mexico (Hillsboro).

ENSTATITE

$16[MgSiO_3]$ Magnesium silicate. Orthorhombic. Crystals usually coarsely crystalline aggregates; sometimes individuals. Greyish, greenish, yellowish or bronze-brown to almost black. Vitreous to silky lustre. Colourless to greyish streak. Uneven fracture. Perfect prismatic cleavage. Hardness 5·5–6·0. SG 3·2–3·9 (SG 3·26–3·28 for gemstones). F 6·0. Almost infusible. Insoluble in hydrochloric acid. Gemstone use.
Occurrence: USA—Colorado, Maryland, New York (Brewster), Pennsylvania. South Africa—diamond fields, especially near Kimberley. Austria — Styria, Tyrol. Czechoslovakia—Moravia. France —Vosges. Germany—Bavaria, Eifel district, Harz Mountains. Norway—Kjörrestad. Poland—Silesia. Sri Lanka. Upper Burma—Mogok Stone Tract. In igneous rocks and common constituent of *meteorites*.
A *pyroxene*. Varieties are *bronzite* and *chladnite*. *Clinoenstatite* is polymorph. statite is polymorph.

ENSTATITE-DIOPSIDE or ENDIOPSIDE

$8[(Mg,Ca)SiO_3]$ with ratio of magnesium to calcium slightly greater than 1:1. Magnesium rich variety of *diopside*.
Occurrence: USA—Montana (Stillwater Complex), Pennsylvania (Harrisburg), Washington (Northern Whatcom County). Mexico—Chihuahua (Camargo).

ENSTATITFELS or ENSTATOLITE

Igneous rock of the *ultrabasic* clan. A yellowish or greenish-grey coarse-grained plutonic rock forming a bladed mass. Contains essentially *enstatite* with *chromite* as main accessory.
Occurrence: USA—North Carolina. Norway—Salten district.

ENSTATOLITE

Synonym of *enstatitfels*.

EOSITE

Type of *quartzite*. Bluish-white with veins or splotches of brownish-red. Fine-grained texture. Contains small crystals of *pyrites*. Used for ornamental carvings.
Occurrence: Germany—Idar.

Name is also given to vanadate and molybdate of lead forming deep aurora-red, imperfectly developed tetragonal octahedra.
Occurrence: Scotland — Lanarkshire (Leadhills). With *cerussite* and *pyromorphite*.

EOSPHORITE

$8[(Mn,Fe)AlPO_4(OH)_2.2H_2O]$ Basic hydrated aluminium manganese and ferrous iron phosphate with ratio of ferrous to manganese 1 : 3·2. Orthorhombic. Long to short prismatic crystals with prism zone striated; often radial groupings of distinct crystals grading into botryoidal masses or crusts with coarse, fibrous structure. Pink or rose-red. Vitreous to resinous lustre. White streak. Subconchoidal to uneven fracture. Poor cleavage in one direction. Hardness 5·0. SG 3·06. When heated, swells up and fuses to black magnetic bead. Soluble in acids. Yields water when heated in closed tube.
Occurrence: USA—Connecticut (Fairfield County), Maine, New Hampshire (Palermo Mine near North Groton). Germany—Bavaria (Hagendorf near Pleystein). In phosphate-rich *pegmatites*. Isomorphous with *childrenite*.

EPIDIDYMITE

$8[NaBeSi_3O_7OH]$ Basic sodium beryllium silicate. Orthorhombic. Colourless crystals, deeply striated and elongated parallel to macrodiagonal. Very perfect cleavage parallel to base and macropinacoid.
Occurrence: Greenland.
Isomorphous with *eudidymite*.

EPIDIORITE

High grade contact metamorphic rock. A type of *schist*. Contains *augite* altered to *hornblende*, *plagioclase* less basic than *labradorite*, and minor accessories.
Occurrence: Scotland—South West Highlands.

EPIDOSITE

High grade contact metamorphic rock. Contains almost entirely granular *epidote* and some *quartz* with secondary *uralite* and *chlorite*.
Occurrence: England—Cornwall (Lizard). Antarctica—Adélie Land.

EPIDOTE or PISTACITE

$2[Ca_2(Al,Fe)_3(SiO_4)_3(OH)]$ Basic calcium aluminium ferric iron silicate. Monoclinic. Long, slender, grooved prismatic crystals; also thin crusts of small crystals, greenish films or massive, fine-granular. Pistachio-green to blackish-green, brown or light yellowish-brown. Vitreous to pearly lustre on cleavage. Colourless to greyish streak.

Uneven fracture. Perfect basal cleavage. Hardness 6·0–7·0. SG 3·4–3·5. F 3·0–3·5. Fuses with bubbling to dull, black, usually magnetic, *glass*. Gelatinises with hydrochloric acid when previously ignited. Gemstone use.
Occurrence: Worldwide. In metamorphic rocks, contact metamorphic *limestones*, altered igneous rocks, *pegmatites*, in trap-rocks with *zeolites* or shrinkage seams in *granite*. Member of *epidote group*. Varieties are *arendalite*, *fouquéite*, *okkolite* *tawmawite* and *withamite*.

EPIDOTE Group

Complex basic silicates with similar atomic structure. General formula is $R^{2+}R^{3+}_3(SiO_4)_3(OH)$ where $R^{2+}=$ calcium and ferrous iron and $R^{3+}=$ mainly aluminium, manganese, cerium and ferric iron. Subdivided according to crystal system:
Orthorhombic:
zoisite—$Ca_2Al_3(SiO_4)_3(OH)$
Monoclinic:
clinozoisite—$Ca_2Al_3(SiO_4)_3(OH)$
epidote—$Ca_2(Al,Fe)_3(SiO_4)_3(OH)$
piedmontite—$Ca_2(Al, Fe, Mn)_3(SiO_4)_3(OH)$
allanite—$(Ca, Fe)_2(Al, Fe, Ce)_3(SiO_4)_3(OH)_2$

EPINATROLITE

Synonym of *natrolite*.

EPISTILBITE

$CaAl_2Si_6O_{16}.5H_2O$ Hydrated calcium aluminium silicate. Monoclinic. Crystals uniformly twins with prismatic habit and rounded, brilliant faces; also granular or radiated spherical aggregates. Uneven fracture. Brittle. Perfect cleavage in one direction. Hardness 4·0–4·5; 3·5 in plane of cleavage. SG 2·25. Soluble with difficulty in concentrated hydrochloric acid without gelatinising. In flame intumesces and forms vesicular enamel.
Occurrence: Scotland—Skye. USA—New Jersey (Bergen Hill). Canada—Nova Scotia (Margaretville). Iceland—Berufiord. Switzerland—Valais (Viesch). India—Poona. North Atlantic—Faeroes Islands.
Member of *zeolite* family.

EPISTOLITE

Sodium titanium silicate and niobate (columbate) with some fluoride. Monoclinic. Large, thin, rectangular plates. Silver-white on large face. Strong pearly lustre. Perfect cleavage parallel to large face.
Occurrence: South Greenland.

EPSOMITE

$MgSO_4.7H_2O$ Magnesium sulphate heptahydrate. Orthorhombic. Hairlike or cottony efflorescences, botryoidal masses or small prismatic crystals. White. Silky, vitreous to earthy lustre. Conchoidal fracture. Brittle to cottony. Perfect cleavage in one direction, less perfect in two others. Hardness 2·0–2·5. SG 1·7. Bitter taste. Very water soluble. Fuses in flame to infusible mass of magnesium sulphate.
Occurrence: England—Surrey (Epsom). USA—California, Indiana, Kentucky, Nevada, New Mexico, Tennessee, Utah, Washington, Wyoming. Canada—British Columbia (near Ashcroft). South Africa—arid regions. Czechoslovakia—especially Bohemia. France—Isère (Peychagnard). Germany—Saxony (Anhalt, Stassfurt). Italy—Mount Vesuvius. Spain—Zaragoza. As efflorescence on cave walls or deposited by salt springs.

EQUANT

Crystal form in which all the faces are of equal dimensions.

EQUIGRANULAR

Dimensionally similar grains.

EQUIMOLAR

With same 'molarity', or chemical concentration.

EREMEEVITE

$12[AlBO_3]$ Aluminium borate. Hexagonal with monoclinic, six-sectored core. Hexagonal prisms with modified prism faces and rounded, irregular or indented terminations. Composite crystals with hexagonal outer zone and monoclinic inner zone. Colourless to pale yellowish-brown. Vitreous lustre. Conchoidal fracture. No cleavage. Hardness 6·5. SG 3·28. In flame turns white and opaque, colours flame green but does not fuse. After ignition, easily soluble in hot sulphuric acid or concentrated potassium hydroxide.
Occurrence: USSR—Eastern Siberia (Adun-Chilon Range in Dauria in the Nertschinsk district). As single crystals in granitic debris.

ERINITE (of Thomson)

Calcium ferrous iron aluminosilicate. Yellowish-red. Resinous lustre. SG 2·04. Unctuous to touch.
Occurrence: Northern Ireland—County Antrim (Giant's Causeway).
A *clay mineral*.

EROSION

Process of wearing of a surface by water or wind, particularly when containing debris. Can also be caused by animals.

ERUBESCITE

Synonym of *bornite*.

ERYTHRITE (of Beudant)

$2[(Co,Ni)_3(AsO_4)_2.8H_2O)]$ Hydrated cobalt arsenate with some nickel. Monoclinic. Prismatic to acicular and flattened crystals with striated faces; radiated or stel-

late groups; globular or reniform masses with drusy surfaces; massive columnar, coarse fibrous, earthy and pulverulent. Crimson or peach-red, becoming paler with increasing nickel content. Vitreous to pearly lustre. Streak paler than colour. Fracture not significant. Perfect micaceous cleavage, parallel to side pinacoid. Hardness 1·5–2·5. SG 2·9. Flexible, sectile laminae. Fuses to flattened grey mass in flame with arsenical smell. If *borax* added, flame turns blue.
Occurrence: England—Cornwall, Cumberland (Alston Moor). USA—Arizona, California, Idaho, Nevada, New Mexico. Canada—Ontario. Australia—Queensland. Austria — Salzburg, Styria. Czechoslovakia—Bohemia. France—Alsace, Isere. Germany—Baden, Hesse-Nassau, Saxony, Thuringia. Sweden—Tunaberg. Switzerland—Wallis. Chile. Mexico. Morocco. In upper parts of veins with *smaltite* and *cobaltite*.
Formed from weathering of cobalt areas.

ERYTHRITE (of Thomson)

Variety of *orthoclase*. Flesh-red colour.
Occurrence: Scotland—Dunbartonshire and Stirlingshire (Kilpatrick Hills).

ESCHYNITE

(Ce,Th)(Nb,Ti)$_2$(O,OH)$_6$ Rare earths titanium thorium niobate (columbate). Orthorhombic. Crystals sometimes large, prismatic to short prismatic; sometimes tabular and striated; also massive. Black to various shades of yellow or brown. Submetallic to resinous to waxy lustre; brilliant or dulled. Almost black to brown streak. Conchoidal fracture. Brittle. Traces of cleavage in one direction. Hardness 5·0–6·0. SG 5·19. Swells and turns to rusty-brown in flame. As fine powder, partly decomposes in sulphuric acid, completely decomposes in hydrofluoric acid.
Occurrence: Norway—Hitterö. Poland—Silesia (Döbschütz). USSR—Ilmen Mountains (Miask), Southern Ural Mountains (*gold* sands of Orenburg district). In *nepheline-syenite* or *miaskite*.

ESKEBORNITE

Iron copper selenide. Hexagonal, pseudocubic. Resembles *pyrrhotite* but much softer. Has directional magnetic properties.
Occurrence: Germany—Harz Mountains (Tilkerode).

ESMERALDITE or NORTHFIELDITE

Igneous rock of the *granite* clan. A white to grey (when pure), various shades of yellow or brown (with addition of *feldspar* or ferrous oxide), hypautomorphic-granular hypabyssal rock.
Occurrence: USA—Massachusetts (Pelham), Nevada (Esmeralda County). Australia—Tasmania (Mount Bischoff). South Africa—Transvaal.

ESPICHELLITE

Igneous rock of the *gabbro* clan. A porphyritic hypabyssal rock. Contains phenocrysts of *hornblende, augite, olivine, magnetite* and *pyrite* in black, compact groundmass of *magnetite, hornblende, augite, mica* and *labradorite* with *orthoclase* rims.
Occurrence: Portugal—Cape Espichel.

ESSEXITE

Igneous rock of the *gabbro* clan. A coarse to fine-grained plutonic rock. Contains *plagioclase* near *labradorite* with small amounts of alkali-*feldspar* and *nepheline*, plus *titanaugite, olivine, apatite, ilmenite* and *analcite*.
Occurrence: Scotland. USA—Massachusetts (Essex County). Canada—Quebec (Mount Yamaska). Czechoslovakia—Bohemia (Rongstock). Norway.

ESSONITE

Synonym of *hessonite*.

ESTERELLITE

Igneous rock of the *diorite* clan. A bluish holocrystalline hypabyssal rock. Type of *dacite*. Contains rare phenocrysts of *quartz* with *andesine, hornblende* and accessory *magnetite* in a groundmass predominantly *andesine, quartz* and *magnetite* with accessory *pyroxene, orthoclase* and *biotite*.
Occurrence: France—Var (1') Esterel.

ETINDITE

Igneous rock of the *svenite* clan. A dark grey to green porphyritic extrusive rock. Contains phenocrysts of *augite* in dense holocrystalline, granular groundmass of *leucite, nepheline* and *augite* with accessory *perovskite, titanite, apatite* and *magnetite*.
Occurrence: Cameroon—Etinde volcano.

EUCLASE

4[BeAlSiO$_4$OH] Basic aluminium beryllium silicate. Monoclinic. Prismatic crystals with two faces vertically striated. Colourless or pale green passing into blue and white. Vitreous lustre, somewhat pearly on cleavage surface. Colourless streak. Perfect cleavage in one direction. Hardness 7·5. SG 3·1. F 5·5. Fuses to white enamel. Not acted on by acids.
Occurrence: Austria—the Alps (Glossglockner region). Germany—Bavaria. Brazil—Minas Gerais. USSR—Ural Mountains (Orenburg district). Tanzania—Morogoro.

EUCOLITE

Variety of *eudialyte*. Prismatic crystals with large or small basal plane. Distinct cleavage in one direction, imperfect in two others. SG 3·0–3·1. Softer than *eudialyte* and different optical properties.
Occurrence: USA—Arkansas (Magnet Cove in the Ozark Mountains). Norway—Langesund Fiord.

EUCRITE

Igneous rock of the *gabbro* clan. A dark green, coarse to medium-grained hypabyssal rock. Contains mainly *anorthite* with *augite*.
Occurrence: Scotland—Argyllshire (Ardnamurchan). Northern Ireland—Carlingford (Grange Irish). Czechoslovakia—Moravia (Gümbelberg).

EUCRYPTITE

3[LiAlSiO$_4$] Lithium aluminium silicate. Hexagonal. Symmetrically arranged crystals. Colourless or white. Basal cleavage. SG 2·667. Gelatinises with hydrochloric acid.
Occurrence: USA—Connecticut (Branchville). Embedded in *albite*.
Alteration product of *spodumene*.

Illustration over page

Above: eucryptite

EUDIALYTE

$4[(Ca,Na,Ce)_5 (Zr,Fe^{2+})_2 Si_6 (O,OH,Cl)_{20}]$ Basic calcium sodium cerium zirconium and ferrous iron chlorosilicate. Hexagonal. Highly modified tabular or rhombohedral crystals; massive in embedded grains, sometimes reniform. Rose-red, bluish-red, brownish-red or chestnut-brown. Vitreous lustre. Colourless streak. Subconchoidal, splintery fracture. Brittle. Perfect cleavage in one direction. Hardness 5·0–5·5. SG 2·91–2·93. F 2·5. Fuses to light green opaque *glass*, colouring flame yellow. Gelatinises with hydrochloric acid.

Occurrence: Greenland—Kangerdluarsuk. USSR—Beloye More Sea (Sedlovaty Island), Kola Peninsula.
Variety is *eucolite*.

EUDIDYMITE

$8[NaBeSi_3O_7OH]$ Basic sodium beryllium silicate. Monoclinic. Twinned, tabular or pyramidal crystals, striated parallel to intersection edges. White. Vitreous lustre on crystalline faces, pearly on basal plane, silky on fracture surfaces. Perfect basal cleavage, imperfect in second direction. Hardness 6·0. SG 2·553. Easily fuses to colourless *glass* in flame. Dissolves with difficulty and incompletely in acids.
Occurrence: Norway—Langesund Fiord. In *zircon-syenite*.

EUDNOPHITE

Variety of *analcite*. Orthorhombic. Six-sided prismatic crystals terminated by a macrodome; commonly massive, cleavable. White, greyish or brownish. Weak lustre, a little pearly on cleavage faces. White streak. Perfect cleavage in one direction, less perfect in second direction. Hardness 5·5. SG 2·27. Cloudy appearance.
Occurrence: Norway—Langesund Fiord (Låven Island). With *catapleiite and leucophanite.*
Possibly only *analcite* with very strong double refraction.

EUHEDRAL

With fully developed crystal grains.

EUKTOLITE, EUKTOLITH or VENANZITE

Feldspathoid igneous rock. A light ash-grey porphyritic extrusive rock. Contains phenocrysts of *olivine, melilite* and *mica* in rough, holocrystalline groundmass of *olivine, melilite, leucite, phlogopite* and *magnetite*. Shows numerous fractures along flow direction.
Occurrence: Italy—Umbria (San Venanzo).

EUKTOLITH

Synonym of *euktolite.*

EULITE

$16[(Fe,Mg)SiO_3)]$ Iron magnesium silicate with molar proportion of ferrous silicate between 70–90 per cent. Variety of *ortho-ferrosilite.* Orthorhombic. SG 3·74–3·86.
Occurrence: USA—California (Riverside). Canada—North West Territories (Baffin Island). Sweden—Mansjö Mountains. China—Manchuria (Je-ho-shen). Sudan—Madial.
An *orthopyroxene.*

EULYSITE

High grade regional metamorphic rock. A fine-grained rock. Contains manganese-*fayalite, diopside,* iron-manganese *garnet* and *amphibole* with accessory *quartz, biotite, apatite* and *iron* ore.
Occurrence: Sweden—Hagermansdalen, Mansjö Mountains, Tunaberg.
Type of metamorphosed *limestone.* Resembles an igneous rock of the *ultrabasic* clan.

EULYTINE or EULYTITE

$4[Bi_4Si_3O_{12}]$ Bismuth silicate. Cubic. Minute, semi-tetragonal crystals, often in groups with rounded edges; also in spherical forms. Colourless, straw-yellow, greyish-white, yellowish-grey or dark brown. Resinous or adamantine lustre. Yellowish-grey or colourless streak. Uneven fracture. Rather brittle. Very imperfect dodecahedral cleavage. Hardness 4·5. SG 6·106. Fuses to dark yellow mass in flame and gives out colourless fumes.
Occurrence: Germany—Saxony (Johanngeorgenstadt, near Schneeberg).

EULYTITE

Synonym of *eulytine.*

EUSTRATITE

Igneous rock of the *diorite* clan. A compact textured hypabyssal rock. Contains rare phenocrysts of *olivine* and corroded *hornblende* in groundmass of automorphic *augite* and titaniferous *augite* in base of *feldspar, mica* and *glass.*
Occurrence: United Arab Republic.

EUTHALITE

Variety of *analcite.* Compact, often nodular, form with concentric structure. Successive layers greenish or greyish-white in colour.
Occurrence: Norway—Langesund Fiord (Lille Arö, Sigtesö Islands).

EUXENITE

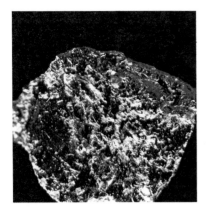

$(Yx,Er,Ce,La,U)(Nb,Ti,Ta)_2(O,OH)_6$ Hydrated niobate (columbate) and titanate-tantalate of uranium and several rare earth elements with some thorium or calcium. Orthorhombic. Stout prismatic crystals with striated faces; often parallel or subparallel and slightly radial aggregates of crystals; also massive. Black, sometimes with greenish or brownish tint. Often brilliant, submetallic lustre, sometimes greasy or vitreous. Yellowish, greyish or reddish-brown streak. Subconchoidal to conchoidal fracture. Hardness 5·5–6·5. SG 4·7–5·0. Infusible in flame, glows markedly and becomes lighter. Nearly completely decomposes in hot concentrated hydrochloric, hydrofluoric or sulphuric acids.
Occurrence: USA—Pennsylvania (Delaware County). Canada—Ontario (Nipissing district, Renfrew County). Western Australia—Cooglegong, Woodstock. South Africa—Cape Province (Namaqualand). Finland—Huntila. Greenland. Norway. Sweden—Alsheda parish. Brazil—Minas Gerais. Congo—Kivu. Madagascar. In *granite pegmatites* or as detrital mineral in areas of granitic rocks.

EVANSITE

$Al_3PO_4(OH)_6 \cdot 6H_2O$ Basic hydrated aluminium phosphate with some ferric oxide.

Amorphous. Massive, as opaline, botryoidal or reniform coatings, sometimes with concentric, colloform structure; also stalactitic. Colourless to milky-white, sometimes tinged blue-green or yellow; varieties with much ferric oxide brown, reddish-brown or red. Vitreous, inclining to resinous or waxy lustre. White or weakly tinted streak. Conchoidal fracture. Very brittle. Hardness 3·0–4·0. SG 1·8–2·2, usually *c* 1·9. Easily soluble in acids. When heated in closed tube yields neutral water, decrepitates and falls into powder.
Occurrence: England—Cheshire (near Macclesfield). USA—Alabama (Coalville, Columbiana), Idaho (Custer County). Australia—Tasmania (Mount Zeehan). Czechoslovakia—Bohemia, Com Gömör, Moravia. France—Marne (Epernay). Hungary. Spain—near Vigo (Teis). Eastern Madagascar—Vatoinandry district.
Secondary mineral found with *limonite* and *allophane*.

EVAPORITE or SALT DEPOSITS

Sedimentary rock. Deposits of physico-chemical origin, mainly involving precipitation and crystallisation from saturated solutions. Forms lenticular masses (if of terrestrial origin) or well-bedded, occasionally regularly interstratified, with fragmental or organic sediment (if laid down in lakes or the sea). Deposits mainly from evaporation processes from lakes, seas or evaporation of ground water drawn up through soil by capillary action. Chemical deposition from springs leads to formation of *calcareous sinters*.
Occurrence: Worldwide, as deposits of chlorides, sulphates, alkali or calcium carbonates, borates, nitrates, silica or *iron*.

EXFOLIATES

See exfoliation.

EXFOLIATION

Process in which surface layers of a rock are removed in thin sheets or 'leaves' as a result of uneven heating and/or cooling.

EXTRUSION

Process of flowing out from the crust over the surface. See intrusion.

EXTRUSIVE

See extrusion.

EYE AGATE

Variegated variety of *chalcedony*. The delicate parallel lines of banded colours can be white, wax-like, pale and dark brown and black, blue or some other shade. Lines usually follow wavy, sometimes straight,

occasionally concentric, circular, courses.
Occurrence: Worldwide. Forms nodules.

FACIES

The sum total of features such as sedimentary rock type, mineral content, sedimentary structures, fossil content etc which characterise a sediment having been deposited in a specified environment.

FAIRCHILDITE

$K_2Ca(CO_3)_2$ Calcium potassium carbonate. Hexagonal. Microscopic, flattened hexagonal plates. Good cleavage in one direction.
Occurrence: USA—Western States. With *calcite* as clinkers formed by fusion of wood ash in partly burned trees, especially hemlock and fir.

FALSE TOPAZ

Synonym of *citrine*.

FAMATINITE

Cu_3SbS_4 Copper antimonide. Cubic. Crusts of minute crystals; usually massive, granular to dense, sometimes reniform. Grey with tinge of copper-red. Uneven fracture. Rather brittle. Hardness 3·5. SG 4·52. When heated on charcoal, fuses to black, brittle, metallic globule. When heated in closed tube, decrepitates and gives off *sulphur* and some *arsenic* sulphide on stronger heating.
Occurrence: USA—Alaska (Kennecott), California (Darwin and Loope districts), Nevada (Goldfield). South West Africa—Tsumeb. Hungary—Matrabánya. Argentina—La Rioja (Sierra de Famatina). Bolivia—Laurani. Peru. Philippine Islands—Luzon (Mancayan). With *enargite*, often intimately intergrown.
Variety is *luzonite*.

FARÖELITE

Silica rich variety of *thomsonite*. Spherical concretions of lamina. Pearly lustre on cleavage surfaces.

Above: faroelite

Occurrence: Denmark—Faeroe Islands. With *mesolite* and *apophyllite*.

FARRISITE

Feldspathoid igneous rock. A chocolate-brown, fine-granular to porphyritic hypabyssal rock. Contains 33 per cent *melilite*-like mineral, *barkevikite, pyroxene* and rare *biotite* and altered *olivine* with accessory *magnetite, pyrite* and *apatite*.
Occurrence: Norway—Lake Farris near Oslo.

FASINITE

Feldspathoid igneous rock. A coarse-grained to porphyritic under saturated plutonic rock. Contains *augite, nepheline* and subsidiary *olivine* and *biotite*.
Occurrence: USA—Arkansas (Magnet Cove in the Ozark Mountains). Germany—Odenwald (Katzenbuckel). Mexico—Tamaulipas. Madagascar—near Ambaliha.

FASSAITE (of Werner)

$(Ca,Mg,Fe,Al)(Si,Al,Fe)O_3$ Calcium magnesium aluminium and ferrous iron silicate with considerable ferric and aluminium oxides, proportion of calcium greater than 0·5 and proportion of magnesium greater than ferrous iron. Variety of *augite* rich in

calcium. Monoclinic. Crystals resemble *epidote* or *diopside*. Pale to dark, sometimes deep, green or pistachio-green. SG 2·979.
Occurrence: Austria—Tyrol (Fassathal). Italy—Piedmont (Traversella), Mount Vesuvius. Norway—Arendal. Also at many *augite* localities.
Mineral of contact metamorphism. A *clinopyroxene*.

FAUJASITE

$cNa_2CaAl_4Si_{10}O_{28} \cdot 20H_2O$ Hydrated sodium calcium aluminium silicate. Cubic. Octahedra crystals. Colourless or white, tarnishing brown externally. Vitreous, sometimes adamantine lustre. Uneven fracture. Fragile. Distinct cleavage in one direction. Hardness 5·0. SG 1·923. Fuses in flame with intumescence to white, blebby enamel. Decomposes in hydrochloric acid without gelatinisation.
Occurrence: Germany—Annarode, Baden (Kaiserstuhl), Eisenach, near Marburg (Stempel). Associated with *augite*.
Member of *zeolite* family.

FAUSTITE

$ZnAl_6(PO_4)_4(OH)_8 \cdot 5H_2O$ Basic hydrated zinc aluminium phosphate. Crystal system uncertain. Vein filling masses. Apple-green. Conchoidal to smooth fracture. Hardness 5·5. SG 2·92.
Occurrence: USA—Nevada (Copper King Mine in Eureka County).
Zinc rich form of *turquoise*.

FAVAS

Water-worn, bean-shaped pebbles of *tourmaline, kyanite, perovskite, rutile, chrysoberyl* and *anatase*.
Occurrence: Brazil—Minas Gerais. In river beds. Usually found associated with *diamonds*.

FAYALITE

$4[Fe_2SiO_4]$ Ferrous iron silicate. Orthorhombic. Massive or minute, tabular crystals. Light yellow, tarnishing dark brown to black by oxidation. Metallic lustre, somewhat resinous on fracture. Imperfect, conchoidal fracture. Brittle.

Below: feldspar

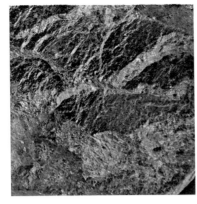

Distinct cleavage in one direction, less so in second direction. Hardness 6·5. SG 4·0–4·14. Gelatinises with acids. Fuses and reduces to black magnetic globule in flame.
Occurrence: Northern Ireland—Mountains of Mourne (near Bryansford). USA—Colorado (Cheyenne Mountain), Wyoming (Obsidian Cliff in Yellowstone National Park). Iceland. Mexico—Cerro de las Navajas. Azores—Fayal. in *pegmatite* or in extrusive igneous rocks.
Member of the *olivine* family. Forms a solid solution with *forsterite*. Variety is *talasskite*.

FELDSPAR Family

Potassium, sodium, calcium or barium aluminous silicates. Isomorphous mixtures of the minerals *orthoclase* $KAlSi_3O_8$, *albite*

$NalSi_3O_8$, *anorthite* $CaAl_2Si_2O_8$, and *celsian* $BaAl_2Si_2O_8$. Isomorphous substitution of one alkali metal with another gives rise to varieties of *feldspar*, as when sodium replaces potassium in *orthoclase* to form soda-*orthoclase*. *Plagioclase* series consists of mixtures of *albite* and *anorthite* in all proportions; replacement of the sodium and silicon with calcium and aluminium results in complete gradation between two end-members. *Albite, oligoclase, andesine, labradorite, bytownite* and *anorthite* are *plagioclase feldspars*. *Alkalifeldspars* are *orthoclase, microcline, soda-orthoclase* and *anorthoclase*. *Bariumfeldspars* are *celsian* and *hyalophane* (both rare).

Occurrence: Worldwide as dominant or important components of most igneous rocks; alkali-*feldspars* in acid rocks, lime-*feldspars* in more basic rocks.

FELDSPATHOID Family

Minerals similar to *feldspars*, especially in chemical composition, but deficient in silica. Aluminium, sodium and/or potassium silicates, generally co-ordinated to a calcium or sodium compound. Series comprises
leucite $KAlSi_2O_6$
nepheline $(Na,K)AlSiO_4$
cancrinite $4(NaAlSiO_4).CaCO_3.H_2O$
sodalite $3(NaAlSiO_4).NaCl$
hauyne $3(NaAlSiO_4).CaSO_4$
nosean $3(NaAlSiO_4).Na_2SO_4$
lazurite $3(NaAlSiO_4).Na_2S$
Occurrence: Worldwide in igneous rocks low in silica and rich in alkalis.

FELDSPATHOID IGNEOUS ROCKS

Type of igneous rock undersaturated in silica. Contains *feldspathoids* (or their equivalents) in place of *feldspars* and no free *quartz*. Includes undersaturated equivalents of the *syenite, diorite* and *gabbro* clans.

FELSITE

Igneous rock of the *granite* clan. A grey or pinkish, compact, aphanitic, hypabyssal rock. Consists of cryptocrystalline aggregates, essentially *quartz* and *orthoclase*. Satellitic to *granite* masses or may be spherulitic.
Occurrence: Worldwide.
Term includes non-porphyritic *rhyolites, trachytes, phonolites, latites* and *andesites*.

FEMAGHASTINGSITE

Variety of *hastingsite*. Contains ratio of ferrous iron to magnesium from $0.5-2.0$. Chemically identical with *ferrohastingsite* except magnesium replaces ferrous iron. SG 3.518.
Occurrence: USA—Vermont (Cuttingsville). Canada— Quebec (Mount Johnston in Montreal). In *åkerites* or *essexite*.
A clino-*amphibole*.

FENCATITE

Synonym of *predazzite*.

FENITE

Igneous rock of the *syenite* clan. A leucocratic, coarse-grained plutonic rock. Contains 70–90 per cent alkali-*feldspar* (usually *orthoclase* or *microcline*), 5–25 per cent *aegirine* or *aegirine-augite* with subordinate alkali-*hornblende* and accessory *titanite* and *apatite*.
Occurrence: Norway—West of Melteig (Fen region).

FERBERITE

$2[FeWO_4]$ Iron tungstate with up to 20 per cent manganese tungstate. Monoclinic. Elongated, somewhat flattened and striated crystals, often with wedge-shaped appearance; also massive or in bladed groups. Black. Brownish-black to black streak. Hardness $c4.5$. SG 7.51. F $2.5-3.0$. Fuses to non-magnetic globule.
Occurrence: USA—Arizona, Colorado (Boulder County), Idaho, New Mexico, South Dakota. Australia—New South Wales, Queensland. France—Haute-Vienne. Germany—Saxony, Vogtland.

Greenland—Ivigtut. Spain—Sierra Almagrera. Bolivia—Colquiri.
Member of the *wolframite* group.

FERGUSITE

Undersaturated igneous rock of the *syenite* clan. A light-grey, medium to coarse-grained granular plutonic rock. Contains mostly *leucite* with subordinate *augite, biotite*, rare *olivine, iron* ore and *apatite*.
Occurrence: USA—Montana (Shonkin Creek in the Highwood Mountains). Italy—South East of Rome.
Related to *missourite*.

FERGUSONITE

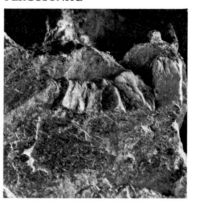

$8[(Y,Er)(Ta,Nb)O_4]$ Erbium yttrium niobate (columbate)-tantalate with proportion of yttrium greater than erbium, proportion of niobium (columbium) greater than tantallum, and small amounts of uranium. Tetragonal. Prismatic to pyramidal crystals, often hemihedral, occasionally long prismatic; also as irregular masses and grains. Grey, yellow, brown or dark brown; brownish to velvety-black on fracture surfaces. Brown, yellowish-brown or greenish-grey streak. Subconchoidal fracture. Brittle. Traces of cleavage in one direction. Hardness $5.5-6.5$. SG $5.6-5.8$, decreasing with hydration and increasing with tantalum content. Infusible. Partly decomposes in sulphuric and hydrochloric acids with separation of niobium (columbium) and tantalum oxide.
Occurrence: USA—Connecticut, Massachusetts, North Carolina, South Carolina, Texas, Virginia. Western Australia. Rhodesia—Fort Victoria. South Africa—Cape Province. Finland. Greenland—Julianehaab district. Poland—Silesia. Southern Norway—Iveland district. USSR—Ilmen Mountains. Japan. Sri Lanka. Madagascar. In *granite pegmatites*. Forms series with *formanite*.

FERRIAN WAD

Synonym of *skemmatite*.

FERRICRETE

Sedimentary rock. *Conglomerates* formed from cementation of gravels by oxidation of percolating solutions of *iron* salts.
Occurrence: Worldwide, especially in arid regions.

FERRIERITE

$(Na,K)_4Mg_2Al_6Si_{30}O_{73} \cdot 20H_2O$ Hydrated sodium potassium magnesium and aluminium silicate. Related to *mordenite* with magnesium replacing calcium. Orthorhombic. Spherical aggregates. White. Pearly lustre.
Occurrence: Canada—British Columbia (Kamloops Lake). In *basalt* with *chalcedony* and *calcite*.
Member of the *zeolite* family.

FERRIMOLYBDITE

$cFe_2Mo_3O_{12} \cdot 8H_2O$ Hydrated ferric molybdate with some ferrous iron. Probably orthorhombic. Massive as fibrous crusts and tufted or radial fibrous aggregates; also subfibrous and as earthy powder or coating. Shades of yellow. Silky or earthy lustre. Pale yellow streak. Hardness 1–2. SG 2·99. Easily soluble in hydrochloric acid or slowly in ammonium hydroxide with separation of ferric hydroxide.
Occurrence: USA. Australia—New South Wales, Tasmania, Western Australia. Italy—Calabria (Bivongi). USSR—Lake Ladoga, Ural Mountains (Minosinsk dis-

trict). Formed from alteration of *molybdenite* and found associated with *limonite*.

FERRIRICHTERITE

Synonym of *juddite*.

FERRI-SICKLERITE

$(Fe^{3+},Li,Mn^{2+})PO_4$ Lithium manganese ferric iron phosphate. Variety of *sicklerite* with proportion of ferric iron greater than manganese. Properties identical to *sicklerite*.
Occurrence: USA—Maine (Peru, Stoneham), New England, New Hampshire (Centre Strafford, North Groton, Rochester). Australia—New South Wales (Euriowie Range). Finland—Tammela. France—Haute-Vienne. Germany—Bavaria. Sweden—Varuträsk. In *pegmatite* as alteration product of *triphylite* and *lithiophilite*.

FERRITE (of Heddle)

Alteration product of *chrysolite* (*of Wallerius*). *Iron*-rich variety of *olivine*. Deep red to chocolate-brown. Prominent cleavage in two directions. Very soft.
Occurrence: Scotland—near Glasgow (between Gleniffar and Boyleston).

FERRITUNGSTITE

$cFe_2WO_6 \cdot 6H_2O$ Basic hydrated iron tungstate. Hexagonal. Earthy crusts. Pale yellow to brownish-yellow. SG 5·57. Decomposes in acids with separation of yellow tungstic oxide.
Occurrence: USA—Washington (Germania Mine in Deer Trail district). Argentina—Jujuy (Cerro Liquinaste). As an alteration product of *wolframite*.

FERROACTINOLITE

Synonym of *ferrotremolite*.

FERRO-ANTHOPHYLLITE

$4[Fe_7Si_8O_{22}(OH)_2]$ Hydrated iron silicate. End-member of *anthophyllite* series, or *iron*-rich *anthophyllite*. Orthorhombic. Asbestiform fibres resemble *chrysotile* in appearance. Greyish-green. SG 3·24.
Occurrence: USA—Idaho (Tamarock Mine in Coeur d'Alene district). North Eastern Japan—Kitakami. Associated with *galena*.
An ortho-*amphibole*.

FERROAUGITE

$8[(Ca,Mg,Fe^{2+})SiO_3]$ Calcium magnesium ferrous iron silicate with or without some ferric oxide or aluminium ferric oxide. Molar proportion of calcium silicate from 25–45 per cent and proportion of ferrous iron greater than magnesium. Iron-rich

variety of *augite*. SG $c3·45$–$3·49$.
Occurrence: Northern Ireland—Portrush. USA—Minnesota (Beaver Bay), New Hampshire (Pilot Range), New York (Adirondack Mountains). South Africa—East Griqualand. South West Africa—Okonjeje. Eastern Greenland—Kangerdlugssuaq. Western Norway—Hitterö. USSR—Karamazar. India—Madras. Japan. In ferro-*gabbros* and *iron rich dolerites* and their *pegmatites*.
A *clinopyroxene*.

FERROBRUCITE

$[(Mg,Fe)(OH)_2]$ Magnesium ferrous iron hydroxide with up to 10 per cent ferrous oxide. Variety of *brucite* (*of Beudant*). Fibrous. Colourless, but turns brown on exposure.
Occurrence: Scotland—Inverness-shire (Muck). Canada—Quebec (Asbestos).

FERROHASTINGSITE

$NaCa_2Fe_4^{2+}(Al,Fe^{3+})(Si_6Al_2O_{22})(OH,F)_2$ Basic sodium calcium aluminium ferric and ferrous iron fluoride and aluminosilicate. *Iron*-rich member of *hastingsite-pargasite* series. Monoclinic. Dark green, black, yellow or brownish-green. No cleavage but partings visible. Hardness 5·0–6·0. SG 3·50.
Occurrence: Northern Ireland—Mountains of Mourne (Slieve Donard). Finland—Salmi (Uuksunjoki). Norway—Tysfiord. USSR—Southern Ural Mountains, Transcaucasia. Japan—Kyushu (Obira Mine). In *syenite*, *granite*, *schist* and *amphibolites*. Type of *hornblende*.

FERROHEDENBERGITE

$8[(Fe,Ca)SiO_3]$ Iron calcium silicate with proportion of iron greater than calcium. Variety of *hedenbergite*. SG 3·48–3·65.
Occurrence: Scotland—Skye (Meall Dearg). USA—New York. South Africa—East Griqualand. East Greenland—Kangerdlugssuaq.
A *clinopyroxene*.

FERROHORTONOLITE

$(Fe,Mg)_2SiO_4$ Iron magnesium silicate with molar proportion of ferric silicate 70–90 per cent. SG 4·0–4·4. Variety of *hortonolite*.
Occurrence: USA—Minnesota. East Greenland—Kangerdlugssuaq.
Member of the *olivine* family.

FERROJOHANNSENITE

$4[Ca(Mn,Fe)Si_2O_6]$ Calcium manganese ferrous iron silicate. Iron-rich mineral with manganese replaced by ferrous iron. SG $c3·55$.
Occurrence: USA—New Mexico (Vana-

dium). Australia—New South Wales (Broken Hill). In metasomatised *limestone* and as a vein material, associated with *bustamite*.

FERROMAGNESIUM MINERALS

Rock-forming silicate minerals which contain essential magnesium and/or iron.

FERROPERICLASE

Synonym of *magnesiowüstite*.

FERROSELITE

$FeSe_2$ Iron selenide with traces of silver, thorium, vanadium and titanium. Orthorhombic. Small prismatic crystals resembling *marcasite*. Steel-grey to tin-white with pinkish tinge. Black powder. Metallic lustre. Very brittle. Hardness $6 \cdot 0$–$6 \cdot 15$. SG $7 \cdot 214$. F $2 \cdot 5$–$3 \cdot 0$. When in closed tube gives dark red sublimate. Dissolves with effervescence in cold nitric acid.
Occurrence: USSR—Siberia (Tuva). Forms solid solution series with *pyrite*, and mixed crystals occur with uranium minerals in *sandstone*.

FERROSILITE

$FeSiO_3$ Iron silicate. Group name to include *orthoferrosilite* and *clinoferrosilite*.

FERROTREMOLITE or FERROACTINOLITE

$2[Ca_2Fe^{2+}{}_5Si_8O_{22}(OH)_2]$ Hydrated calcium ferrous iron silicate. Variety of *actinolite* with molar proportion of ferrous greater than 90 per cent. Fibrous. SG $3 \cdot 34$–$3 \cdot 38$.
Occurrence: USA—Idaho (Tamarack Mine), Minnesota (Eastern Mesabi district).
A clino-*amphibole*. End-member of *actinolite-tremolite* series.

FERRUGINOUS

Containing oxides of iron, and thus often having a brownish or reddish colouration.

FERVANITE

$FeVO_4 \cdot 1\frac{1}{4}H_2O$ Hydrated ferric vanadate. Probably monoclinic. Parallel fibrous aggregates. Golden brown. Brilliant lustre. No cleavage. Insoluble in water.
Occurrence: USA—numerous localities in the uranium-vanadium districts of South West Colorado and South East Utah. Associated with *carnotite* (*of Friedel and Cumenge*), *gypsum* and various black vanadium minerals.

FIASCONITE

Igneous rock of the *gabbro* clan. A dark grey, compact porphyritic extrusive rock. Contains phenocrysts of *augite* and *olivine* in aphanitic groundmass of *augite, leucite, anorthite, olivine, biotite, magnetite* and *apatite*.
Occurrence: Italy—vicinity of Monte Fiascone.

FIBRE

A structure resembling a thread.

FIBROLITE or BUCHOLZITE

Hydrated variety of *sillimanite*. Massive, fibrous or fine columnar, firm and compact, sometimes radiating. Greyish-white to pale brown, pale olive-green or greenish-grey. SG $3 \cdot 24$.
Occurrence: India—Madras.

FIBROUS

Having a structure of fine threads (or fibres).

FIBROUS ZEOLITES

Members of the *zeolite* family with framework structures of silicate-tetrahedra arranged in groups of five, with aluminium replacing part of the silicon. Minerals monoclinic or orthorhombic. Series comprises *natrolite* $Na_2(Al_2Si_3O_{10}) \cdot H_2O$ *mesolite* $Na_2Ca_2(A_2Si_3O_{10})_3 \cdot 8H_2O$ *scolecite* $Ca(Al_2Si_3O_{10}) \cdot 3H_2O$ *thomsonite* $NaCa_2((Al,Si)_5O_{10}) \cdot 6H_2O$ *gonnardite* $Na_2Ca((Al,Si)_5O_{10}) \cdot 6H_2O$ *edingtonite* $Ba(Al_2Si_3O_{10}) \cdot 4H_2O$

FICHTELITE

$cC_{18}H_{32}$ Crystalline hydrocarbon. Monoclinic. Small tabular crystals and aggregates. White. Pearly, somewhat greasy lustre. Brittle. Melting point 46 degrees C. No taste or smell. Easily soluble in ether, less so in alcohol. Soluble in cold nitric acid.
Occurrence: Germany—Bavaria (Fichtelgebirge, near Redwitz, near Rosenheim). As incrustations on wood in *peat* beds.

FILIFORM

Thread-like or fibrous.

FINANDRANITE

Igneous rock of the *syenite* clan. A coarse-grained plutonic rock. Contains *microcline* and *amphibole* with some *biotite, ilmenite* and *apatite*.
Occurrence: Madagascar.

FIRECLAY

Sedimentary rock. A *clay* consisting of dominant *kaolinite* with *illite, chlorite* and *chlorite-kaolinite*. Deficient in *iron* and alkalis. Grey or white. Unbedded, sometimes nodular and pisolitic, structure. Brittle, plastic to hard, form.
Occurrence: England—Hampshire, Yorkshire. In shallow-water, plant-rich rock successions. Widespread as seat earth associated with many *coal* beds.

FIRE OPAL

Variety of *opal*. Transparent to translucent with red to orange-red body colour which may or may not show a play of colour.
Occurrence: Mexico—Hidalgo (Zimapan).

FISSILE

Readily split or cleaved. Certain heavy elements are also described as fissile—they may undergo radioactive fission, as in nuclear power applications.
See cleavage.

FLAME TEST

See bead test.

FLASER GABBRO

Metamorphosed igneous rock. Dynamically metamorphosed *gabbro* which has been crushed and sheared into lenticular masses separated by wavy ribbons and streaks of finely granulated and recrystallised material. Original pseudo-porphyritic igneous texture retained. Rock has not become schistose. Contains large *plagioclases*, often crushed around peripheries and *pyroxenes* showing uralitisation around borders.
Occurrence: Worldwide.
Variety is *zobtenite*.

FLASER-GNEISS

Low grade dynamic metamorphic rock. Coarse-grained. Typically *gneiss* assemblage of minerals. Contains lenticular masses of parts of the original rock in finely crystalline, foliated groundmass.
Occurrence: Worldwide.

FLÈCHES D'AMOUR

Consists of *quartz* crystals enclosing long, hair-like needles of red or golden-coloured *rutile*.
Occurrence: USA—North Carolina (Alexander County), Vermont (West Hartford). Switzerland—Grisons, Val Tavetsch. Brazil—Minas Gerais. Madagascar.

FLINT

Variety of *chalcedony*. Contains mainly granular *chalcedony* with small proportion *opal* silica and sometimes organic remains,

chiefly sponge spicules. Grey, smoky-brown or brownish-black. Exterior often whitened from mixture with *chalk*. Barely glistening, subvitreous lustre. Conchoidal fracture, giving sharp cutting edge. Tough. Very compact, microcrystalline.

Occurrence: *Chalk* formations of England (Kent, Lincolnshire, Sussex, Yorkshire) and Western Europe. Forms nodules, irregular concretionary layers and vein-like masses in lime or *chalk* deposits.

FLINTY CRUSH-ROCK

Medium to high grade dynamic metamorphic rock. Partly fused variety of *mylonite* with flinty composition and black colour. Usually structureless, occasionally showing traces of crystallisation.

Occurrence: Scotland—North West Highlands.

FLOCCULENT

Woolly. Specifically the result of flocculation, a process in which the material carried in a colloid clumps together and settles to the bottom.

FLORENCITE

$CeAl_3(PO_4)_2(OH)_6$ Basic cerium earths aluminium phosphate. Hexagonal. Rhombohedral crystals; sometimes pseudocubical. Clear pale yellow. Greasy to resinous lustre. Splintery to subconchoidal fracture. Good cleavage in one direction, in traces in second direction. Hardness $5 \cdot 1 – 6 \cdot 0$. SG $3 \cdot 586$. Infusible. Partly soluble in hydrochloric acid. Gives off acid water when heated in closed tube.

Occurrence: South West Africa—Klein Spitzkopje. Brazil—near Diamantina (Matta dos Creoulos), Minas Gerais (near Ouro Preto). In *pegmatite* in South West Africa and *diamond*-bearing sands in Brazil.

FLOS-FERRI

Variety of *aragonite*. Forms groups of delicately interlacing and coalescing stems. Snow–white. Like coral in appearance.

Occurrence: Austria—Carinthia (Hüttenberg-Lölling), Styria (Erzberg near Eisenerz). In *siderite* deposits, often coating *hematite*.

FLOW BRECCIA

Igneous rock. Lava flow, usually siliceous, in which fragments of solidified or partly solidified lava produced by explosion of flowage have become welded together or cemented by the still fluid parts of the same flow.

Occurrence: Worldwide in regions associated with volcanic activity.

FLUELLITE

$8[3AlF_3.4H_2O]$ Hydrated aluminium

Below: flos ferri

fluoride with hydroxide substituting in part for fluorine. Orthorhombic. Dipyramidal, often modified crystals. Colourless to white; sometimes faint yellow. Vitreous lustre. Cleavage observed in one direction, indistinct in second direction. Hardness 3·0. SG 2·17.

Occurrence: England—Cornwall (Stenna Gwyn). Czechoslovakia—Bohemia (near Marienbad). Germany—Bavaria (in the Oberpfalz). Spain—Estremadura (Montaña de Cáceres).

FLUOBORITE

$2[Mg_3BO_3(F,OH)_3]$ Magnesium fluorborate with hydroxide substituting in part for fluorine. Hexagonal. Acicular, hexagonal prisms, sometimes in fan-shaped or stellate groups; also fluffy, belted aggregates. Colourless to white. Indistinct cleavage in one direction. Hardness $c3·5$. SG 2·98.

Occurrence: USA—Nevada (Lincoln County), New Jersey (Sterling Hill). Sweden—Norberg (Tallgruvan Mine). Korea—Hol Kol Mine. Malaysia—Selibin (Beatrice Mine). Hydrothermal mineral found as product of contact metamorphism.

FLUOCERITE (of Weibull and Tedin)

$c(Ce,La$ etc$)_2OF_4$ Cerium lanthanum group of rare earths fluoride-oxide. Hexagonal. Prismatic or tabular crystals; also massive, coarsely granular. Pale wax-yellow when fresh, changing yellowish and reddish-brown. Vitreous to resinous lustre; pearly on cleavage surfaces. Almost white streak. Subconchoidal to uneven and splintery fracture. Brittle. Distinct cleavage in one direction, indistinct in second direction. Hardness 4·0–5·0. SG 6·14. Infusible in flame, darkening in colour. Soluble in sulphuric acid but almost unaffected by hydrochloric and nitric acids.

Occurrence: USA—Colorado (Pikes' Peak district in El Paso County). Sweden—Dalarne (Broddbo, Finbo, Österby). In *pegmatite*.

FLUOR

Synonym of *fluorite*.

FLUOR-APATITE

$Ca_5(PO_4)_3(F,Cl,OH)$ with proportion of fluorine greater than chlorine or hydroxide. Variety of *apatite*. SG 3·1–3·2.

Occurrence: England—Cornwall, Devonshire. USA—Maine, New England, New Jersey, New York. Canada—Ontario, Quebec. Austria—Salzburg, Tyrol, Untersulzbachtal. Czechoslovakia. France—Villeder. Germany—Bavaria, Rhine district, Saxony. Italy—Piedmont. Spain—Estremadura, Murcia. Sweden—Nordmark. Mexico. USSR—Kola Peninsula, Ural Mountains.

Varieties are *francolite* and *saamite*.

FLUORESCENCE

Phenomenon in which a substance absorbs electromagnetic radiation of one wavelength (for example in the ultraviolet) and re-emits it at once at a longer wavelength (often visible). Certain species of minerals fluoresce characteristically.

FLUORESCENT

See fluorescence.

FLUORITE, FLUORSPAR or FLUOR

CaF_2 Calcium fluoride. Cubic. Cubic crystals, sometimes octahedra, occasionally in complex formations or twinned; also massive, fine-grained. Colourless, black, white, brown and all spectral and pastel intermediates. Vitreous lustre. White streak. Conchoidal fracture. Brittle. Perfect octahedral cleavage. Hardness 4·0. SG 3·0–3·3. Often fluorescent and thermoluminescent. Powder mixed with sulphuric acid and boiled in glass tube etches

144

the glass surfaces to just above solution.
Fuses on charcoal with difficulty. Used for

carving but too soft for jewellery.
Occurrence: Worldwide. In sedimentary rocks, ore veins and *pegmatites*.
Varieties are blue john, chlorophane, yttro-fluorite and antozonite.

Below: fluorite

FLUORSPAR

Synonym of *fluorite*.

FLUOTARAMITE

$2[(Na, K)_{2.5}(Mg, Fe^{2+}, Fe^{3+}, Ca)_5Si_8O_{22}(OH, F)_2]$ Basic sodium potassium magnesium calcium ferric and ferrous iron fluorsilicate; there may be a moderate amount of aluminium. Monoclinic. Long acicular crystals. Greenish-black.
Occurrence: USSR—Ukraine (Mariupol district). In dykes of *syenite-pegmatite*.
A clino-*amphibole*. Not closely related to *taramite*, but shows analogies to *crossite*.

FLUTED

Having a channelled, or grooved, structure or texture.

FLUVIAL

Applied to sediments formed by rivers. See alluvial.

FOLD

A flexure in rocks; a change in the amount of dip in a bed.

FOLIATED

Having a layered 'onion skin' form. Inter-foliated structures are interleaved.

FONTAINBLEAU SANDSTONE

Sandstone with a cement of crystalline *calcite*.
Occurrence: France—Fontainbleau.

FOOL'S GOLD

Synonym of *pyrite*.

FORELLENSTEIN

Synonym of *troctolite*.

FORESITE or COOKEITE (of D'Achiardi)

Variety of stilbite, forming crystalline crusts. Pearly lustre on cleavage. Distinct cleavage in one direction. SG 2·405. Decomposes with difficulty in hydrochloric acid even after ignition. Expands and melts in flame. Other properties as *stilbite*.
Occurrence: Italy—Island of Elba (San Piero in Campo). On *tourmaline* or *heulandite* lining cavities in *granite*.

FORMANITE

$8[(Y,Er)(Ta,Nb)O_4]$ Yttrium erbium niobate (columbate)-tantalate with proportion of yttrium greater than erbium and propor-

145

tion of tantalum greater than niobium (columbium). Metamict mineral cyrstallising on heating in the tetragonal system. Same properties as *fergusonite*. Is the nearly pure tantalum end-member of a solid-solution series with *fergusonite*. SG greater than 5·8.
Occurrence: Western Australia—Cooglegong. With *cassiterite, monazite, euxenite* and *gadolinite*.

FORSTERITE

$4[Mg_2SiO_4]$ Magnesium silicate. Orthorhombic. Prismatic crystals, also massive. White, yellowish-white, wax-yellow, greyish, bluish-grey or greenish; sometimes becomes yellow on exposure.Vitreous lustre. Colourless streak. Subconchoidal to uneven fracture. Distinct cleavage in one direction, less so in second. Hardness 6·0–7·0. SG 3·21–3·33. Infusible and unaltered in flame. Decomposes in hydrochloric acid with separation of gelatinous silica.
Occurrence: Germany—Kaiserstuhl (Schelinge Matte). Italy—Albani Mountains (Baccano crater), Mount Vesuvius. Norway—Snarum. USSR—Ural Mountains (Zlatoust).
Member of the *olivine* family. Forms solid solution with *fayalite*.

FORTUNITE

Igneous rock of the *diorite* clan. A brownish-grey, fine-grained porphyritic extrusive rock. Contains phenocrysts of *enstatite* and *bronzite* with ortho-*pyroxene* and clino-*pyroxene*, mica, *feldspar* and some *glass*. Variety of *verite*.
Occurrence: Spain—Marcia Province (Fortuna).

FOSHAGITE

$Ca_5Si_3O_{10}(OH)_2.2H_2O$ Hydrous calcium silicate. Orthorhombic. Compact, fibrous. White. Silky lustre. Hardness 3·0. SG 2·36–2·67. Soft and brittle fibres. Readily soluble in hydrochloric acid with gelatinisation. Becomes incandescent and con-

verts to infusible, vitrified mass in flame.
Occurrence: USA—California (Crestmore). Fills veins in *idocrase*.

FOSHALLASSITE

$Ca_3Si_2O_7.2H_2O$ Hydrated calcium silicate. Scaly, sometimes as spheroidal aggregates. Snow-white. Pearly lustre on cleavage surfaces. Perfect tabular cleavage. Hardness 2·5–3·0. SG 2·5.
Occurrence: USSR—Kola Peninsular (Yukspor Mountain). In veinlets associated with *calcite* and *mesolite*.

FOSSIL

Animal or plant remains preserved in a rock.

FOUQUÉITE

$(Ca,Fe^{2+})_2(Al,Fe^{3+})_3Si_3O_{12}OH$ Basic calcium aluminium ferric and ferrous iron silicate, with ferric oxide up to six per cent and ferrous oxide up to four per cent. Variety of *epidote*. Monoclinic. Elongated crystals, usually with rounded outlines. Yellow or white. Good cleavage in one direction. SG 3·24–3·31. Infusible in flame.
Occurrence: USA—Massachusetts (Salem). Sri Lanka—Kandy. In *anorthitegneiss* with *epidote, scapolite, garnet, amphibole* and *pyroxene*.

FOURCHITE

Feldspathoid igneous rock. A melanocratic, porphyritic hypabyssal rock. Contains 75 per cent *pyroxene* as phenocrysts of *augite* in greatly altered holocrystalline groundmass which may have been *analcite*, together with accessory iron oxide.
Occurrence: USA—Arkansas (Fourche Mountains).

FOURMARIERITE

$PbU_4O_{13}.5H_2O$ Hydrated uranium lead oxide. Orthorhombic. Tabular and usually elongated and striated crystals. Red to golden-red and brown. Adamantine lustre. Perfect cleavage in one direction. Hardness 3·0–4·0. SG 6·046. Blackens in flame but does not fuse. Easily soluble in acids.
Occurrence: Germany—Bavaria (Wölsendorf). Congo—Katanga (Kasolo). Secondary mineral found with *torbernite*, and *curite* as alteration product of *uraninite*.

FOWLERITE

$10[(Mn,Zn)SiO_3]$ Zinc manganese silicate with up to 10 per cent zinc oxide. Zinc-bearing variety of *rhodonite*. Forms crystals up to 12·5mm to 25mm through or massive, foliated or compact. Reddish.
Occurrence: USA—New Jersey (Franklin Furnace). In *calcite*.

FOYAITE

Undersaturated igneous rock of the *syenite* clan. A coarse to fine-grained plutonic rock, although term foyaite has been sometimes restricted to those rocks with a trachytoid texture. Contains *orthoclase, nepheline* and *hornblende* with accessory *pyroxene, mica, iron* ore and *sodalite, nosean* and *zircon*.
Occurrence: Scotland—Assynt. USA—Arkansas (Fourche Mountains, Magnet Cove in the Ozark Mountains), Montana (Crazy Mountains). Canada—British Columbia. South Africa—Transvaal. Norway. Portugal—Serra de Monchique (Foya, Picota). Sweden—Island of Alnö. Mexico—Tamaulipas.

FRACTURE

Breaking behaviour of mineral, not associated with cleavage directions.

FRAMESITE BORT or BLACK DIAMOND

Variety of *diamond*. A grey to black, granular to cryptocrystalline stone with minute brilliant *diamond* points. More difficult to cut than *bort*.

FRANCKEÏTE

$Pb_5Sn_3Sb_2S_{14}$ Lead tin antimonide-sulphide. Orthorhombic. Thin tabular, elongated and striated crystals, often warped or bent. Usually massive, radial or foliated. Greyish-black, tarnishing iridescent. Metallic lustre, Greyish-black streak. Flexible but not elastic and slightly malleable. Perfect cleavage in one direction. Hardness 2·5–3·0. SG 5·90. F 1·0. Decomposes in nitric acid with separation of tin and antimony oxides.
Occurrence: Bolivia—widespread as common mineral in *silver*-tin veins, occurring with other lead, *arsenic* and *antimony* minerals.

FRANCOLITE or STAFFELITE

$c2[Ca_5(PO_4,CO_3,OH)_3(F,OH)]$ Variety of *fluor-apatite* containing considerable amounts of carbon dioxide. Stalactitic masses.

Occurrence: England—Devonshire (Tavistock).

FRANKLINITE

$8[(Zn,Mn,Fe^{2+})(Fe^{3+},Mn^{3+})_2O_4]$ Zinc, divalent and trivalent manganese and ferric and ferrous iron oxide; mainly zinc and ferric iron. Cubic. Octahedral crystals, alone or with dodecahedra; edges often rounded; massive compact, granular or as rounded grains. Iron-black. Metallic lustre. Black or brownish-black, reddish-brown or dark brown streak. Conchoidal fracture. Brittle. Indistinct octahedral cleavage. Hardness 5·5–6·5. SG 5·07–5·22. Borax bead colours amethyst in oxidising flame; bottle-green in reducing flame. Powder may be slightly magnetic.
Occurrence: USA—New Jersey (Franklin, Sterling Hill). Forms thick beds in crystalline *limestone*, sometimes admixed with various amounts of *zincite, willemite, rhodonite* and *tephroite*.
Member of *spinel* family.

FREIBERGITE

$(Cu,Ag)_3SbS_3$ Copper silver antimonide-sulphide, with silver up to 18 per cent. Argentine variety of *tetrahedrite*.
Occurrence: Germany—Freiberg.

FREIESLEBENITE

$Ag_5Pb_3Sb_5S_{12}$ Lead silver antimony sul-phide. Monoclinic. Prismatic and striated crystals. Colour and streak light steel-grey, inclining to silver-white or lead-grey. Metallic lustre. Subconchoidal to uneven fracture. Rather brittle. Imperfect cleavage in one direction. Hardness 2·0–2·5. SG 6·04–6·23. F 1·0. In flame gives coating of lead and *antimony*, leaving globule of *silver*.
Occurrence: Czechoslovakia—Bohemia (Příbram). Germany—Saxony (Himmel-fürst Mine at Freiberg). Rumania – Felsöbanya, Kapnik. Spain—Hiendelaen-cina.

FRENCH CHALK

Synonym of *steatite*.

FRENZELITE

Synonym of *guanajuatite*.

FRIABLE

May readily be crumbled into a powder.

FRIEDELITE

$6[(Mn,Fe)_8Si_6O_{15}(OH,Cl)_{10}]$ Hydrous manganese iron chloride-silicate. Hexagonal. Tabular, striated crystals; massive with saccharoidal structure and distinct cleavage to close compact with indistinct cleavage. Rose-red or powder-pale rose. Perfect cleavage in one direction. Hardness 4·0–5·0. SG 3·07. Fuses easily in flame to black *glass*. Dissolves in hydrochloric acid with gelatinisation.
Occurrence: USA—New Jersey (Franklin Furnace). Austria—Styria (Veitsch). France—Hautes Pyrénées (Vallée du Louron). Sweden—Wermland (Harstig Mine near Pajsberg).
Bementite is member of family.

FRONDELITE

$4[(Mn^{2+}Fe^{3+})_4(PO_4)_3(OH)_5]$ Basic divalent manganese ferric iron phosphate with proportion of manganous greater than ferric iron. Orthorhombic. Botryoidal masses and crusts with a radial, elongated, fibrous or fine-columnar structure. Dark green or olive-green to greenish-black in fresh material, becoming brownish-green or dark reddish-brown on oxidation. Vitreous to dull lustre. Uneven fracture. Brittle. Excellent cleavage in one direction, good in second direction, fair in third direction. Hardness 4·5. SG 3·3–3·49. Easily fusible to magnetic globule in flame. Soluble in hydrochloric acid but not in nitric acid or sulphuric acid.
Occurrence: USA—New Hampshire (North Groton). Brazil—Minas Gerais (Conselheira Pena). Secondary mineral found in *limonite* deposits or as alteration product of manganese-iron phosphates in *pegmatites*.

FRUSTULE

Silica-based shell of a diatom (minute plant).

FUCHSITE or CHROME-MICA

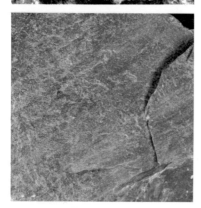

$4[K(Al,Cr)_3Si_3O_{10}(OH)_2]$ Hydrated potassium aluminium chromium silicate with up to 5 per cent chromic oxide. Chrome-rich variety of *muscovite*. Green. SG 2·78–2·88.
Occurrence: USA—Maryland (Montgomery County). Canada—Lake Huron (Aird Island). Austria — Zillertal (Schwarzenstein). Brazil—Ouro Preto. USSR—Ural Mountains. India—many localities with *quartz* as type of *aventurine quartz* used in jewellery.

FULGURITE

Type of natural *glass*. Thin tubes of glassy rock from fusion of sand or loose or compact rocks caused by the intense heat set up when a lightning flash strikes the sands of a desert or an exposed crag.
Occurrence: Worldwide in arid or mountainous regions.

FULLER'S EARTH

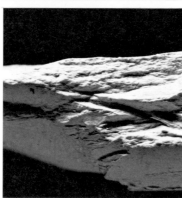

Sedimentary rock. A *clay*. Contains mainly *montmorillonite* with other hydrated aluminium silicates and accessory minerals such as *quartz*, *mica* and *glauconite*. Greenish-brown, greenish-grey, bluish or yellowish. Soft, earthy and soapy texture. Yields to finger with shining streak. Adheres to tongue. In flame fuses to porous slag and ultimately to white, blistery *glass*. Forms powder in water. Strong absorbent properties which are utilised commercially.
Occurrence: England. USA—Arkansas, California, Eastern Coastal Plain from Florida to New York, South Dakota. Germany. As sedimentary beds.

FUMAROLE

Hot volcanic spring round which material often builds up by crystallisation.

FUSAIN

Constituent of *bituminous coal*. Charcoal-like substance consisting of *fusinite*. Extremely friable and easily reduced to fine powder. Consists of fibrous strands forming patches and wedges somewhat flattened parallel to bedding plane of the *coal*. Formed from alteration of wood in oxidising environment.

FUSE

Melt. Often different substances will mix together on fusion.

FUSINITE

Major constituent of *fusain*. Consists of wood of which very little is left but woody conducting cells or thick-walled elements so highly carbonised as to contain only traces of ulmins.

GABBRO

Clan of igneous rocks containing essentially *plagioclase*; more calcic than *anorthite* associated with *clinopyroxene*. *Orthopyroxene*, *olivine*, *apatite* and *iron* ore may or may not be present. Melanocratic. Silica content from 45 to 55 per cent. Term generally refers to plutonic rocks of coarse-grained, granular texture, but clan definition covers all modes and textures.

GABBRO-NORITE

Synonym of *hyperite*.

GABBROPHYRE

Synonym of *odinite*.

GABBROPHYRITE or GABBRO-PORPHYRY

Igneous rock of the *gabbro* clan. A porphyritic plutonic rock. Contains phenocrysts of basic *plagioclase* and sometimes *pyroxene* and/or *olivine* in fine-grained, equigranular holocrystalline groundmass of basic *plagioclase*, *pyroxene* and *magnetite* with accessory *biotite*, *hornblende* and *iron* ores.

Occurrence: Germany—Odenwald (Frankenstein).

GABBRO-PORPHYRY

Synonym of *gabbrophyrite*.

GADOLINITE

$2[Be_2Fe^{2+}Y_2Si_2O_{10}]$ Beryllium yttrium ferrous iron silicate with small amounts of rare earths replacing yttrium. Monoclinic. Rough and coarse, commonly prismatic crystals; also massive. Black, greenish-black or brown. Vitreous to greasy lustre. Greenish-grey streak. Conchoidal or splintery fracture. Brittle. No cleavage. Hardness 6·5–7·0. SG 4·0–4·5. Decomposes with gelatinisation in hydrochloric acid,

but not if previously ignited. Borax bead gives reactions for iron.

Occurrence: Northern Ireland — Mountains of Mourne, Newcastle. Republic of Ireland—near Galway. USA—Colorado (Douglas County), Texas (Llano County). Germany—Harz Mountains. Italy—Baveno. Norway—Hitterö, Malö. Poland—Silesia. Sweden—especially Dalarne. Principally in *pegmatite* veins, often associated with *allanite*.

GAGEITE

$(Mn,Mg,Zn)_8Si_3O_{14}.(2-3)H_2O$ Hydrated zinc magnesium manganese silicate with 10 per cent magnesium oxide and 4 per cent zinc oxide. Orthorhombic. Radiating aggregates of acicular crystals. Colourless to pinkish. SG 3·584.

Occurrence: USA—New Jersey (Franklin). Associated with *zincite*, *willemite* and *calcite*.

GAHNITE (of da Silveira)

Synonym of *idocrase*.

GAHNITE (of Von Moll)

$8[ZnAl_2O_4]$ Aluminium zinc oxide, often with some magnesium. Cubic Octahedral crystals. Dark green, bluish-green, sometimes yellow or brown. Grey streak. SG 4·62 (calc.). Lighter green-coloured stones used as gemstones.

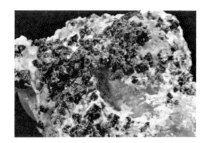

Occurrence: USA—Colorado, Connecticut, Georgia, Maryland, Massachusetts, North Carolina, New Jersey, Pennsylvania. Western Australia. Finland. Germany—Riesengebirge. Italy—Calabria. Poland—Silesia. Sweden—especially Fahlun. Brazil—Minas Gerais. India—Madras. Madagascar—Ambatofotsikely. In zinc deposits, crystalline *schists*, *granite pegmatites* or contact metamorphised *limestones*.

Varieties are *dysluite*, *kreittonite* and *limaite*. Member of *spinel* family.

GAHNOSPINEL

$8[Mg,Zn)Al_2O_4]$ Magnesium zinc aluminium oxide. Zinc-rich variety of *spinel*. Pale-blue to dark blue. SG 3·58–4·06. Gemstone.

Occurrence: Sri Lanka—gem gravels

Below: galena (*see over page*)

Grades into *gahnite* (*of Von Moll*) with increasing zinc content.

GALAXITE

8[$MnAl_2O_4$] Manganese aluminium oxide containing considerable ferrous oxide. Cubic, Massive, coarse granular to compact and as irregular or rounded embedded grains. Black reddish-brown streak. SG 6·04(calc.).
Occurrence: USA—North Carolina (Bald Knob in Alleghany County). As vein with *alleghanyite, calcite, spessartite, rhodonite* and *tephroite*.
Member of *spinel* family.

GALENA or BLEISCHWEIF

4[PbS] Lead sulphide. Cubic. Cubic, cubo-octahedral or tabular crystals; also skeletal crystals; also massive, cleavable, coarse to fine-granular, fibrous or plumose. Lead-grey. Metallic lustre. Lead-grey streak. Flat sub-conchoidal fracture. Perfect cleavage in one direction. Hardness 2·5–2·75. SG 7·58. F 2·0. Decomposes in sulphuric acid with separation of some sulphur and formation of lead sulphide.

Occurrence: Worldwide. In ore veins, igneous and sedimentary rocks and disseminated through sediments. Commonly associated with *blende, pyrite,* and *chalcopyrite* with *quartz, siderite, dolomite, fluorite, calcite* and *borite* as gangue materials.

GALENOBISMUTITE

4[$PbBi_2S_4$] Lead bismuth sulphide with antimony substituting in small amounts for bismuth. Orthorhombic. Lathlike, elongated, flattened and striated crystals; also as needles or extremely thin plates; also massive, columnar to fibrous, compact. Light grey to tin-white or lead-grey, sometimes tarnished yellow or iridescent. Metallic lustre. Black streak. Good cleavage in one direction. Hardness 2·5–3·5. SG 7·04. Crystals frequently flexible, bent and twisted. F 1·0. Soluble in hot hydrochloric acid.
Occurrence: USA—Idaho (Quartzburg district), Montana (Jefferson County). Canada—British Columbia (Cariboo district). Italy—Lipari Islands. Sweden—Gladhammer, Nordmark.

GANGUE

The mineral part of an ore rock which is not the objective of the mining operation in question, i.e. the first mineral waste product.

GANISTER

Sedimentary rock. A highly siliceous *sandstone.* Consists of medium to fine *quartz* grains cemented together with silica. Compact form with fine and even granular texture. Highly refractory.
Occurrence: Northern and Central England. Eastern Scotland. In *coal* fields under *coal* beds.

GANOMALITE

$Ca_2Pb_3Si_3O_{11}$ Lead calcium silicate. Tetragonal. Prismatic crystals or granular massive. Colourless to grey. Resinous to vitreous lustre. Uneven fracture. Very brittle. Distinct cleavage in two directions. Hardness 3·0. SG 5·74. Easily soluble in nitric acid with separation of gelatinous silica. Easily fuses in flame to clear *glass* which is coloured black by reduced lead in reducing flame.
Occurrence: Sweden—Wermland (Jacobsberg, Långban, Nordmark). Mainly with *calcite* and *jakobsite* and/or *manganophyllite* and *tephroite.*

GANOPHYLLITE

$Mn_7Al_2Si_8O_{26}.6H_2O$ Manganese or aluminosilicate. Monoclinic. Short prismatic crystals with dull, striated faces up to 25mm; also foliated, micaceous. Brown. Vitreous, brilliant lustre. Perfect basal cleavage. Hardness 4·0–4·5. SG 2·84. Readily dissolves in strong acids but becomes nearly insoluble after ignition.
Occurrence: Sweden—Wermland (Harstig Mine near Pajsberg). Embedded in *calcite* or implanted on *rhodonite.*

GARÉWAITE

Igneous rock of the *ultrabasic* clan. A dark-coloured porphyritic hypabyssal rock. Contains phenocrysts of corroded *diopside* in fine-grained, holocrystalline groundmass of *olivine, pyroxene, chromite* and *magnetite* with accessory *spinel* and *labradorite.*
Occurrence: USSR—Northern Ural Mountains (Garewaia River).

GARNET Group

Silicates of various divalent and trivalent metals. General formula $R^{2+}_3R^{3+}_2(SiO_4)_3$ where R^{2+} is calcium, magnesium, iron or manganese and R^{3+} is iron, aluminium, chromium or titanium. Cubic. Mainly dodecahedral or tetragonal trisoctahedral crystals; granular, massive, compact, lamellar, disseminated crystals or sand. Red, brown, white, yelow, apple-green or black. Vitreous to resinous lustre. White streak. Subconchoidal to uneven fracture. Brittle. Sometimes friable when granular massive. Sometimes rather distinct cleavage in one direction. Hardness 6·5–7·5. SG 3·15–4·3.
Occurrence: Worldwide. Common in *gneisses, schists, crystalline limestones* and metamorphosed igneous rocks. Sometimes primary minerals in igneous rocks, especially *syenites.* Common in heavy detrital residues in sediments.
Series comprises *grossular, pyrope, almandine, spessartite* and *andradite* (gemstones) and *uvarovite* (non-gemstone).

GARNETOID

Group name for minerals which are not primarily silicates, but with structures similar to *garnet.* Series includes *griphite, berzeliite* (*of Kühn*), *hydrogarnet, plazolite* and *hibschite.*

GARNIERITE

$(Ni,Mg)_3Si_2(OH)_4$ Hydrated nickel magnesium silicate with ratio of nickel to magnesium very variable. Amorphous. Bright apple-green or pale green to nearly white. Dull lustre. SG 2·3–2·8. Partly unctuous. Sometimes adheres to tongue. Soft and friable. When heated in closed tube yields water and blackens. Borax bead gives nickel reactions in flame.
Occurrence: USA—North Carolina (Webster), Oregon (Riddle). New Caledonia—especially Noumea. Spain—

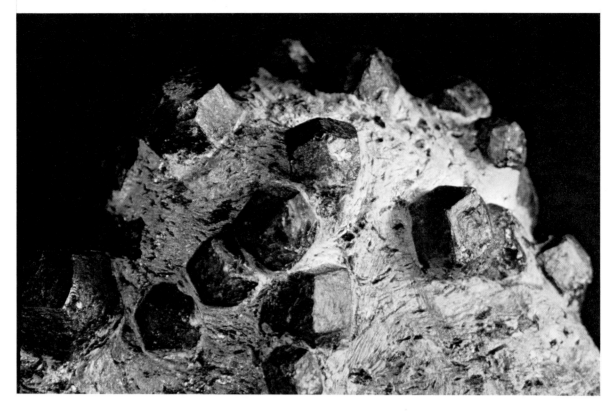

Above: garnet
Below: garnierite

Malaga. USSR—Ural Mountains (Revda). With *serpentine*.
An alteration product of *serpentine, chromite* or *talc*. Member of *serpentine* family. Varieties are *genthite* and *noumeite*. Possible variety is *connarite*.

GASTRALDITE

Synonym of *glaucophane*.

GAUSSBERGITE

Igneous rock of the *syenite* clan. A porphyritic extrusive rock. Contains phenocrysts of *leucite, augite* and *olivine* in glassy groundmass potentially *sanidine*.
Occurrence: Antarctic—Kaiser Wilhelm II Land (Gaussberg volcano).

GAUTEITE

Igneous rock of the *diorite* clan. A light grey to greenish-grey with brown weathered crusts, porphyritic hypabyssal rock. Contains phenocrysts of *hornblende, augite*, occasional *biotite* and lime-soda *feldspar* in groundmass of mainly *feldspar* strips plus small amounts of *magnetite, augite, hornblende* and *biotite* and commonly a little *analcite*. Type of *trachyandesite*.
Occurrence: Czechoslovakia (near Gauté).

GAY-LUSSITE

$Na_2Ca(CO_3)_2.5H_2O$ Hydrated sodium calcium carbonate. Monoclinic. Crystals often elongated or flattened and wedge-shaped; usually rough and striated surfaces. Colourless to yellowish-white, greyish-white or white. Vitreous lustre. Colourless to greyish-white streak. Conchoidal fracture. Very brittle. Perfect cleavage in one direction. Hardness 2·5–3·0. SG 1·99. Effloresces slowly in dry air. Easily soluble in acids with effervescence. Easily fusible in flame to white enamel.
Occurrence: USA—California (Mono Lake, Searles Lake), Nevada (Carson Desert), Wyoming (Independence Rock). Venezuela—near Merida. Mongolia— Eastern Gobi Desert. As a deposit from soda lakes.

GEDRITE or BIDALOTITE

$4[Mg,Fe^{2+},Al)_7(Si,Al)_8O_{22}(OH)_2]$ Hydrated magnesium aluminium ferrous iron aluminosilicate with silicon as low as 6·5, ratio of magnesium to ferrous iron varying widely and aluminium and ferrous iron present in larger amounts than *anthophyllite*. Variety of *anthophyllite*. SG 2·98– 3·23. Orthorhombic.
Occurrence: USA. Finland—near Helsingfors (Stansvik). France—near Gèdres (Valley of Héas). Greenland—Fiskernäs. Norway—Bamle, Hilsen, Snarum.
An ortho-*amphibole*.

GEHLENITE or VELARDEÑITE

$4[Ca_2Al_2SiO_7]$ Calcium aluminium silicate with ferric iron replacing aluminium and magnesium replacing calcium. Tetragonal. Short, square prismatic crystals, sometimes tabular, often resembling cubooctahedra. Different shades of greyish-green to liver-brown. Resinous lustre, inclining to vitreous. White to greyish-white streak. Uneven to splintery fracture. Brittle. Imperfect cleavage in one direction, in traces in second direction. Hardness 5·5– 6·5. SG 2·9–3·07. In flame with borax fuses slowly to glass coloured by iron. Gelatinises in hydrochloric acid to solution containing ferric and ferrous oxides.

151

Occurrence: Austria—Tyrol (Fassathal, Fleimsthal). Rumania—Banat (Orawitza). In contact zone in *limestone* or artificially in slags.
Member of *melilite* group.

GEIKIELITE

2[MgTiO$_3$] Magnesium titanate. Hexagonal. Brownish-black. Black streak. Rhombohedral cleavage. SG 4·05. Other properties as *ilmenite*.
Occurrence: USA—California (in *marble* of Riverside area). Sri Lanka—gem gravels of Balangoda and Rakwana districts.
Member of *ilmenite* series with magnesium replacing iron.

GELATINISATION

Formation of a gelatin, or colloid, as the result of chemical action.

GENÉVITE

Synonym of *idocrase*.

GENTHELVITE

((Zn,Fe,Mn)$_8$Be$_6$Si$_6$O$_{24}$S$_2$) Zinc manganese ferrous iron and beryllium sulphate-silicate; the proportion of zinc end member is greater than 85 per cent. Cubic. Purplish-pink to reddish-brown. Poor cleavage in one direction. Hardness 6·5. SG 3·44–3·70. Decomposes in hydrochloric acid with the evolution of hydrogen sulphide.
Occurrence: England—Cornwall (Treburland). USA—Colorado (El Paso County). Northern Nigeria—Jos. In *granite, granite pegmatites* and contact metamorphosed rocks.
Member of the *helvine* family.

GENTHITE

(Ni,Mg)$_x$Si$_2$(OH)$_4$ Hydrated nickel magnesium silicate. Ratio of nickel to magnesium very variable. Variety of *garnierite*. Amorphous. Massive with delicately hemispherical or stalactitic surface; also incrusting. Pale apple-green or yellowish.

Resinous lustre. Greenish-white streak. Soft and friable. Hardness 3·0–4·0 (sometimes so soft as to be polished under the nail and falling to pieces in water). SG 2·2–2·8. Infusible. Borax bead is violet in oxidising, and grey in reducing, flame. Decomposes in hydrochloric acid without gelatinising.
Occurrence: USA—Lake Superior (Michipicoten Island), Pennsylvania (Lancaster County). Switzerland—Upper Valais (Saasthal).

GEOCRONITE

[Pb$_{27}$(As,Sb)$_{12}$S$_{46}$] Lead arsenide-antimonide-sulphide with ratio of arsenic to antimonide from 1:2 to 1:1. Orthorhombic. Rare, tabular crystals; usually massive, granular and earthy. Light lead-grey to greyish-blue colour and streak. Metallic lustre. Uneven fracture. Distinct cleavage in one direction, less so in second direction. Hardness 2·5. SG 6·4. F 1·0. When heated in closed tube gives faint sublimate of sulphur and antimony sulphide.
Occurrence: Republic of Ireland—County Clare (Kilbricken Mine). USA—California (Inyo County, Mono County), Utah (Tintic district), Virginia (Louisa County). Italy—Val di Castello. Spain—Asturias (Meredo). Sweden—Orebro, Sala.
Isomorphous with *jordanite*.

GEODE

A void in a rock, subsequently filled or partly filled with a mass of crystals radiating in towards the centre.

GEOLOGICAL TIME

Geological time is subdivided into a number of systems, incorporating all the rocks developed on the Earth since its formation approximately 4000 million years ago.

These systems are:

Pre-Cambrian—from the formation of the Earth's crust to approximately 550 million years ago. Areas of metamorphic rocks greatly affected by orogenesis, forming the shield areas of Fennoscandia and North America and many minor areas. Remains of life very sparse.
Cambrian—550 million to 500 million years ago. Mainly sedimentary rocks, often later metamorphosed. Abundant marine invertebrates.
Ordovician—500 million to 435 million years ago. Extensive sedimentary and extrusive igneous rocks, often later metamorphosed. Characterised by graptolite fauna.
Silurian—435 million to 395 million years ago. Various rock types developed. Extensive marine invertebrate fauna.
Devonian—395 million to 345 million

years ago. Extensive continental (Old Red Sandstone) and marine sediments developed; also extrusive igneous activity. Fish fauna characterises the Old Red Sandstone.
Carboniferous—345 million to 280 million years ago. Extensive marine limestones and deltaic/freshwater deposits developed, the latter being characterised by coal deposits.
Permian—280 million to 225 million years ago. Mainly continental desert deposits with associated igneous activity. Fauna and flora very varied.
Triassic—225 million to 195 million years ago. Mainly continental deposits with later marine sediments. Fauna and flora very sparse.
Jurassic—195 million to 135 million years ago. Characterised by extensive marine and continental sediments with ammonite fauna and dinosaur remains.
Cretaceous—135 million to 64 million years ago. Mainly marine deposits including extensive limestones (chalk) with very varied fauna.
Eocene, Oligocene, Miocene and Pliocene—64 million to one million years ago. Sedimentary and igneous rocks developed. Fauna and flora similar to present day, showing the evolution of the mammals.
Pleistocene—one million years ago to the present day, deposits representative of the Ice Age.
Holocene—recent post-glacial deposits.

These systems may be simplified by grouping into major Eras of geological time:

Azoic—the Pre-Cambrian.
Palaeozoic—the Cambrian, Ordovician, Silurian, Devonian, Carboniferous and Permian.
Mesozoic—the Triassic, Jurassic and Cretaceous.
Cainozoic—the Eocene, Oligocene, Miocene, Pliocene, Pleistocene and Holocene.

The Cainozoic Era is further subdivided into:

Tertiary—the Eocene, Oligocene, Miocene and Pliocene.
Quaternary—the Pleistocene and Holocene.

GERHARDTITE

Cu$_2$NO$_3$(OH)$_3$ Basic copper nitrate. Orthorhombic. Thick tabular crystals, the pyramid zone strongly striated and faces often in oscillatory combination. Dark green to emerald-green. Light green streak. Perfect cleavage in one direction, good in second direction. Hardness 2·0. SG 3·40–3·43. Flexible crystals with separation in one plane. When heated in closed tube gives nitrous fumes and acid water. Insoluble in water, soluble in dilute acids.
Occurrence: USA—Arizona (Clifton-

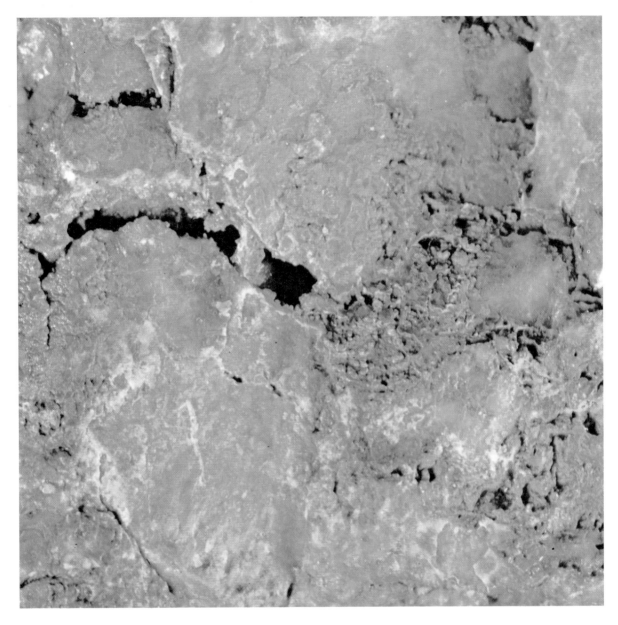

Above: genthite
Right: germanite

Morenci district, Jerome). Congo—Katanga (Likasi). Secondary mineral found with other *copper* minerals in cavities in *cuprite*.

GERMANITE

8[Cu$_3$(Ge,Ga,Fe,Zn)(As,S)$_4$] Copper germanium sulphide with germanium from 6–10 per cent and zinc, gallium, arsenic and ferrous iron in smaller amounts, some as impurities. Cubic. Dark reddish-grey. Metallic lustre. Dark grey to black streak. Brittle. No cleavage. Hardness 4·0. SG 4·46–4·59. Soluble in nitric acid. Decrepitates on heating.

Occurrence: South West Africa—Tsumeb. No crystals found but occurs associated with *pyrite, tennantite, enargite, galena* and *blende*.

GERSDORFFITE

4[NiAsS] Nickel sulphide-arsenide. Cubic. Octahedral, cubo-octahedral, pyritohedral, sometimes striated crystals; also massive, granular and lamellar. Silver-white to steel-grey, often tarnishing grey or greyish-black. Metallic lustre. Greyish-black streak. Uneven fracture. Brittle. Perfect cleavage in one direction. Hardness 5·5. SG 5·9. F 2·0. Decrepitates when heated in closed tube and gives yellowish-brown sublimate of arsenic sulphide.

Occurrence: USA—Connecticut (Chatham), Pennsylvania (Phoenixville). Canada—Ontario (Cobalt, Sudbury dis-

trict). Rhodesia—Selukwe Peak. Austria—Carinthia, Styria, Thuringia. Czechoslovakia—Bohemia (Dobschau). Germany—Harz Mountains, Westphalia. Sweden—Helsingland. With other nickel minerals in vein deposits.

GEYSERITE

Variety of *siliceous sinter*. Porous, stalactitic, filamentous or cauliflower-like; also massive, compact or scaly. White or greyish. Hardness 5·0. Sometimes falls to powder when drying in air.
Occurrence: USA—Wyoming (Yellowstone National Park). New Zealand. Iceland. Constitutes concretionary deposits around geysers.

GHIZITE

Igneous rock of the *diorite* clan. A dark grey, dense, compact, aphanitic extrusive rock. Contains large phenocrysts of *biotite* and smaller ones of *augite* and *olivine* in groundmass of *augite*, *olivine*, *analcite* and *magnetite* in a mesostasis of abundant *glass* (potentially *andesine*).
Occurrence: Italy—Sardinia (Monte Ferru).

GIANNETTITE

$cNa_3Ca_3Mn(Zr,Fe)TiSi_6O_{21}Cl$ Calcium sodium manganese chloro-zircono-titano-silicate. Triclinic. Minute crystals. Cleavage observed in one direction.
Occurrence: Brazil—Poços de Caldas. Alteration product of *aegirine* occurring intimately intergrown with *aegirine* in *nepheline* rocks.

GIBBSITE (of Torrey), HYDRARGILLITE (of Cleaveland) or WAVELLITE (of Dewey)

$8[Al(OH)_3]$ Aluminium hydroxide. Monoclinic. Tabular crystals with hexagonal aspect; occasionally spheroidal concretions, stalactitic, small mammillary and incrusting; also compact earthy and as enamel-like coatings. White or greyish, greenish or reddish-white; reddish-yellow when impure. Pearly lustre on cleavage surfaces, vitreous on other surfaces. White streak. Rough fracture. Perfect cleavage in one direction. Hardness 2·5–3·5. SG 2·4 (2·3–2·4 when massive). Infusible in flame, whitens, gives off light and becomes very hard.
Occurrence: USA—Arizona, Arkansas, Massachusetts, New York, Pennsylvania. Australia—Tasmania. France. Germany—Hesse. Hungary—*bauxite* deposits. Italy—Mount Vesuvius. Norway—Langesund Fiord. Turkey—Ephesus. Brazil—Minas Gerais. French Guiana. Guyana. USSR—Ural Mountains. India. Ghana. Madagascar. Seychelles Island. Secondary product from alteration aluminous minerals found chiefly in *bauxite* deposits.

GIBELITE

Igneous rock of the *syenite* clan. A light grey, porphyritic extrusive rock. Contains large and abundant phenocrysts of soda-*microcline*, sometimes *microperthite* and a few small prisms of *augite* in groundmass of soda-*microcline* and accessory *augite* and *amphibole*.
Occurrence: Italy—Pantellaria (Montagna Grande).

GIESECKITE

Magnesium potassium aluminosilicate, sometimes with appreciable ferrous oxide. Formula uncertain. Alteration product of *nepheline*, differing in that it contains water. Six-sided prisms. Greenish-grey. Greasy lustre. Perfect hexagonal cleavage.
Occurrence: USA—New York (Diana). Greenland—Akulliardsuk, Kangerdluarsuk. In compact *feldspar*.

Member of the *mica* family, closely related to *muscovite*.

GILBERTITE

Variety of *muscovite*. A secondary *mica* resulting from alteration of numerous rock-forming minerals such as *feldspar* and *andalusite*.
Occurrence: Widespread in igneous and metamorphic rocks as alteration product of *feldspar*, *topaz* or *andalusite*. Fine silky scales or fibres.

GILLESPITE

$4[BaFeSi_4O_{10}]$ Barium ferrous iron silicate. Tetragonal. *Mica*-like scales. Red. Pink streak. Good basal cleavage, poor in second direction. Hardness 4·0. SG 3·33. Fuses readily to black bead in flame. Easily decomposes in hydrochloric acid leaving glistening scales of silica.
Occurrence: USA—Alaska (Alaska Range), California (Mariposa County). With *diopside* and barium-*feldspar*.

GISMONDINE

$c8[CaAl_2Si_2O_8 . 4H_2O]$ Hydrated calcium aluminium silicate with some replacement of calcium by potassium. Monoclinic. Square octahedral crystals with terminal angle, faces rough and composite. Colourless, white, bluish-white, greyish or reddish. Vitreous lustre. Conchoidal fracture. Hardness 4·5. SG 2·265. Easily soluble in acids with gelatinisation. At 100 degrees yields one third of its water and becomes opaque.

Occurrence: Czechoslovakia—Bohemia (Salesl). Germany—near Fulda, near Giessen, Westphalia (near Bühne). Italy—Sicily (Val di Noto), South East of Rome (Capo di Bove). Poland—Silesia (near Görlitz). Switzerland—near Zermatt.
Member of the *zeolite* family.

GIUMARRITE

Igneous rock. A variety of *monchiquite*. Contains much *amphibole*.
Occurrence: Italy—Sicily (Giumarra).

GLACIAL

Produced by, or relating to, ice action.

GLADKAITE

Igneous rock of the *diorite* clan. A fine-grained hypabyssal rock. Contains *plagioclase* (*oligoclase* to *andesine*), abundant *quartz*, *hornblende* and a little *biotite* with accessory *magnetite* and *apatite* and secondary *muscovite* and *epidote*.
Occurrence: USSR—Northern Ural Mountains (Gladkaia Spoka).

GLASERITE

$c[K_3Na(SO_4)_2]$ Sodium potassium sulphate. Variety of *aphthitalite*, the pure end-member with 3:1 ratio of potassium to sodium.
Occurrence: Italy—lavas of Mount Vesuvius. Poland—Stebnick.

GLASS

Igneous rock constituent, the cryptocrystalline or amorphous material which represents the final phase of magma consolidation. Occurs as part of the groundmass. May also form extrusive igneous rocks when rapid chilling occurs (*obsidian*).

GLASSY

Glassy, or vitreous, substances are solids with absolutely no crystalline structure. Rare geologically but can be formed by a variety of processes.

GLASSY TUFF

Consolidated volcanic dust with porous texture. When unaltered has shreds of *glass* in interstitial mass of fine *glass* dust, iron oxide and secondary *chalcedony* or hydrous silica. Usually altered by solution of substance or addition of material with cementation and consolidation. *Glass* usually becomes devitrified, rocks can become hard and dense resembling *hornfels*, minerals can be introduced or secondary minerals can form.
Occurrence: Worldwide in regions associated with extrusive volcanic activity.

GLAUBERITE

$4[Na_2Ca(SO_4)_2]$ Calcium sodium sulphate. Monoclinic. Tabular, prismatic, dipyramidal, often striated crystals. Grey or yellowish, sometimes colourless; often reddish from inclusion of iron oxide. Vitreous to slightly waxy lustre, pearly on cleavage. White streak. Conchoidal fracture. Brittle. Perfect cleavage in one direction, indistinct in second direction. Hardness 2·5–3·0. SG 2·75–2·85. Slightly saline taste. Whitens in water from leach-

ing of sodium sulphate and deposition of *gypsum*. F 1·5. Completely soluble in hydrochloric acid. Decrepitates, turns white and fuses to white enamel in flame.
Occurrence: Worldwide as constituent of salt deposits of oceanic and lacustrine origin, as fumarole deposit and with nitrate deposits in extremely arid regions. Isolated crystals in clastic sediments formed under arid conditions or in cavities in basic extrusive rocks.

GLAUCOCHROITE

$4[(Mn,Ca)_2SiO_4]$ Calcium manganese silicate with calcium and manganese in equimolar proportions. Orthorhombic. Columnar, prismatic crystals. Bluish-green resembling *beryl*. Hardness 6·0. SG 3·407.
Occurrence: USA—New Jersey (Franklin Furnace).
Member of the *olivine* family.

GLAUCODOT

$8[(Co,Fe)AsS]$ Cobalt-iron arsenide-sulphide with ratio of cobalt to iron from 0·5–0·6. Orthorhombic. Prismatic crystals, striated in prism zone; also massive. Greyish tin-white to reddish silver-white. Metallic lustre. Black streak. Uneven fracture. Brittle. Perfect cleavage in one direction, less perfect in second direction. Hardness 5·0. SG 6·04. F 2·0–3·0. Decomposes in nitric acid with separation of sulphur and formation of pink solution.
Occurrence: USA—Oregon (Sumpter). Australia—Tasmania (Mount Wellington, North East Dundas). Norway—Skutterud. Rumania—Oravicza. Sweden—Hakansbo. Chile—Atacama region (around Huasco). Mainly with *cobaltite*, *pyrite*, and *chalcopyrite*.
Member of *arsenopyrite* family.

GLAUCONITE

$c2[K_{1.5}(Fe^{3+}, Mg, Al, Fe^{2+})_{4-6}(Si, Al)_8O_{22}(OH)_4]$ Hydrated magnesium aluminium potassium ferric and ferrous iron aluminosilicate, with ferric iron $c2$, ferrous minium potassium ferric and ferrous iron

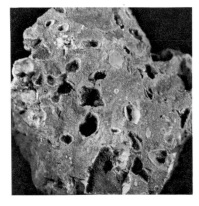

aluminosilicate, with ferric iron $c2$, ferrous $c0.5$, silicon $c7$ and calcium present. Monoclinic. Earthy masses. Olive-green, blackish-green, yellowish-green and greyish-green. Dull or glistening lustre. Hardness 2·0. SG 2·2–2·4. Fuses easily in flame to dark magnetic glass. Yields water when heated in closed tube.

Occurrence: England—Cheviot Hills, Greensands of Kent and Sussex. Wales. USA—Missouri (Red Bird), New Jersey, Pennsylvania (French Creek). South Africa—Agulhas Bank. Belgium—Anvers, Havre. USSR—Grondno Valley, Ontika, Svir River. India. Either in cavities in rocks or loosely granular massive.

GLAUCOPHANE, GASTRALDITE or GREEN RHODONITE

$2[Na_2(Mg,Fe^{2+})_3Al_2Si_8O_{22}(OH)_2]$ Hydrated sodium magnesium aluminium and ferrous iron silicate with proportion of magnesium greater than ferrous and including the iron-free end-member. Monoclinic. Thin, indistinct, prismatic crystals; usually massive, fibrous or columnar to granular. Azure-blue, lavender-blue, bluish-black or greyish. Vitreous to pearly lustre. Greyish-blue streak. Subconchoidal to uneven fracture. Brittle. Perfect cleavage in one direction. Hardness 6·0–6·5. SG 3·103–3·113.

Occurrence: USA—California (coast ranges). New Caledonia—Balade Mine. France—Groix Island on North West coast. Greece—Kikládhes Islands. Italy—Corsica. Switzerland—widespread throughout the Alps. Yugoslavia—Croatia (in the Fruška Gora). Japan—Shikoku. As hornblendic constituent of *glaucophane schists* but prominent in most types of metamorphic rocks.

A clino-*amphibole*. Varieties are *holmquistite* and *rhodusite*.

GLAUCOPHANE SCHISTS

High grade regional metamorphic rocks. Consist of *glaucophane* with *lawsonite*, *jadeite-quartz*, *aegirine* and *pumpellyite*. Ordinary schistose texture. Formed as derivatives of sediments and basic lavas, *tuffs* and intrusives.

Occurrence: USA—the Rocky Mountains. Europe—the Alps. South America—the Andes.

GLENMUIRITE

Igneous rock of the *diorite* clan. A mesocratic, coarse to medium-grained plutonic rock. Contains *orthoclase*, *labradorite*, *analcite*, *augite* and *olivine* with accessory *apatite*, *biotite* and *iron* ore.

Occurrence: Scotland—Ayrshire (Glenmuir Water), Fife.

GLIMMERGABBRO

Igneous rock of the *gabbro* clan. A coarse-grained plutonic rock. Consists of usual gabbroic composition with addition of *biotite* and optional accessories *quartz*, *orthoclase*, *hornblende*, *magnetite*, *apatite* and *pyrite*.

Occurrence: Sweden—Småland.

GLIMMERITE

Synonym of *biotitite*.

GLIMMERTON

Synonym of *illite*.

GLISTENING

Rather faint.

GLUCINITE

Synonym of *herderite*.

GMELINITE or SARCOLITE (of Vauquelin)

$c(Na_2,Ca)Al_2Si_4O_{12}.6H_2O$ Hydrated sodium calcium aluminosilicate with small amounts of potassium. Hexagonal. Rhombohedral, often horizontally striated crystals. Colourless, yellowish-white, greenish-white, reddish-white or flesh-red. Vitreous lustre. Uneven fracture. Brittle. Cleavage easy in one direction. Hardness 4·5. SG 2·04–2·17. F 2·5–3·0. Fuses in flame to white enamel. Decomposes in hydrochloric acid with separation of silica.

Occurrence: Scotland—Skye (Taliser). Northern Ireland—County Antrim (Glen-arm, near Larne, Portrush). USA—New Jersey (Bergen Hill). Canada—Nova Scotia (Cape Blomidon, Five Islands, Two Islands). Cyprus—near Pyrgo. Rumania—Transylvania.

GNEISS

High grade regional metamorphic rock. A coarse-grained granular rock with irregular banding and poor schistosity. Contains quartzo-feldspathic and ferromagnesian-minerals.

Occurrence: Worldwide. Metamorphism of *granites*, *syenites* or *diorites*.

GNEISSIC GRANITE

Type of *granite* with a gneissic structure.

GNEISSOID GRANITE

Igneous rock of the *granite* clan. A granitic gneissoid textured plutonic rock. Constituents as bands or parallel arrangements; gneissoid structure due to constrained movements of a viscous magma during crystallisation.

Occurrence: Widespread in areas of regional metamorphism.

GOETHITE, α GOETHITE, PYRRHOSIDERITE or XANTHOSIDERITE (of Schmid)

$4[FeO.OH]$ Hydrous ferrous oxide.

Orthorhombic. Commonly as slender, flattened plates, velvety surfaces of needles, occasionally brilliant rosettes of radiating plates and rarely small, shiny, equidimen-

Above, below left and over page: goethite

sional crystals; also fibrous, massive with reniform surfaces, compact or earthy. Brilliant black to brownish-black crystals; brown to yellow fibres. Lustre of crystals imperfectly adamantine-metallic, sometimes dull; lustre of fibres often silky. Brownish-yellow, orange or ochre-yellow streak. Uneven fracture. Brittle. Perfect side pinacoid cleavage. Hardness 5·0–5·5. SG 3·3–4·3. When heated in closed tube gives off water and turns to *hematite*. Practically infusible on charcoal but becomes magnetic.

Occurrence: England. USA. Austria. Czechoslovakia. Germany. Chile. USSR. Constitutes an important ore of iron. Found in

secondary oxidised deposits, sometimes in low-temperature veins.

GOLD

4[Au] Native gold. Cubic. Octahedral, dodecahedral or cubic crystals; often elongated and flattened, as parallel groups, reticulated, dendritic, arborescent, filiform or spongy, also massive, rounded fragments, flattened grains or scales. Colour and

streak gold-yellow when pure; silver-white to orange-red when impure. Metallic lustre. Jagged fracture. No cleavage. Hardness 2·5–3·0. SG 19·3. Very malleable and ductile. Insoluble in acids except aqua regia. Readily fuses on charcoal, drawing into golden-button.
Occurrence: Worldwide in *quartz* veins and stream deposits.

GOLDEN BERYL

Synonym of *heliodor*.

GONDITE

High grade regional metamorphic rock. A fine-grained granular rock. Contains *spessartite*, *quartz* and *apatite*.
Occurrence: India—Bombay Province (Narukot), Central Province (Nagpur-Balaghut area). Metamorphism of manganese rich sediments.

GONNARDITE

$c2[Na_2CaAl_4Si_6O_{20} . 7H_2O]$ Hydrated sodium calcium aluminium silicate. Some replacement of sodium and silicon content with calcium and aluminium. Orthorhombic. Finely fibrous. SG 2·256–2·357.
Occurrence: France—Puy-de-Dôme (Gignat).

Above left, left, below and below right: gold

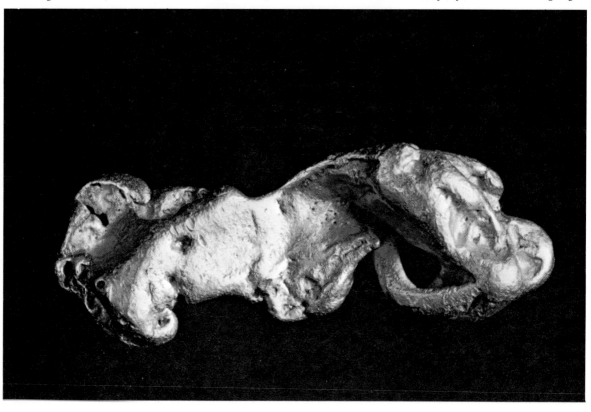

A *fibrous zeolite* closely related to *thomsonite*.

GONYERITE

$c(Mn^{2+}, Mg, Fe^{3+})_{5.91}(Si, Fe^{3+}, Al)_4(O, OH)_{18}$ Hydrated magnesium divalent manganese aluminium ferric and ferrous iron silicate. Rich in manganese oxide (33·83 per cent) and poor in aluminium oxide (0·58 per cent). Orthorhombic. Small, radial aggregates of laths and plates. SG 3·01. Dark brown.
Occurrence: Sweden—Långban. In hydrothermal veinlets cutting *skarn*.
Member of *chlorite group*.

GOODERITE

Igneous rock of the *diorite* clan. A light grey, occasionally spotted with pink, medium to fine-grained plutonic rock. Contains mainly *albite* with *biotite, nepheline, microperthite* and *calcite*.
Occurrence: Canada—Ontario (Gooderham).

GORCEIXITE

$39[Al_3P_2O_7(OH)_7]$ Basic hydrated aluminium phosphate. Hexagonal. Grains and pebbles in part microcrystalline. Brown, sometimes mottled. Vitreous to dull lustre. Porcelaneous fracture. Hardness 6·0. SG 3·036–3·185.
Occurrence: Brazil—as *favas* in diamanti-

ferous sands. Guyana—Mozaruni district. Ghana, Rhodesia, Sierra Leone—as *favas*.

GORDUMITE

Igneous rock of the *ultrabasic* clan. A coarse-grained porphyritic plutonic rock. Massive with indistinct schistosity. Contains pseudo-phenocrysts of *pyrope* in groundmass of *olivine* and *diopside* with accessory *picotite, magnetite* and *rutile* needles. A garnet-bearing *wehrlite*.
Occurrence: Switzerland—Tessin (Bellinzona).

GOSHENITE

Variety of *beryl*. Colourless or white. Rare. Gemstone use.
Occurrence: USA—Massachusetts (Goshen in Hampshire County).

GOSLARITE

$4[ZnSO_4 . 7H_2O]$ Hydrated zinc sulphate. Orthorhombic. Crystals rare, normally efflorescent crusts or stalactitic masses. Colourless or white. Vitreous or silky lustre. Brittle fracture. Perfect cleavage in one direction. Hardness 2·0–2·5. SG c 2·0. Astringent taste. Soluble in water.
Occurrence: USA—Arizona (Globe district of Gila County), California (Island Mountain in Trinity County), Montana (Gagman Mine near Butte). France—Lyons (Sain-Bel). Germany—Harz Moun-

tains (Rammelsberg Mine). Alteration product of *blende*.

GOSSAN

The 'iron hat' of a decomposed and leached rock at the surface of an outcrop of a sulphide vein, generally containing hydrated iron oxides.

GRAHAMITE

A variety of solid *bitumen* derived from petroleum-bearing rocks.

GRAIN

Mineral particle in rock, the rock being described on basis of grain type and var-

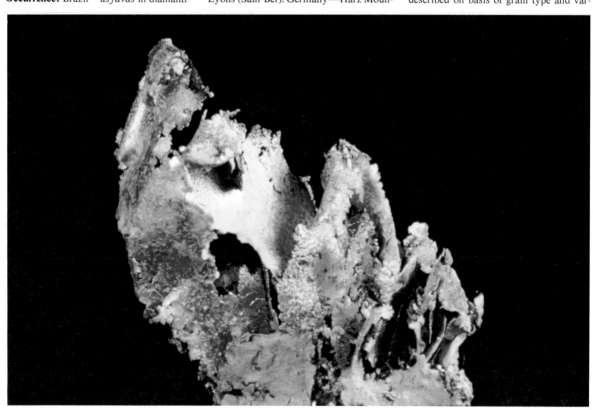

iety, size and range. Grain size may be approximated to following scale:

microscopic less than 0·1mm
fine   0·1mm–1mm
medium   1mm–5mm   ⎱ 'granular'
coarse   more than 5mm ⎰ material

Scale of grain or particle size in sedimentary rocks:

clay   less than 0·005mm
silt   0·005–0·05mm
sand   0·05–2mm
gravel 2–4mm
pebble 4–65mm
cobble 65–250mm
boulder more than 250mm

GRANITE

Igneous rock of the *granite* clan. A coarse-grained granular plutonic rock. Contains *quartz*, alkali-*feldspars*, *biotite*, *muscovite* and *plagioclase*.
Occurrence: Worldwide.

GRANITE APLITE

Type of *aplite* with a granitic composition.

GRANITE CLAN

Igneous rocks containing at least 66 per cent silica and an average of 30 per cent *quartz*. Also contain alkali-*feldspar* and *plagioclase* with numerous ferromagnesian minerals. Subdivided using feldspar composition and environment of formation. Alkali-*feldspar* over 60 per cent of total *feldspar: granite* (plutonic), *microgranite* (hypabyssal) and *rhyolite* (extrusive). Alkali-*feldspar* and *plagioclase* in equal amounts: *adamellite* (plutonic), *micro-adamellite* (hypabyssal) and *toscanite* (extrusive). *Plagioclase* over 60 per cent of total *feldspar: granodiorite* (plutonic), *microgranodiorite* (hypabyssal) and *dacite* (extrusive).

GRANITE GREISEN

Igneous rock of the *granite* clan. A coarse-grained granular plutonic rock. Contains *quartz* (50 per cent), *orthoclase*, *microcline*, *oligoclase* and *muscovite*.
Occurrence: USA—California (Kernville), New Hampshire (Troy).

GRANITE PEGMATITE

Type of *pegmatite* with a granitic composition.

GRANITE PORPHYRY

Igneous rock of the *granite* clan. A medium to fine-grained porphyritic hypabyssal rock. Contains phenocrysts of *plagioclase* and *orthoclase* in a groundmass of *quartz*, *orthoclase* and *muscovite*.
Occurrence: USA—Montana (Wolf Butte in the Little Belt Mountains). Sweden—Medelpad (Rödön).

GRANITISATION

Suspected but unproven process in which rocks are decomposed into *granite* by the action of fluids produced deep in the crust.

GRANITITE

Igneous rock of the *granite* clan. A coarse-grained granular plutonic rock. Contains *quartz*, *orthoclase*, *oligoclase* and *biotite*.
Occurrence: Worldwide.

GRANOBLASTIC

Metamorphic equivalent of equigranular.

GRANODIORITE

Igneous rock of the *granite* clan. A coarse-grained granular plutonic rock. Contains *quartz*, *plagioclase*, *orthoclase*, *biotite*, *hornblende* and *pyroxene*.
Occurrence: Worldwide.

GRANOGABBRO

Type of *granodiorite* containing *plagioclase* in the range *labradorite-bytownite*.

GRANOPHYRE

Igneous rock of the *granite* clan. A fine-grained porphyritic hypabyssal or extrusive rock. Contains phenocrysts of *feldspar*, *augite* and *biotite* in a groundmass of intergrown *quartz* and alkali-*feldspar*.
Occurrence: Worldwide.

GRANOPHYRIC

A rock with a micro*granite* texture such as *granophyre*.

GRANULAR

Grainy texture.
See grain.

GRANULITE, HÄLLEFLINTGNEISS or LEPTITE

High grade regional metamorphic rock. A granular rock containing *quartz*, *feldspar*, *pyroxene* and *garnet*.
Occurrence: Worldwide.

GRAPESTONE

Descriptive term for accumulations of fine-grained carbonates in a back reef environment. Aggregates of pellets form botryoidal clusters.
Occurrence: Bahamas.

GRAPHIC GRANITE

Type of *granite* composed of a coarse intergrowth of *quartz* and *microcline*. Intergrowth resembles runic characters.
Occurrence: USA—Connecticut (Andrews Quarry near Portland), New York (Kinkles Quarry near Bedford). Norway—Arendal. Mozambique—Fort Jagaia.

GRAPHITE

4[C] Carbon. Hexagonal. Tabular crystals; also massive. Iron-black to steel-grey.

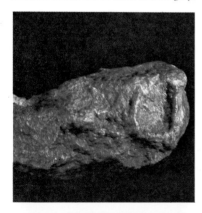

Metallic to dull lustre. Black or dark grey streak. Perfect basal cleavage. Hardness 1·0–2·0. SG 2·09–2·23. Feels greasy.
Occurrence: England—Cumberland (Borrowdale near Keswick). USA—New Jersey (Sterling Hill), New York (Ticonderoga in the Adirondack Mountains). Canada—Quebec (Buckingham). Finland—Pargas.
Metamorphism of carbonaceous material in sediments.

GRATONITE

$Pb_9As_4S_{15}$ Lead arsenic sulphide. Hexagonal. Prismatic crystals; also massive. Dark grey. Metallic lustre. Black streak. Brittle fracture. No cleavage. Hardness 2·5. SG c6·2.
Occurrence: Peru—Cerro de Pasco (Excelsior Mine).

GRAVEL

Unconsolidated grains consisting of water worn particles between 2mm and 4mm in diameter.

GRAVITY, SPECIFIC

See specific gravity.

GREENALITE

A hydrous iron silicate which is an important ore of *iron*.

GREEN CALAMINE

Synonym of *aurichalcite*.

GREEN MUD

Oceanic *argillaceous* deposit containing detrital material enriched with *glauconite* and *chlorite*.

GREENOCKITE

2[CdS] Cadmium sulphide. Hexagonal. Pyramidal crystals; also earthy masses. Shades of yellow or orange. Adamantine to resinous lustre. Orangish-yellow to brick-red streak. Conchoidal fracture. Good cleavage in one direction. Hardness 3·0–3·5. SG c5·0. Soluble in concentrated hydrochloric acid with the evolution of hydrogen sulphide.
Occurrence: Scotland — Dumfriesshire (Wanlockhead), Renfrewshire (Bishopton). USA—California (Topaz in Mono County), New Jersey (Franklin). Greece—Laurium. Bolivia—Llallagua.

GREENOVITE

A red or pinkish variety of *sphene* containing manganese oxide.

GREEN RHODONITE

Synonym of *glaucophane*.

GREENSAND

Green marine *arenaceous rock*. Contains detrital *quartz* and *feldspar* and grains of *glauconite*.
Occurrence: England—the Lower and Upper Greensand of Kent and Sussex.

GREEN SCHIST

High grade regional metamorphic rock. A green schistose rock containing *actinolite*, *chlorite*, *epidote* or *hornblende*.
Occurrence: Widespread. Metamorphism of basic igneous rocks.

GREISEN

Pneumatolytically altered igneous rock of the *granite* clan. A fine to coarse-grained granular plutonic rock. Contains *quartz, zinnwaldite, cassiterite* and minor amounts of *orthoclase, tourmaline, fluorite* and *topaz.*
Occurrence: England—Cornwall (St Michael's Mount), Cumberland (Cambe Height on Carrock Fell). Germany—Saxony (Altenberg).

GREYWACKE

Arenaceous rock. Contains *quartz, feldspar* and rock fragments in a matrix of *sericite, illite* and *chlorite.* Shows graded bedding with a coarse-grained base and a fine-grained upper surface.
Occurrence: Worldwide.

GRIPHITE

Basic manganese calcium aluminium sodium and iron phosphate. Formula uncertain. Cubic. Massive. Dark brown to brownish-black. Resinous to vitreous lustre. Uneven to conchoidal fracture. No cleavage. Hardness 5·5. SG 3·4. Easily soluble in hydrochloric acid.
Occurrence: USA—South Dakota (Riverton Lode near Harney City in Pennington County). Australia—Northern Territory (Mount Ida).

GRIQUAITE

Intergrowth of *augite* and *garnet.*

GRIT

Type of coarse-grained *sandstone* with angular grains.

GRITTY

Resembling or containing grit.

GROCHAUITE

Magnesium rich variety of *sheridanite.*

GRÖNLANDITE

Igneous rock of the *ultrabasic* clan. A coarse-grained granular plutonic rock. Contains *hornblende* and *hypersthene.*
Occurrence: Greenland—Upernivik Island.

GRORUDITE

Igneous rock of the *granite* clan. A fine to medium-grained porphyritic hypabyssal rock. Contains phenocrysts of *microcline, microcline-microperthite, aegirine* and *hornblende* in a groundmass of *quartz, microcline* and *aegirine.*
Occurrence: Norway—Grorud (Gussletten), Kallerud, Varingskollen.

GROSSULAR or WILUITE

$8[Ca_3Al_2Si_3O_{12}]$ Calcium aluminium sili-

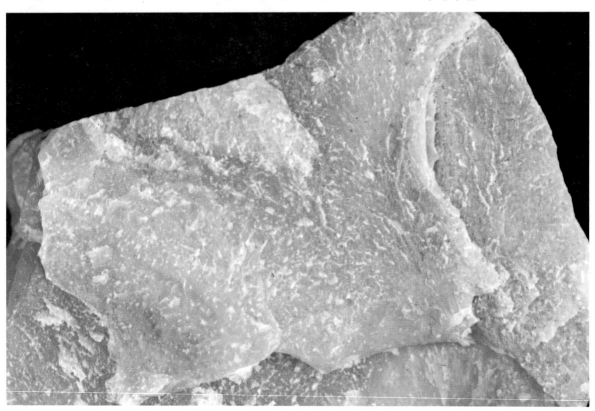

cate. Cubic. Rhombdodecahedral crystals. Greenish-white to olive-green. Vitreous lustre. White streak. Subconchoidal to uneven fracture. No cleavage. Hardness 7·5. SG 3·5.
Occurrence: Worldwide.
A *garnet*. Varieties are *hessonite, landerite, South African jade* and *succinite*.

GROSSULAROID

Synonym of *hydrogarnet*.

GROTHITE

Synonym of *sphene*.

GROUNDMASS

In igneous rocks, fine grained crystalline body in which are contained large crystals forming phenocrysts.

GROUTITE

4[MnO.OH] Basic manganese oxide. Orthorhombic. Pyramidal crystals. Black. Submetallic to adamantine lustre. Dark brown streak. Perfect cleavage in one direction. SG 4·14.
Occurrence: USA—Minnesota (Cuyuna Range).
A *diaspore* mineral.

GRUNERITE

$2[(Fe,Mg)_7Si_8O_{22}(OH)_2]$ Iron magnesium silicate. Monoclinic. Fibrous or lamellar masses. Brown. Silky lustre. Brown streak. Perfect cleavage in two directions. Hardness 5·0–6·0. SG c3·5.
Occurrence: USA—Massachusetts (Rockport).
An *amphibole*; forms a solid solution series with *cummingtonite*.

GUANAJUATITE or FRENZELITE

$4[Bi_2Se_3]$ Bismuth selenide. Orthorhombic. Acicular crystals; also massive. Bluish-grey. Metallic lustre. Grey streak.

Left: grossular

Good cleavage in one direction. Hardness 2·5–3·5. SG 6·25–6·98. Soluble when heated in concentrated sulphuric acid.
Occurrence: USA—Idaho (Salmon in Lemhi County). Germany—Harz Mountains (Andreasberg). Sweden—Fahlun. Mexico—Guanajuato (Santa Catarina Mine near Sierra de Santa Rosa).

GUANO

Phosphatic deposit. Greyish-brown. Contains calcium phosphate and various nitrates and carbonates. Compact or friable.
Occurrence: Rocky Islands of the West Indies and Eastern Pacific Ocean. Deposit of sea birds in an arid environment.

GUMMITE

Hydrated uranium lead thorium and rare earths oxide. Formula and crystal system uncertain. Massive. Yellow, red, brown or black. Greasy to vitreous lustre. Yellow or olive-green streak. Conchoidal to uneven fracture. No cleavage. Hardness 2·5–5·0. SG 3·9–6·4.
Occurrence: Widespread.
Alteration product of *uraninite*. Related to *limonite*.

GYMNITE

Synonym of *deweylite*.

GYPSITE

A variety of *gypsum* containing sand and grit.

GYPSUM

$8[CaSO_4.2H_2O]$ Hydrated calcium sulphate. Monoclinic. Tabular or prismatic crystals; also massive. Colourless, white, grey or yellowish-brown. Subvitreous to pearly lustre. White streak. Conchoidal fracture. Good cleavage in two directions. Hardness c2·0. SG c2·3. Slightly soluble in water. Soluble in hydrochloric acid.
Occurrence: Worldwide.
Varieties are *gypsite, selenite* (crystalline),

satin spar (fibrous) and *alabaster* (massive).

HACKLY

Irregular surface fracture with sharp angular projections.

HACKMANITE

Calcium rich variety of *sodalite*.

HAILSTONE BORT

Variety of *diamond*. A grey or greyish-black, rounded stone. Composed of concentric shells of clouded *diamond* and cement-like material.

HALITE or ROCK SALT

4[NaCl] Sodium chloride. Cubic. Cubic crystals; also massive. Colourless, white, yellow, red, blue or purple. Vitreous lustre. Colourless or white streak. Conchoidal fracture. Perfect basal cleavage. Hardness 2·0. SG c2·2. Tastes salty. Soluble in water.

Above and below: halite

Occurrence: Worldwide.
An *evaporite*. Variety is *hydrohalite*.

HALLOYSITE

2[Al$_2$Si$_2$O$_5$(OH)$_4$.2H$_2$O] Basic hydrated

aluminium silicate. Monoclinic or triclinic. *Clay*-like masses. White, often tinted grey, green, yellow, red or blue. Pearly to dull lustre. Conchoidal fracture. Hardness 1·0–2·0. SG 2·0–2·2. Decomposes in hydrochloric acid.

Occurrence: Widespread.

A *kaolinite clay mineral* produced by the alteration of ore veins, *granites* or aluminous sediments.

HAMRONGITE

Igneous rock of the *granite* clan. A fine-grained porphyritic hypabyssal rock. Contains phenocrysts of *biotite* in a ground-mass of *quartz, biotite* and *andesine*.

Occurrence: Sweden—Hamrånge parish (Gävle).

HANCOCKITE

(Pb, Ca, Sr)$_2$(Al, Fe)$_3$Si$_3$O$_{12}$OH Basic lead aluminium iron silicate. Monoclinic. Lath shaped crystals; also massive. Brownish-red to yellowish-brown. Vitreous lustre. Hardness 6·5–7·0. SG 4·03.

Occurrence: USA—New Jersey (North Mine Hill at Franklin Furnace).

HANLÉITE

8[Mg$_3$Cr$_2$Si$_3$O$_{12}$] Magnesium chromium silicate. Cubic. Cubic crystals. Green.

Right: hancockite

Vitreous lustre. Conchoidal to uneven fracture. No cleavage. Hardness 6·5–7·5. SG 3·5–4·3.

Occurrence: India—Kashmir (Hanlé Monastery near Rupshu).

A *garnet*.

HAPLITE

Synonym of *aplite*.

HARDNESS

Prime property of rocks and minerals, measure of which is of major importance. Various scales exist; the following is that produced by Mohs, materials with higher values being able to scratch those of lower values.

1·0	talc	6·0	orthoclase
2·0	gypsum	7·0	quartz
3·0	calcite	8·0	topaz
4·0	fluorite	9·0	corundum
5·0	apatite	10·0	diamond

Note the following useful values.

2·3	fingernail	6·0	glass
5·0	tooth, 'copper' coin	7·0	knife

Note also that this is not a linear scale and that most rocks and minerals vary to some extent in hardness depending on source of sample and direction of scratch.

HARDYSTONITE

$2|Ca_2ZnSi_2O_7|$ Calcium zinc silicate. Tetragonal. Granular masses. White. Vitreous lustre. Good cleavage in three directions. Hardness 3·0–4·0. SG c4·0.
Occurrence: USA—New Jersey (Franklin Furnace).

HARMOTOME or HERCYNITE

$2|BaAl_2Si_6O_{16}.6H_2O|$ Hydrated barium aluminium silicate. Monoclinic. Square prismatic crystals. White, grey, yellow, red or brown. Vitreous lustre. White streak. Uneven to subconchoidal fracture. Good cleavage in one direction. Hardness 4·5. SG 2·44–2·50. Decomposes in hydrochloric acid.
Occurrence: Scotland — Argyllshire

(Strontian). Canada—Ontario (Rabbit Mountain near Port Arthur). Germany—Harz Mountains (Andreasberg).

HARRINGTONITE

Amorphous chalky-white mixture of *mesolite* and *thomsonite* from County Antrim in Northern Ireland.

HARRISITE

Igneous rock of the *gabbro* clan. A coarse-grained granular hypabyssal rock. Contains *olivine*, *anorthite*, *enstatite* and *chromite*.
Occurrence: Scotland—Inverness-shire (Isle of Rhum).

HARZBURGITE

Igneous rock of the *ultrabasic* clan. A medium-grained granular plutonic rock. Contains *olivine* and orthorhombic *pyroxene* with *serpentine*, *magnesite* and *calcite*.
Occurrence: Austria—Tyrol (Gotthard Tunnel). Germany—Harz (Baste). USSR—Ural Mountains (Naravet Ken).

HASTINGSITE

Alkali enriched *hornblende*. Varieties are *femaghastingsite* and *magnesiohastingsite*.

HATCHETTITE

Paraffin wax. Amorphous. Massive. Yellowish white or shades of yellow (tarnishes black). Pearly lustre. SG 0·92. Soluble in ether. Wax-like greasy feel.
Occurrence: Scotland—Argyllshire (Loch Fyne). Wales—Glamorganshire (Perth).

HATCHETTOLITE

$8|(Ca,Fe,U)(Nb,Ta,Ti)_2(O,OH,F)_7|$ Basic calcium iron uranium niobium (columbate) and tantalate. Cubic. Octahedral crystals; also massive. Black or yellowish-brown. Resinous lustre. Subconchoidal fracture. Hardness 4·0–5·0. SG 4·5–4·9.
Occurrence: USA—North Carolina (the Mica Mines of Mitchell County).
Related to *pyrochlore*.

HAUERITE

$4|MnS_2|$ Manganese sulphide. Cubic. Octahedral crystals; also globular masses. Reddish-brown to brownish-black. Metallic adamantine lustre. Brownish-red streak. Uneven to subconchoidal fracture. Perfect basal cleavage. Hardness 4·0. SG c3·4. Soluble in warm concentrated hydrochloric acid with the evolution of hydrogen sulphide and the separation of sulphur.
Occurrence: USA—Texas (Big Hill in Matagorda County). Italy—Sicily (Raddusa near Catania).

HAUGHTONITE

Type of *biotite* rich in iron.

HAUSMANNITE or BLACK MANGANESE

8[Mn_3O_4] Manganese oxide. Tetragonal. Pyramidal crystals; also granular masses. Brownish-black. Submetallic lustre. Chestnut-brown streak. Uneven fracture. Good basal cleavage. Hardness 5·5. SG 4·84. Soluble in hot hydrochloric acid with the evolution of chlorine.

Occurrence: England — Cumberland (Whitehaven). Scotland—Aberdeenshire (Granan). USA—California (Prefumo Canyon district of San Luis Obispo County), New Jersey (Franklin). Sweden—Wermland (Pajsberg Mine). In contact metamorphic rocks and high temperature hydrothermal veins.

Variety is *hydrohausmannite*.

HAÜYNE, HAÜYNITE or ROEBLINGITE

$(Na,Ca)_{4-8}Al_6Si_6O_{24}(SO_4)_{1-2}$ Sodium calcium aluminosilicate and sulphate. Cubic. Rounded grains. Shades of blue, green, red or yellow. Vitreous to greasy lustre. Colourless or pale blue streak. Conchoidal to uneven fracture. Good cleavage in three directions. Hardness 5·5–6·0. SG 2·4–2·5. Decomposes in hydrochloric acid with the separation of gelatinous silica.

Occurrence: Widespread. In extrusive igneous rocks.

HAÜYNITE

Synonym of *haüyne*.

HAÜYNOLITH

Name given to any igneous rock containing 100 per cent *haüyne*.

HAWAIITE

Name given to *andesine basalts* due to the wide occurrence of these rocks on the Hawaii Islands.

HAYDENITE

Yellow crystalline variety of *chabazite*.

HEAVY SPAR

Synonym of *barytes*.

HEBRONITE

Synonym of *amblygonite*.

HECTORITE

Altered hydrated green *pyroxene* occurring in *serpentinite* in the Dun Mountains of New Zealand.

Also magnesium rich variety of *montmorillonite*.

HEDENBERGITE

4[$CaFeSi_2O_6$] Calcium iron silicate. Monoclinic. Prismatic crystals; also lamellar masses. Black. Vitreous lustre. Conchoidal fracture. Perfect cleavage in two directions. Hardness 6·0. SG 3·5–3·58.

Occurrence: Norway—Arendal. Sweden—Tunaberg.

A *pyroxene*. Variety is *ferrohederbergite*.

HEDRUMITE

Igneous rock of the *syenite* clan. A fine to medium-grained porphyritic hypabyssal rock. Contains phenocrysts of *microperthite* and *lepidomelane* in a groundmass of *microperthite, nepheline, analcime, aegirine, lepidomelane* and *hornblende*.

Occurrence: Norway—Gran (Skirstad-

Below: hedenbergite

Kjern), Hedrum (Longenthal), Kristiania-fjord (Island of Osto).

HELIDOR or GOLDEN BERYL

Golden-yellow, gem quality variety of *beryl*.

HELIODORE

Synonym of *chrysoberyl* (*of Werner*).

HELIOTROPE or BLOOD STONE

Variety of green *plasma* with inclusions of red *jasper*.

HELSINKITE

Igneous rock of the *diorite* clan. A medium-grained granular hypabyssal rock. Contains *sodaclase, epidote, biotite* and *apatite*.
Occurrence: Finland—Gulf of Finland (Suursaari Island).

HELVINE

$(Mn,Fe,Zn)_8Be_6Si_6O_{24}S_2$ Manganese iron zinc sulphide and berylliosilicate. Cubic. Tetrahedral crystals; also massive. Honey-yellow. Vitreous to resinous lustre. Colourless streak. Uneven to conchoidal fracture. Indistinct cleavage. Hardness $6·0-6·5$. SG $3·16-3·36$. Decomposes in hydrochloric acid with the evolution of hydrogen sulphide and the separation of gelatinous silica.
Occurrence: USA—Virginia (the Mica Mines near Amelia Court House in Amelia County). Germany—Saxony (Schwarzenberg).
Family member is *donalite*.

HEMATITE

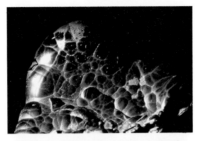

$2[Fe_2O_3]$ Ferric oxide. Hexagonal. Tabular crystals; also massive. Steel-grey or red. Metallic or dull lustre. Reddish-brown streak. Subconchoidal to uneven fracture. No cleavage. Hardness $5·0-6·0$. SG $5·26$. Soluble in concentrated hydrochloric acid.
Occurrence: Worldwide.
Important ore of iron. Varieties are *kidney ore* (reniform masses), *micaceous hematite* (foliated masses), *red ochre* (earthy masses), *specular hematite* (crystals with a splendant metallic lustre and *titano-hematite* (titanium rich).

HEMIMORPHITE or ELECTRIC CALAMINE

$2[Zn_4Si_2O_7(OH)_2.H_2O]$ Basic hydrated zinc silicate. Orthorhombic. Pyramidal crystals; also massive. White. Vitreous to adamantine lustre. White streak. Uneven to subconchoidal fracture. Perfect cleavage in one direction. Hardness $4·5-5·0$. SG $3·4-3·5$. Gelatinises in acetic acid.

Occurrence: England — Cumberland (Roughten Gill near Caldbeck), Derbyshire (Rutland Mine near Matlock). Scotland—Lanarkshire (Leadhills). USA—New Jersey (Sterling Hill), Virginia (Austin Mine in Wythe County). Belgium—Moresnet.

HEPTORITE

Igneous rock of the *gabbro* clan. A fine-grained porphyritic hypabyssal rock. Contains phenocrysts of *augite, hornblende, haüyne* and *labradorite* in a groundmass of *labradorite, olivine, magnetite* and *glass*.
Occurrence: Germany—Rhineland (Rhöndorfer Tal near Siebengebirge).

HERCYNITE

$8[FeAl_2O_4]$ Iron aluminium oxide. Cubic. Granular masses. Black. Vitreous lustre. Dark greyish-green to dark green streak. Conchoidal fracture. No cleavage. Hardness $7·5-8·0$. SG $4·39$. Soluble with difficulty in concentrated sulphuric acid.
Occurrence: USA—New York (Peekskill in Westchester County), Virginia (Whittles in Pittsylvania County). Australia—Tasmania (Moorina). Germany—Black Forest (Schenkenzell).
Variety is *chromohercynite*.
Also synonym of *harmotome*.

HERDERITE or GLUCINITE

$4[CaBePO_4(OH,F)]$ Basic calcium beryllium phosphate with fluorine. Monoclinic. Prismatic or tabular crystals; also spherical aggregates. Colourless, pale yellow or

greenish-white. Vitreous to subvitreous lustre. Subconchoidal fracture. Indistinct cleavage. Hardness 5·0–5·5. SG $c2·9$. Soluble in hydrochloric acid.
Occurrence: USA—Maine (Stoneham in Oxford County), New Hampshire (Fletcher Mine near North Groton in Groton County). Germany—Bavaria (Fichtelgebirge).

HERKIMER DIAMOND

Name given to gem quality *quartz* crystals from Herkimer County in New York State, USA.

HERONITE

Igneous rock of the *syenite* clan. A fine-grained granular hypabyssal rock. Contains *analcite, orthoclase, labradorite* and *aegirine*.
Occurrence: Canada—Ontario (Heron Bay on Lake Superior).

HERSCHELITE

Synonym of *seebachite*.

HERZENBERGITE

4[SnS] Tin sulphide. Crystal system uncertain. Irregular grains. Black. Metallic lustre. Black streak. Easily soluble in hydrochloric acid with the evolution of hydrogen sulphide.
Occurrence: Bolivia—Oruro (Maria-Teresa Mine near Huari).

HESSITE

48[Ag$_2$Te] Silver telluride. Cubic. Massive crystals. Lead-grey. Metallic lustre. Hardness 2·5. SG 8·4. When powdered and heated with strong-sulphuric acid gives a reddish-violet solution. When heated on charcoal gives *silver* bead.
Occurrence: USA—California. Canada—Ontario (Porcupine mining area). Western Australia—Kalgoorlie. Chile. Mexico.

HESSONITE, CINNAMON STONE or ESSONITE

Yellow variety of *grossular*.

HETAEROLITE

4[ZnMn$_2$O$_4$] Zinc manganese oxide. Tetragonal. Octahedral crystals; also massive. Black. Submetallic lustre. Dark brown streak. Uneven fracture. Indistinct cleavage. Hardness 6·0. SG $c5·2$.
Occurrence: USA—New Jersey (Franklin, Sterling Hill).

HETEROGENITE

CoO·OH Hydrous cobalt oxide. Crystal system uncertain. Globular or reniform masses. Black, brownish-black or reddish-brown. Dull to vitreous lustre. Dark brown streak. Conchoidal fracture. No cleavage. Hardness 3·0–4·0. SG 3·44.

Occurrence: Germany—Saxony (Schneeberg).
Alteration product of *smaltite*.

HETEROSITE

4[(Fe,Mn)PO$_4$] Iron manganese phosphate. Orthorhombic. Massive. Deep rose to reddish-purple. Satiny lustre. Pale red or purple streak. Uneven fracture. Good cleavage in one direction. Hardness 4·0–4·5. SG 3·2–3·4. Easily soluble in hydrochloric acid.
Occurrence: USA—Maine (Newry in Oxford County), New Hampshire (Palermo Mine near North Groton in Groton County), South Dakota (Hill City in Pennington County). Germany—Bavaria (Rabenstein).
Alteration product of *triphylite*.

HEULANDITE or STILBITE

(Na,Ca)$_{4-6}$Al$_6$(Al,Si)$_4$Si$_{26}$O$_{72}$·24H$_2$O Hydrated sodium calcium aluminosilicate. Monoclinic. Tabular crystals; also massive. Colour variable. Vitreous to pearly lustre. White streak. Subconchoidal to uneven fracture. Perfect cleavage in one direction. Hardness 3·5–4·0. SG 2·18–2·22.
Occurrence: Widespread in *basalt* and *gneiss*.
A *zeolite*. Silicon-rich variety is *clinoptilolite*. Barium-bearing variety is *beaumontite (of Lévy)*.

HEUMITE

Igneous rock of the syenite clan. A fine-grained granular hypabyssal rock. Contains soda-*orthoclase*, soda-*microcline*, *barkevikite, biotite, nepheline* and *sodalite*.
Occurrence: Norway—Brathagen, Heum.

HEXAGONAL

See *crystal*.

HEXAGONITE

Pink manganese rich variety of *tremolite*.

Above: heulandite
Below: hexagonite

HEXAHEDRITE

Type of *meteorite*.

HIBSCHITE

$8[Ca_3Al_2Si_2O_8(OH)_4]$ Hydrous calcium aluminium silicate. Cubic. Octahedral crystals. Colourless. Hardness 6·0. SG 3·05. Soluble in hydrochloric acid.
Occurrence: Czechoslovakia—Bohemia (Aussig). France—Ardèche (Aubenas). In contact metamorphic rocks.
A *hydrogarnet.*

HIDDENITE

Green variety of *spodumene.*

HIGH GRADE METAMORPHISM

Intense level of metamorphism (regional or contact) resulting in the growth of new minerals and the development of completely new textures. e.g. *gneiss* or *hornfels.*

HIGHWOODITE

Igneous rock of the *syenite* clan. A coarse-grained granular plutonic rock. Contains soda-*orthoclase, labradorite, diopside, biotite* and *apatite.*
Occurrence: USA—Montana (Highwood Peak in the Highwood Mountains).

HILAIRITE

Igneous rock of the *syenite* clan. A coarse-grained porphyritic plutonic rock. Contains phenocrysts of *sodaclase, nepheline*

and *lavenite* in a groundmass of *orthoclase, aegirine* and *sodalite.*
Occurrence: Canada—Quebec (Mount St Hilaire).

HIRNANTITE

Igneous rock of the *diorite* clan. A fine-grained granular plutonic rock. Contains *albite, chlorite* and *augite.*
Occurrence: Wales—Berwyn Hills (Craig-ddu near Hirnant).

HOEGBOMITE

$Mg(Al,Fe,Ti)_4O_7$ Magnesium aluminium iron and titanium oxide. Hexagonal. Crystals rare, normally irregular grains. Black. Metallic adamantine lustre. Grey streak. Conchoidal fracture. Imperfect cleavage. Hardness 6·5. SG *c*3·8.
Occurrence: USA—New York (Peekskill in Westchester County), Virginia (Whittles in Pittsylvania County). Norway—Söndmöre (Rödstand).

HOKUTOLITE or ANGLESOBARYTES

Lead rich variety of *barytes.*

HOLLANDITE

$(Ba,Na,K)Mn_2^{2+}Mn_6^{4+}O_{16}H_2O$ Hydrated manganese barium oxide and alkalis. Tetragonal. Prismatic crystals; also mas-

169

sive. Silver grey, greyish-black or black. Metallic lustre. Black streak. Brittle fracture. Good cleavage in two directions. Hardness c6·0. SG 4·95.
Occurrence: India—Jhabua State (Kajlidongri). Madagascar—Ambatomiady.

HOLMITE

Igneous rock of the *gabbro* clan. A finegrained porphyritic hypabyssal rock. Contains phenocrysts of *augite* and *olivine* (altered to *serpentine*, *calcite* and *magnetite*) in a groundmass of *melilite*, *augite* and *biotite*.
Occurrence: Scotland—Orkney Islands (Holm).
Also synonym of *clintonite*.

HOLMQUISTITE

Lithium rich variety of *glaucophane*.

HOLOCRYSTALLINE

Having 100 per cent degree of crystallinity. Hemicrystalline rocks have glassy and crystalline components.

HOLYOKEITE

Igneous rock of the *diorite* clan. A white, fine-grained granular hypabyssal rock. Contains *sodaclase* and *calcite*.
Occurrence: USA — Massachusetts (Mount Tom)

HORNBLENDE, LAMPROBOLITE, OXYHORNBLENDE or SIDERITE

$2[(Ca,Mg,Fe,Na,Al)_{7-8}(Al,Si)_8O_{22}(OH)_2]$
Calcium magnesium iron sodium and aluminium aluminosilicate. Monoclinic. Prismatic crystals; also massive. Black or greenish-black. Vitreous lustre. Subconchoidal to uneven fracture. Perfect cleavage in two directions. Hardness 5·0–6·0. SG 3·0–3·4.
Occurrence: Worldwide.
Iron-free variety is *edenite*. Alkali variety is *anophorite*. Other varieties are *carinthine* and *kaersutite*.

Above: hornblendite
Right: hortonolite

HORNBLENDE GABBRO

Synonym of *bojite*.

HORNBLENDE (Labrador)

Synonym of *hypersthene*.

HORNBLENDITE

Igneous rock of the *ultrabasic* clan. A coarse-grained granular hypabyssal or plutonic rock. Contains *hornblende*, *biotite* and *magnetite*.
Occurrence: USA—Idaho (the North fork of the Clearwater River near Ahsahka).

HORNFELS or HÄLLEFLINTA

High grade contact or rarely regional metamorphic rock. A fine-grained granular rock. Contains *quartz*, *micas* and *feldspar*.
Occurrence: Widespread. Metamorphism of many rock types.
Varieties are *astike*, *aviolite* and *edolite*.

HORN SILVER

Synonym of *cerargyrite*.

HORTONOLITE

$4[(Fe,Mg)_2SiO_4]$ Iron magnesium silicate. Orthorhombic. Crystalline masses. Yellow or dark yellowish-green, tarnishing black. Vitreous to resinous lustre. Uneven fracture. Good cleavage in two directions. Hardness 6·5. SG 3·9.
Occurrence: USA—New York (Monroe in Orange County).
An *olivine* variety is *ferrohortonolite*.

HOUSEHOLD COAL

Synonym of *bituminous coal*.

HOWIEITE

$Na(Fe^{2+}, Mn)_{11}(Fe^{3+}, Al)_2(Si, Ti)_{12}O_{31}(OH)_{13}$ Basic titanosilicate of sodium and iron. Triclinic. Bladed crystals. Dark green or black. Good cleavage in one direction.
Occurrence: USA—California (Laytonville district of Mendocino County). In metamorphosed *shales*, *ironstones* and *limestones*.

HOWLITE

$Ca_2B_5SiO_9(OH)_5$ Basic calcium boron silicate. Orthorhombic. Prismatic crystals. White. Subvitreous lustre. Smooth frac-

ture. No cleavage. Hardness 3·5. SG c 2·55.
Occurrence: Canada—Nova Scotia (Brookville in Hawks County).

HÜBNERITE

Synonym of *huebnerite*.

HUDSONITE

Synonym of *cortlandtite*.

HUEBNERITE or HÜBNERITE

$2[MnWO_4]$ Manganese tungstate. Monoclinic. Prismatic crystals; also radiating masses. Yellowish-brown, reddish-brown or brownish-black. Submetallic to resinous lustre. Yellow to reddish-brown streak. Uneven fracture. Perfect cleavage in one direction. Hardness 4·0–4·5. SG c 7·12. Decomposes in hot concentrated hydrochloric acid.
Occurrence: Widespread. Associated with *wolframite*.

HULLITE

Iron magnesium calcium and alkalis aluminosilicate. Crystal system uncertain. Massive. Velvet-black. Waxy to dull lustre. Hardness 2·0.
Occurrence: Scotland—Fifeshire (Kinkell). Northern Ireland—Belfast (Carnmoney Hill). Fills cavities in *basalt*.
A *chlorite*.

HUMBOLDTILITE

Synonym of *melilite*.

HUMIC COAL

Banded and jointed woody *coal*. Contains a similar range of plant tissues but varying volatiles. Main types: *anthracite*, *bituminous coal*, *lignite* and *peat*.

HUMITE

$4[Mg_7Si_3O_{12}(F,OH)_2]$ Magnesium silicate with fluoride and hydroxide. Orthorhombic. Elongated pyramidal crystals. White, yellowish-white, yellow or chestnut-brown. Vitreous to resinous lustre. Subconchoidal to uneven fracture. Good cleavage in one direction. Hardness 6·0–6·5. SG 3·1–3·2. Gelatinises with hydrochloric acid.
Occurrence: USA—New York (Tilly Foster Iron Mine in Brewster County). Sweden—Wermland (Ladu Mine near Filipstad).

HUNGARITE

Igneous rock of the *diorite* clan. A fine-grained porphyritic extrusive rock. Contains phenocrysts of *hornblende* and *plagioclase* in a groundmass of *plagioclase*.
Occurrence: Widespread.

HUNTITE

$2[Mg_3Ca(Co_3)_4]$ Magnesium calcium carbonate. Orthorhombic. Platy crystals; also earthy masses. White. Earthy lustre. White streak. Smooth to subconchoidal fracture. No cleavage. Hardness 1·0–2·0. SG c 2·7.
Occurrence: USA—Nevada (Currant Creek in White Pine and Nye Counties). Hungary—Dorog (Dorog Mine).

HUSEBYITE

Igneous rock of the *syenite* clan. A coarse-grained granular plutonic rock. Contains *orthoclase*, *andesine*, *nepheline* and *titanaugite*.
Occurrence: Norway—Oslo (Huseby).

HUTTONITE

Monoclinic polymorph of *thorite*.

HYACINTH

Reddish-orange or brownish-orange gem variety of *zircon*. Also synonym of *harmotome, idocrase* and *meionite*.

HYALITE

Colourless variety of *opal* occurring as botryoidal or globular crusts or stalactitic shape. Also synonym of *axinite*.

HYALOANDESITE

Volcanic glass with the composition of *andesite*.

HYALOBASALT

Volcanic glass with the composition of *basalt*.

HYALODIABASE

Volcanic glass with the composition of *diabase*.

HYALOPHANE

$4[(K,Na,Ba)(Al,Si)_4O_8]$ Potassium sodium barium aluminosilicate. Monoclinic. Prismatic crystals; also massive. Colourless, white or flesh red. Vitreous lustre. Conchoidal fracture. Perfect cleavage in one direction. Hardness 6·0–6·5. SG 2·8.
Occurrence: Sweden—Wermland (Jakobsberg). Switzerland—Valais (Binnental). A *feldspar*.

HYALOPILITIC

With a structure of a mass of needle-like

crystals in a glassy substrate.

HYALOSIDERITE

Variety of *chrysolite* (*of Wallerius*) with 30–50 per cent molar proportion of ferric silicate.

HYDRARGILLITE

Synonym of *aluminite, turquoise* and *wavellite*.

HYDRARGILLITE (of Cleaveland)

Synonym of *gibbsite* (*of Torrey*).

HYDRATED

Containing water in chemical combination or as water of crystallisation.

HYDROBIOTITE

Hydrated variety of *biotite*.

HYDROCASTORITE

Amorphous mixture of *heulandite* and *stilbite*.

HYDROCHLORIC ACID

Important testing reactant, solution of hydrogen chloride gas (HCl) in water. Generally used dilute, occasionally used concentrated (fuming).

HYDRODOLOMITE or HYDROMAGNESITE

Mixture of *calcite* and *hydromagnesite*.

HYDROFLUORIC ACID

Hydrogen fluoride (HF) or, more usually, its solution in water. Occasionally used for testing.

HYDROGARNET or GROSSULAROID

Group of minerals closely related to *garnets* but with $(2H_2O)$ replacing silica in the lattice. Main types are *hibschite* and *plazolite*.

HYDROGOETHITE

Alteration product intermediate between *goethite* and *limonite*.

HYDROHALITE

Variety of *halite* with the formula $NaCl.2H_2O$. A product of crystallisation of sea water or saline spring water in an extremely cold climate.

HYDROHEMATITE, TURGITE or TURITE

Massive reddish-black to red alteration product intermediate between *hematite* and *goethite*.

HYDROHETAEROLITE

$4[Zn_2Mn_4O_8.H_2O]$ Hydrated zinc manganese oxide. Tetragonal. Fibrous crusts. lic lustre. Dark brown streak. Cleavage parallel to fibres. Hardness 5·0–6·0. SG 4·6. Easily soluble in hydrochloric acid with the evolution of chlorine.
Occurrence: USA—Colorado (Wolftone Mine near Leadville), New Jersey (Sterling Hill).

HYDROMAGNESITE

$2[Mg_4(OH)_2(CO_3(_3.3H_2O]$ Hydrated magnesium carbonate-hydroxide. Monoclinic. Acicular or bladed crystals; also massive. Colourless or white. Vitreous to earthy lustre. Perfect cleavage in one direction. Hardness 3·5. SG 2·2. Soluble with effervescence in hydrochloric acid.
Occurrence: USA—California (Sulphur Creek area of Colusa County), Pennsylvania (Wood's Chrome Mine near Texas in Lancaster County). Canada—British Columbia (Cariboo district). Germany—Baden (Limburg). Italy—Tyrol (Predazzo).
Also synonym of *hydrodolomite*.

HYDROMAGNETITE

Hydrous variety of *magnetite*.

HYDROMICA

Group name for hydrated minerals with properties and compositions similar to *micas*. Main types are *pinite* and *sericite*.

HYDRONEPHELINE or RANITE

Sodium aluminosilicate. Formula uncertain. Hexagonal. Massive. White or dark grey. Vitreous lustre. Hardness 4·5–6·0.

SG 2·3. Gelatinises with hydrochloric acid.
Occurrence: USA—Maine (Litchfield). Norway—Langesund Fiord (Låven Island).

HYDROPHANE

White variety of *opal* which has a low refractive index and is virtually transparent when placed in water.

HYDRORHONDITE

Manganese magnesium calcium and lithium silicate. Formula uncertain. Triclinic. Massive. Reddish-brown. Vitreous lustre. Brownish-white streak. Good cleavage in one direction. Hardness 5·0–6·0. SG 2·7. Soluble in hydrochloric acid with the separation of silica.
Occurrence: Sweden—Wermland (Långban).

HYDROTACHYLYTE

Type of *tachylyte* containing up to 13 per cent water.

HYDROTALCITE

3[$Mg_6Al_2CO_3(OH)_{16}$·$4H_2O$] Hydrated magnesium aluminium carbonate-hydroxide. Hexagonal. Massive. White. Pearly to waxy lustre. White streak. Perfect basal cleavage. Hardness 2·0. SG 2·06. Feels greasy. Easily soluble with effervescence in hydrochloric acid.
Occurrence: USA—New Jersey (Vernon in Sussex County), New York (Somerville in St Lawrence County). Norway—Snarum.

HYDROTHERMAL

Applied to igneous processes in which very hot chemical fluids play a major part as an important reactant. Such processes lead to alteration effects in certain minerals and to localised types of deposition around an intrusion.

HYDROTITANITE

Synonym of *anatase*.

HYDROUS

Containing water.

HYDROXYL-ANNITE or ANNITE (of Winchell)

Hydrous variety of *lepidomelane*.

HYDROXYL-APATITE

Variety of *apatite* with the formula $Ca_5(PO_4)_3OH$.

HYDROZINCITE or EARTHY CALAMINE

2[$Zn_5(CO_3)_2(OH)_6$] Zinc carbonate-hydroxide. Monoclinic. Massive. White, often tinted various colours. Dull to silky lustre. Brittle fracture. Perfect cleavage in one direction. Hardness 2·0–2·5. SG *c* 3·6. Easily soluble in hydrochloric acid.
Occurrence: Wales—Montgomery (Llanidloes). USA—California (Cerro Gordo Mine in Inyo County), Pennsylvania (Friedensville in Lehigh County). Western Australia—North Kimberley.
Alteration product of *blende*.

HYPABYSSAL

Intermediate between plutonic and volcanic, generally of medium grain size, formed for instance in dykes.

HYPAUTOMORPHIC

With grains showing a slight tendency to crystallinity.

HYPERITE or GABBRO-NORITE

Swedish term for rocks intermediate between *gabbro* and *norite*.

HYPERSTHENE, HORNBLENDE (Labrador) or HYPERITE

16[$(Mg,Fe)SiO_3$] Magnesium iron silicate.

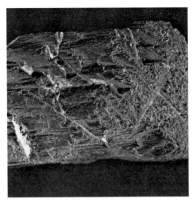

Orthorhombic. Prismatic crystals; also foliated masses. Dark brownish-green, greyish-black or greenish-black. Pearly lustre. Grey or brownish-grey streak. Uneven fracture. Perfect cleavage in one direction. Hardness 5·0–6·0. SG 3·4–3·5.
Occurrence: Worldwide in extrusive rocks of the *diorite* and *gabbro* clans.
A *pyroxene*.

HYPERSTHENFELS

Synonym of *norite*.

HYPERSTHENITE

Igneous rock of the *ultrabasic* clan. A coarse-grained granular plutonic rock composed of *hypersthene*.
Occurrence: Western Australia—Norseman. Variety is *bahiate*.
Also synonym of *norite*.

HYPOSTILBITE

Silica poor variety of *stilbite*.
Also synonym of *laumontite*.

ICELAND SPAR

Synonym of *calcite*.

IDDINGSITE

Iron calcium magnesium silicate. Formula uncertain. Orthorhombic. Pseudomorphs

after *olivine*. Reddish-brown. Good cleavage in three directions. Hardness 3·5. SG 2·5–2·8.
Occurrence: Widespread in basaltic rocks. Metasomatic alteration of *olivine* crystals.

IDOCRASE, DUPARCITE, GAHNITE (of da Silveira), GENEVITE, VESUVIANITE or WILUITE

$4[Ca_{10}(Mg, Fe^{2+}, Fe^{3+})_2Al_4Si_9O_{34}(OH)_4]$ Basic calcium aluminium silicate with magnesium and iron. Tetragonal. Prismatic crystals; also columnar or granular masses. Brown to green. Vitreous lustre. White streak. Subconchoidal to uneven fracture. Indistinct cleavage. Hardness 6·5. SG 3·35–3·45. Partially decomposes in hydrochloric acid.
Occurrence: USA—California (San Carlos in Inyo County), Maine (Phippsburg), New Jersey (Newton). Canada—Quebec (Litchfield in Pontiac County). Czechoslovakia—Bohemia (Eger). Italy—Mount Vesuvius.
Varieties are *californite* (green), *cyprine* (blue) and *xanthite* (yellowish-brown).

IGELSTROMITE or IRON KNEBELITE

Iron rich variety of *knebelite*.

IGNEOUS

Major type of rock, usually viewed as primary. Crystalline or glassy, or both. Classification generally based on mineral composition and texture. Found in extrusions and in intrusions.
See metamorphic and sedimentary.

IGNIMBRITE or WELDED TUFF

Pyroclastic igneous rock. A fine-grained rock containing *glass* and mineral fragments which have been flattened and welded together by impact with the ground.
Occurrence: Worldwide.

IJOLITE

Feldspathoid igneous rock. A medium-grained granular plutonic rock. Contains *nepheline*, *diopside* and *aegirine-augite*.
Occurrence: Canada—British Columbia (Ice River). Norway—Telemark (Meltey).

IJOLITE PEGMATITE

Type of *ijolite* occurring as a *pegmatite*.

ILLITE or GLIMMERTON

Basic potassium aluminium aluminosilicate. Formula uncertain. Monoclinic. Massive. White. Perfect basal cleavage. Hardness 1·0–2·0. SG 2·6–2·9.
Occurrence: Worldwide.
A *clay mineral*. Group member is *glauconite*. Variety is *brammalite*.

ILMENITE or MENACCANITE

$2[FeTiO_3]$ Titanium iron oxide. Hexagonal. Tabular crystals; also massive. Black. Metallic lustre. Black streak. Conchoidal to subconchoidal fracture. No cleavage. Hardness 5·0–6·0. SG c4·75. Slowly soluble in hydrochloric acid.
Occurrence: Widespread. Accessory mineral in *gabbros*, *diorite*, *anortosite* and *pegmatite*. Also in ore veins and *placer deposits*.
Also synonym of *columbite*.

ILMENITITE

Igneous rock of the *ultrabasic* clan. A medium-grained granular hypabyssal rock. Contains *ilmenite* and minor *pyrite*, *chalcopyrite* and *pyrrhotite*.
Occurrence: Southern Norway.

ILMENORUTILE

Columbium rich variety of *rutile*.

ILVAITE, LIEVRITE or YENITE

$[CaFe_2^{2+}Fe^{3+}Si_2O_8.OH]$ Basic calcium iron silicate. Orthorhombic. Prismatic crystals; also columnar masses. Black or dark greyish-black. Submetallic lustre. Black streak. Uneven fracture. Good cleavage in two directions. Hardness 5·5–6·0. SG c4·0. Gelatinises with hydrochloric acid.
Occurrence: USA—Massachusetts (Milk Row Quarry near Somerville), Rhode Island (Cumberland). Germany—Saxony

(Schneeberg). Italy—Island of Elba (Rio la Marina).

IMPREGNATION

Filling of rock pores after formation by liquids such as oil or brine or by mineral material, sometimes with displacement of existing impregnating substances.

IMPSONITE

Sulphur rich variety of *albertite*.

INCANDESCENCE

Light produced by substances raised to high temperatures, often characteristic of the substance and thus relevant to testing. See bead test.

INCRUSTATION

Covering with a hard surfaced layer.

INDIALITE

Granular variety of *anorthite* from India.

INDICOLITE

Blue variety of *tourmaline*.

INDOCHINITES

Type of *tektites* found in Indochina.

INDURATE

See induration.

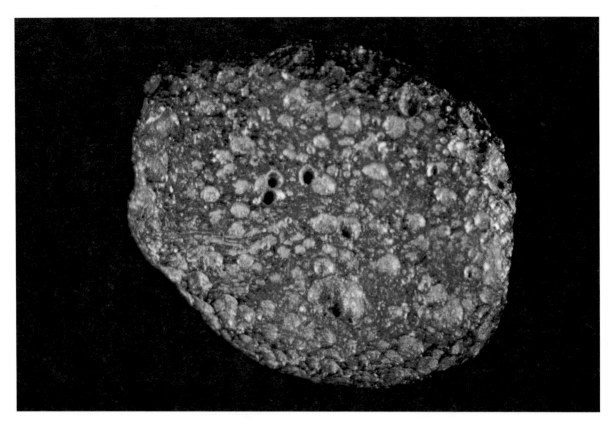

Above: indochinite

INDURATION

The hardening of a sediment by heat, pressure and/or lithification, and as such grading into true metamorphism.

INFUSIBLE

Having a very high melting point.
See fuse.

INJECTION GNEISS

Type of *gneiss* in which the banding is due to injection of a *granite* magma into schistose or fissile rocks.

INSOLUBLE

Substance with negligible solubility (in water unless otherwise stated).

INTERFOLIATED

See foliated.

INTERSTITIAL

Position between the normal sites in the crystal lattice in which additional atoms may be found, thus causing a lattice defect. Interstitial compounds have the atoms of one component in the lattice of the other.

INTRUSION

Igneous formation in which the rock is forced through and into existing rock, perhaps along lines of weakness such as faults and bedding planes.

INTRUSIVES

See intrusion.

INTUMESCENCE

Swelling.

IOLITE

Synonym of *cordierite*.

IRIDESCENT

Many coloured, often the result of interference effects in the light reflected by the surface.

IRIDOSMINE or OSMIRIDUM

Iridium and osmium alloy. Formula variable. Hexagonal. Rhombohedral crystals. Tin-white to steel-grey. Metallic lustre. Hardness 6·0–7·0. SG 19·3–21·12.
Occurrence: Australia—New South Wales (Bingera). Canada. South Africa. Brazil. USSR—Ural Mountains. In *gold* washings and ores.

IRINITE

Basic sodium cerium and thorium niobate (columbate) and titanate. Formula uncertain. Cubic. Cubic crystals. Reddish-brown, brown or brownish-yellow. Greasy lustre. No cleavage. Hardness 5·0–5·5. SG *c*4·5. Partially soluble in hot sulphuric acid.
Occurrence: Rare. In *pegmatites* associated with *columbite*.

IRON

2[Fe]. Cubic. Crystals rare, normally massive. Steel-grey to black. Metallic lustre. Hackly fracture. Poor basal cleavage. Hardness 4·0. SG 7·3–7·9. Easily soluble in hydrochloric acid with the evolution of hydrogen.
Occurrence: USA—Missouri (Cameron in Clinton County). Canada—Ontario (St Josephs Island on Lake Huron). Greenland—Disko Island. Also in *meteorites*.

IRON GLANCE

Synonym of *specular hematite*.

IRON KNEBELITE

Synonym of *igelstromite*.

IRON MONTICELLITE

Iron rich variety of *monticellite*.

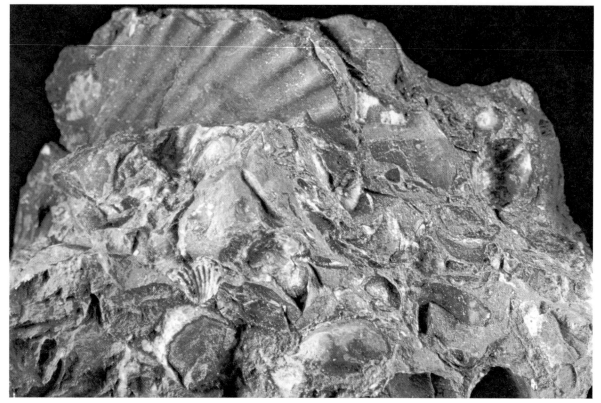

Above: ironstone

IRON OOLITE

Type of *oolite* formed by *iron* minerals.

IRON PYRITES

Synonym of *pyrite*.

IRON RHODONITE

Synonym of *pyroxmanganite*.

IRON ROSE

Type of crystal habit displayed by *hematite* or *ilmenite* where large crystals radiate from a centre in a rose-like pattern.

IRONS

A metal based *meteorite*.

IRONSTONE

Name applied to *argillaceous, arenaceous* or calcareous sedimentary rocks which contain a high percentage of iron minerals. The minerals may be a result of primary deposition or later alteration.

ISHKYLDITE or δ-CHRYSOTILE

$Mg_{15}Si_{11}O_{27}(OH)_{20}$ Basic magnesium silicate. Monoclinic. Fibrous masses. Pale bluish-green. Greasy lustre. Hardness 1·0. SG 2·62. Feels greasy. Variety of *chrysotile*.
Occurrence: USSR—Orskho-Khalilaro region (Ishkyldino).

ISINGLASS

Old synonym of *mica*.

ISOMORPHOUS

Loosely, having the same crystal structure; more strictly. also being able to form a series of solid solutions.

ISOSTRUCTURAL

Loosely equivalent to isomorphous; strictly, having the same crystal structure but not normally able to form a series of solid solutions.

ISSITE

Igneous rock of the *gabbro* clan. A medium-grained granular hypabyssal rock. Contains *hornblende, pyroxene* and *labradorite*.
Occurrence: USSR—Ural Mountains (Isse River near Kamenouchki).

ITABIRITE

Regional metamorphic rock. A schistose rock containing *quartz, hematite,* *oligoclase* and *muscovite.*
Occurrence: Brazil— Itabira.

ITACOLUMITE

Thinly bedded micaceous type of *sandstone* which is flexible in thin laminae.

ITALITE

Feldspathoid igneous rock. A coarse-grained granular plutonic rock. Contains *leucite, augite* and *andradite.*
Occurrence: Italy—Albano Hills (Grotta Ferrata).

ITSINDRITE

Igneous rock of the *syenite* clan. A fine to coarse-grained granular hypabyssal rock. Contains *microcline, nepheline, biotite* and *aegirine.*
Occurrence: Madagascar—Itsindra Valley.

JACOBSITE

8[MnDe$_2$O$_4$] Manganese iron oxide. Cubic. Crystals rare, normally massive. Black or brownish-black. Metallic lustre. Brown streak. No cleavage. Hardness 5·5–6·5. SG 4·76. Soluble in hydrochloric acid with the evolution of chlorine.
Occurrence: Sweden—Wermland (Jacobsberg). Bulgaria—near Tatar Pozardzik (Debarstica).
Related to *magnetite*.

JACUPIRANGITE

Igneous rock of the *ultrabasic* clan. A fine-grained granular plutonic rock which weathers to schistose appearance. Contains *augite*, *magnetite* and *ilmenite*.
Occurrence: Brazil—Sao Paulo (Jacupiranga).

JADE

White to dark green tough compact variety of *jadeite*.

JADEITE

4[NaAlSi$_2$O$_6$] Sodium aluminium silicate. Monoclinic. Granular, columnar, fibrous or compact masses. Shades of green and white. Subvitreous to pearly lustre. Colourless streak. Splintery fracture. Good cleavage in two directions. Hardness 6·5–7·0. SG 3·33–3·35.
Occurrence: Widespread in Eastern Asia. Varieties are *chloromelanite* and *jade* (compact masses).

JAMESONITE

[Pb$_4$FeSb$_6$S$_{14}$] Lead iron antimony sulphide. Monoclinic. Acicular or fibrous crystals; also massive. Greyish-black. Metallic lustre. Greyish-black streak. Good basal

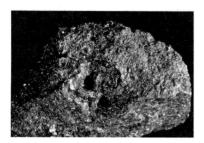

cleavage. Hardness 2·5. SG 5·63. Decomposes in nitric acid with the separation of antimony oxide and lead sulphate.
Occurrence: England—Cornwall (St. Endellion). USA—Idaho (Slate Creek in Custer County). Canada—Ontario (Barrie in Franttenac County).

JARGON or JARGOON

Colourless or smoky variety of *zircon* from Sri Lanka.

JAROSITE

KFe$_3$(SO$_4$)$_2$OH$_6$ Basic potassium iron sulphate. Hexagonal. Tabular crystals; also massive. Yellow to dark brown. Subadamantine to vitreous lustre. Pale yellow streak. Uneven to conchoidal fracture. Good basal cleavage. Hardness 2·5–3·5. SG 2·91–3·26. Soluble in hydrochloric acid.
Occurrence: USA—Arizona (Vulture Mine in Maricopa County), Colorado (Iron Arrow Mine in Chaffee County), Utah (Mammoth Mine in the Tintic district). Greece—Laurium.

JASPER

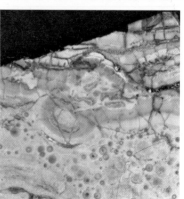

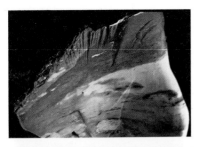

Variety of cryptocrystalline silica. Normally spotted or banded and coloured red or brown by iron oxides.
Variety is *orbicular jasper*.

JAVAITES

Type of *tektites* found in Java.

JEFFERISITE

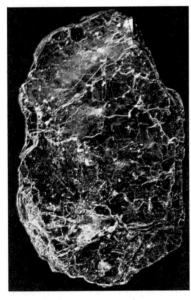

Magnesium aluminium silicate. Formula and crystal system uncertain. Broad crystals or crystalline plates. Dark yellowish-brown to brownish-yellow. Pearly lustre. Perfect basal cleavage. Hardness 1·5. SG 2·3. Decomposes in hydrochloric acid.
Occurrence: England—Lancashire (Walney Island). USA—Pennsylvania (West Chester in Chester County).
In *serpentinite* and *granite*.

JEFFERSONITE

4[Ca(Mn,Zn,Fe)Si$_2$O$_6$] Calcium manganese zinc and iron silicate. Monoclinic. Prismatic crystals; also massive. Greenish-black; weathers chocolate-brown. Vitreous lustre. Uneven to conchoidal fracture. Good cleavage in one direction. Hardness 5·0–6·0. SG 3·36.
Occurrence: USA—New Jersey (Franklin Furnace).
A *pyroxene*.

JET

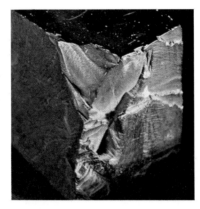

Compact, hard black variety of *lignite*.

JORDANITE

2[Pb$_{14}$As$_7$S$_{24}$] Lead arsenic sulphide. Monoclinic. Tabular crystals; also reniform masses. Lead-grey. Metallic lustre. Black streak. Conchoidal fracture. Perfect cleavage in one direction. Hardness 3·0 SG 6·39. Decomposes in nitric acid with the separation of lead sulphate.
Occurrence: Switzerland—Valais (Binnental). Japan—Aomori (Yunosawa Mine).

JOSEPHINITE

A nickel-iron alloy.

JUANITE

Ca$_{10}$Mg$_4$Al$_2$Si$_{12}$O$_{39}$.4H$_2$O Hydrated calcium magnesium aluminium silicate. Orthorhombic. Fibrous sheaves. White. Hardness 5·5. SG 3·0.

Occurrence: USA—Colorado (Iron Hill). Alteration product of *melilite*.

JUDDITE or FERRIRICHTERITE

Manganese *amphibole* from India.

JUMILLITE

Igneous rock of the *syenite* clan. A brownish-grey, fine-grained porphyritic extrusive rock. Contains phenocrysts of *sanidine*, *olivine* and *phlogopite* in a groundmass of *sanidine*, *diopside*, *aegirine-augite* and *leucite*.
Occurrence: Spain—Murai Province (Jumilla).

JURASSIC

See *geological time*.

JUVITE

Igneous rock of the *syenite* clan. A medium-grained granular plutonic rock. Contains *orthoclase, nepheline, cancrinite, muscovite* and *aegirine*.
Occurrence: Norway—Fen district (Juvet).

KAERSUTITE

Black titanium rich variety of *hornblende*.

KAINITE

16[KClMgSO$_4$3H$_2$O] Hydrous magnesium potassium chloride and sulphate. Monoclinic. Granular. White. Hardness 3·0. SG 2·1.

KAIWEKITE

Igneous rock of the *diorite* clan. A dull green, medium-grained porphyritic extrusive rock. Contains phenocrysts of *anorthoclase*, *aegirine-augite*, *hornblende*, *nepheline* and *serpentine* pseudomorphed after *olivine* in a groundmass of *oligoclase*, *augite* and *magnetite*.
Occurrence: New Zealand—Kaikorai Valley (Long Beach).

KAJANITE

Feldspathoid igneous rock. A fine-grained porphyritic extrusive rock. Contains phenocrysts of *olivine* and *mica* in a groundmass of *leucite, diopside* and *magnetite*.
Occurrence: Borneo—Bawoei Mountains (Oele Kajan).

KAKIRITE

Dynamic metamorphic rock. Contains augens of residual material surrounded by zones of granulation and recrystallisation.

Occurrence: Sweden—Lapland (Lake Kakir).

KAKORTOKITE

Igneous rock of the *syenite* clan. A coarse-grained granular plutonic rock. Contains alkali-*feldspars*, *nepheline*, *eudialyte*, *arfvedsonite* and *aegirine*.
Occurrence: Greenland — Julianehaab (Kringlerne).

KALIALASKITE

Igneous rock of the *granite* clan. A fine to coarse-grained granular plutonic rock. Contains *quartz, orthoclase, microcline* and variable minor ferromagnesian minerals.
Occurrence: USA—Alaska (Terra Cotta Range of the Tordrillo Mountains), Massachusetts (Rockport).
Variety is *birkremite*.

KALIGRANITE

Type of *granite* containing very little or no *plagioclase*.

KALIKERATOPHYRE

Sodaclase free type of *keratophyre*.

KALINITE

$KAl(SO_4)_2 . 11H_2O$ Hydrated potassium

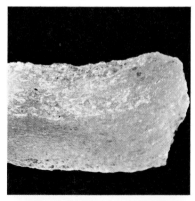

aluminium sulphate. Monoclinic. Fibrous masses.
Occurrence: USA—California (San Bernardino County). Australia—Mount Wingan.

KALIOPHILITE

$54[KAlSiO_4]$ Potassium aluminium silicate. Hexagonal. Bundles of acicular crystals. Colourless. Silky lustre. Perfect basal cleavage. Hardness 6·0. SG 2·49. Gelatinises with hydrochloric acid.
Occurrence: Italy—Monte Somma. In ejected blocks.

KALIRHYOLITE

Plagioclase poor or free type of *rhyolite*.

KALISILITE

High temperature polymorph of *kaliophilite*.

KALISYENITE

Igneous rock of the *syenite* clan. A medium- to coarse-grained granular plutonic rock. Contains *orthoclase, microperthite, anorthoclase, biotite, hornblende, diopside* and minor *quartz*.
Occurrence: Widespread.

KALITHOMSONITE

Synonym of *ashcroftine*.

KALITORDRILLITE

Igneous rock of the *granite* clan. A fine-grained porphyritic extrusive or hypabyssal rock. Contains phenocrysts of *quartz, orthoclase, anorthoclase* and *albite* in a groundmass of intergrown *quartz* and *orthoclase*.
Occurrence: USA—Alaska (Tordrillo Mountains).

KALITRACHYTE

Plagioclase poor or free type of *trachyte*.

KAMACITE

Cubic nickel iron mineral (containing six per cent nickel) found in *meteorites*.

KÄMMERERITE

Red hexagonal variety of *pennine*.

KAMPERITE

Igneous rock of the *syenite* clan. A fine to medium-grained granular hypabyssal rock. Contains *orthoclase, oligoclase* and *biotite*.
Occurrence: Norway—Fen district (Kamperhough Valley).

KANDITE

Group name for *kaolinite, dickite, chamosite, greenalite* and *halloysite*.

KANKAN-ISHI

Japanese synonym of *sanukite*.

KAOLIN

Synonym of *kaolinite*.

KAOLINITE, KAOLIN, CHINA CLAY or NACRITE

$2[Al_2Si_2O_5(OH)_4]$ Hydrous aluminium sili-

cate. Monoclinic. Hexagonal scales or *clay*-like masses. White or grey, often tinted. Pearly to dull lustre. Perfect basal cleavage. Hardness 2·0–2·5. SG 2·6–2·63.
Occurrence: Widespread. Decomposition product of aluminous minerals in *granite* or *gneiss*.
Variety is *lithomarge*.

KARINTHINE

Synonym of *carinthine.*

KASOITE

Potassium rich variety of *celsian* from Japan.

KASSAÏTE

Igneous rock of the *diorite* clan. A fine-grained porphyritic hypabyssal rock. Contains phenocrysts of *haüyne, labradorite, barkevikite* and *augite* in a groundmass of *hastingsite, andesine, oligoclase* and *orthoclase.*
Occurrence: Guinea—Los Islands (Kassa).

KATAPHORITE

A brownish variety of *arfvedsonite.*

KATUNGITE

Igneous rock of the *gabbro* clan. A very fine-grained porphyritic extrusive rock. Contains phenocrysts of *olivine, melilite, leucite, zeolites, fluorite* and *apatite* in a cryptocrystalline groundmass.
Occurrence: Uganda—Katunga volcano.

KATZENBUCKELITE

Feldspathoid igneous rock. A fine-grained porphyritic hypabyssal rock. Contains phenocrysts of *nepheline, biotite, olivine, nosean, leucite* and *apatite* in a groundmass of *nepheline, leucite* and *aegirine.*
Occurrence: Germany—Odenwald (Michelsberg near Katzenbuckel).

KAYSERITE

Synonym of *diaspore.*

KAZANSKITE

Igneous rock of the *gabbro* clan. A black, fine-grained granular hypabyssal rock. Contains *olivine, plagioclase, magnetite* and *spinel.*
Occurrence: USSR—Nicolai-Pawda (Kazansky).

KEDABEKITE

Igneous rock of the *gabbro* clan. A medium-grained granular plutonic rock. Contains *bytownite, andradite* and *hedenbergite.*
Occurrence: USSR—Jelisabetpol (Kedabek).

KEILHAUITE or YTTROTITANITE

Calcium aluminium iron and yttrium metals titanosilicate. Formula uncertain. Monoclinic. Prismatic crystals. Brownish-black to greyish-brown. Vitreous to

resinous lustre. Greyish-brown to pale yellow streak. Good cleavage in one direction. Hardness 6·5. SG 3·52–3·77. Decomposes in hydrochloric acid.
Occurrence: Norway—Arendal, Snarum.

KENTALLENITE

Igneous rock of the *diorite* clan. A medium-grained granular plutonic rock. Contains *olivine, augite, biotite, orthoclase* and *plagioclase* (*labradorite* to *andesine*).
Occurrence: Scotland—Argyllshire (Kentallen near Ballachulish). Sweden—Smålingen.

KENYTE

Olivine bearing type of *phonolite.*

KERALITE

Type of *hornfels* containing *quartz* and *biotite* as essential minerals.

KERARGYRITE

Synonym of *cerargyrite.*

KERATOPHYRE

Altered igneous rock of the *syenite* clan. A brownish-green, fine-grained porphyritic extrusive rock. Contains phenocrysts of *orthoclase, sodaclase, biotite, kataphorite, riebeckite, barkevikite, diopside* and *aegirine* in a groundmass of *sanidine, quartz* and minor ferromagnesian minerals.
Occurrence: Scotland—Lanarkshire (Gavinan Lairs near Glasgow). Germany—Harz (Ostberg bei Ellingerode). Ecuador—Cerro de los Llanganates.
Variety is *kalikeratophyre.*

KERMESITE

8[Sb_2S_2O] Antimony oxysulphide. Monoclinic or orthorhombic. Red needle-shaped crystals. Alteration product of *stibnite* in the oxidized zones of *antimony* deposits.
Occurrence: With *senarmontite* and *valentinite.*

Above: kermesite

KERNITE or RASORITE

4[$Na_2B_4O_7 \cdot 4H_2O$] Hydrated sodium borate. Monoclinic. Elongated wedges; also massive. Colourless or white. Vitreous lustre. White streak. Perfect cleavage in one direction. Hardness 2·5. SG 1·9. Slowly soluble in cold water; readily soluble in hot water.
Occurrence: USA—California (Kramer borate district in Kern County).

KEROGEN

Essential constituent of *oil shale:* decayed carbonaceous matter which gives crude oil when distilled.

KERRITE

Magnesium aluminosilicate. Formula uncertain. Monoclinic. Fine scales. Pale greenish-yellow. Pearly lustre. Hardness 1·0–2·0. SG 2·3. Decomposes in hydrochloric acid with the separation of silica.
Occurrence: USA—North Carolina (Culsagee Mine near Franklin in Macon County). A *vermiculite.*

KERSANTITE

Igneous rock of the *diorite* clan. A fine-grained porphyritic hypabyssal rock. Contains phenocrysts of *oligoclase* (altered to *kaolinite,* white *mica* and carbonates), *biotite* and *hornblende* in a groundmass of *plagioclase* and *diopside.*
Occurrence: France—Kersanton, Markirch.
Variety is *calcikersantite.*

KERSTENITE

Lead selenite. Formula and crystal system uncertain. Botryoidal masses. Yellow. Greasy to vitreous lustre. Fibrous fracture. Good cleavage in one direction. Hardness 3·0–4·0.
Occurrence: Germany—Thuringia (Friedrichsglück Mine near Hildburghausen). Also synonym of *chelevtite.*

KHAGIARITE

Igneous rock of the *granite* clan. A black, very fine-grained glassy extrusive rock. Contains phenocrysts of soda. *Microcline, diopside, aegirine* and *cossyrite* in a glassy groundmass.
Occurrence: Mediterranean Sea—Island of Pantelleria.

KIDNEY ORE

Variety of *hematite* occurring as reniform masses.

KIESERITE

$4[MgSO_4H_2O]$ Hydrous magnesium sulphate. Massive, granular or compact. White.
Occurrence: Germany—Stassfurt. In salt deposits.

KIIRUNAVAARITE

Igneous rock of the *ultrabasic* clan. A coarse-grained granular plutonic rock containing *magnetite* and *apatite*.
Occurrence: Sweden—Lapland (Kiruna).

KILLAS

Cornish term for contorted Devonian *slates*.

KIMBERLITE

Igneous rock of the *ultrabasic* clan. A coarse-grained hypabyssal rock which is normally brecciated and altered. Contains *olivine, serpentine, tremolite, bastite, bronzite*, chrome *diopside, biotite, pyrope* and *calcite*; also contains rock fragments and *diamonds*.
Occurrence: Widespread in the Canadian Shield, Southern and West Africa, and Siberia.

KINZIGITE

Regional metamorphic rock. A coarse-grained granular rock containing *garnet* and *biotite* with *quartz, orthoclase, oligoclase, muscovite, cordierite* or *sillimanite*.
Occurrence: Germany—Black Forest (Kinzig).

KIVITE

Igneous rock of the *gabbro* clan. A fine-grained porphyritic extrusive rock. Contains phenocrysts of *leucite, bytownite, olivine, biotite, augite* and *magnetite* in a groundmass of *labradorite, leucite* and *augite*.
Occurrence: Rwanda—Lake Kivu (Volcano Kisi).

KJELSASITE

Calcium rich, alkali poor type of *larvikite*.

KLEMENTITE

Magnesium rich, iron poor variety of *thuringite*.

KLOCKMANNITE

6[CuSe] Copper selenide. Hexagonal. Granular aggregates. Slate-grey, tarnishing to bluish-black. Metallic lustre. Perfect cleavage in one direction. Hardness 3·0. SG *c*5·0.
Occurrence: Germany—Harz Mountains (Lehrbach). Sweden—Skakerum. Argentina—Sierra de Umango.

KNEBELITE

$4[(Fe,Mn)_2SiO_4]$ Iron manganese silicate. Orthorhombic. Crystalline masses. Grey, red, brown, yellow, green or black. Greasy lustre. Subconchoidal to uneven fracture. Good cleavage in one direction. Hardness 6·5. SG 3·9–4·17. Decomposes in hydrochloric acid with the separation of gelatinous silica.
Occurrence: Sweden — Södermanland (Tunaberg).
Varieties are *igelströmite* and mangan-*knebelite*.

KNOPITE

Black, cerium rich variety of *perovskite*.

KOCHUBEÏTE

Synonym of *kotschubeite*.

KODURÏTE

Igneous rock of the *syenite* clan. A medium-grained granular plutonic rock. Contains *microcline, spandite* and *apatite*.
Occurrence: India—Vizagapatam (Kodur Mine).

KONDRIKOVITE

A variety of *natrolite* from the Kola Peninsular, USSR.

KOPPITE

Red potassium and cerium rich variety of *pyrochlore*.

KORNERUPINE or PRISMATÏNE

$(Mg,Fe^{2+},Fe^{3+},Al)_{40}(Si,B)_{18}O_{86}$ Magnesium iron aluminium borosilicate. Orthorhombic. Crystals rare, normally fibrous or columnar aggregates. Green to brownish-green. Vitreous lustre. Perfect cleavage in two directions. Hardness 6·5. SG *c*3·3.
Occurrence: Greenland—Fiskernös. Sri Lanka—gem gravels.

KOSWITE

Igneous rock of the *ultrabasic* clan. A coarse-grained granular plutonic rock. Contains *diallage, olivine, hornblende* and *magnetite*.
Occurrence: USSR—Ural Mountains (Koswinsky Kamen).

KOTSCHUBEITE or KOCHUBEÏTE

Chromium rich variety of *clinochlore*.

KRAGERÖITE

Igneous rock of the *diorite* clan. A pinkish-grey, fine-grained granular hypabyssal rock. Contains *oligoclase, microcline* and *rutile*.
Occurrence: Norway—Kragerö (Lindvik-kollen).

KREITTONITE

Iron rich variety of *gahnite* (*of Von Moll*).

KTYPÉITE

Synonym of *aragonite*.

KULAITE

Igneous rock of the *diorite* clan. A fine-grained porphyritic hypabyssal rock. Contains phenocrysts of *olivine, diopside* and *hornblende* in a groundmass of *bytownite, orthoclase, nepheline, diopside* and *magnetite*.

KULLAITE

Igneous rock of the *diorite* clan. A medium-grained porphyritic hypabyssal rock. Contains phenocrysts of *plagioclase* (altered to *mica, calcite, epidote, chlorite* and iron oxide) and soda-*orthoclase* in a groundmass of *plagioclase*, soda-*microcline* and *chlorite*.
Occurrence: Sweden—Kullagården, Lund (Dalby).

KUNZITE

Lilac coloured, gem quality variety of *spodumene* from California, USA.

KUPFERNICKEL

Synonym of *niccolite*.

KUPFFERITE

Magnesium rich variety of *anthophyllite*.

KUTNAHORITE

Calcium enriched variety of *rhodochrosite*.

KVELLITE

Igneous rock of the *ultrabasic* clan. A medium-grained porphyritic hypabyssal rock. Contains phenocrysts of *lepidomelane, olivine, barkevikite, apatite, ilmenite* and *magnetite* in a groundmass of *anorthoclase*.
Occurrence: Norway—Oslo (Kvelle near Laurvik).

KYANITE, CYANITE or DISTHENE

$4[Al_2SiO_5]$ Aluminium silicate. Triclinic. Bladed crystals; also columnar or fibrous masses. Blue or white. Vitreous to pearly lustre. Colourless streak. Perfect cleavage in one direction. Hardness $5.0–7.25$. SG $3.56–3.67$.
Occurrence: Widespread in aluminous *gneiss* and *schists*.

Below: kyanite

KYANOPHILITE

$(K,Na)Al_2Si_2O_7(OH)$ Hydrous aluminium alkalis silicate. Triclinic. Laths and nodules. Apple-green. Hardness $2.5–4.0$. SG $2.89–2.9$.
Occurrence: India—Mysore (Marvinhalli). In *schist*. Alteration product of *kyanite* and *sillimanite*.

KYLITE

Igneous rock of the *gabbro* clan. A fine-grained granular hypabyssal rock. Contains *augite, labradorite, olivine, ilmenite* and *nepheline*.
Occurrence: Scotland—Ayrshire (the Kyle district).

LAANILITE

Igneous rock, a form of *pegmatite*. A very coarse-grained rock. Contains *quartz, biotite* and *garnet* with accessory iron ore minerals.
Occurrence: Finland—Laanila.

LABRADITE

Igneous rock of the *gabbro* clan. An *anorthosite*, a grey plutonic rock. Texture coarsely but evenly granular. Composed mainly of *labradorite* and *clinopyroxenes*.
Occurrence: USA—Minnesota, New York (Adirondack Mountains). Canada—Labrador, Quebec. Norway. Forms large plutons.

LABRADORESCENCE

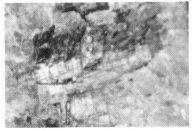

Form of iridescence observed on cleavage planes (especially when wet) with blues, greens and reds. Optical interference caused by minute inclusions in crystal.

LABRADORITE or ANEMOUSITE

A *plagioclase*. Triclinic. Crystals rare, generally granular or cryptocrystalline masses. Usually greyish brown with a vitreous lustre. No streak. Conchoidal fracture. Distinct cleavage in two directions. Hardness 5·0–6·0. SG 2·71. Shows labradorescence when turned in the light. Decomposes with difficulty in hydrochloric acid.
Occurrence: USA—New York (Adirondack Mountains). Canada—Labrador Coast. Rumania—Transylvanian Mountains. Scandinavia. In basic igneous rocks.

LACCOLITH

Dome-like intrusion with great variety of possible forms.

LACUSTRINE

Applied to sediments formed in lakes.

LAKARPITE

Igneous rock of the *syenite* clan. A coarse-grained plutonic rock similar in appearance to *diorite*. Contains *orthoclase* or *microcline* with *albite*, *amphiboles* and minor *aegirine*, *apatite* and altered *nepheline*.
Occurrence: Sweden—around Lakarp. Mexico—San José. USSR—Ural Mountains (Miask).

LAMELLA, -R

Same as lamina, -r.

LAMINA, -R

A thin bed or sheet of consistent material.

LAMPROBOLITE

Synonym of basaltic *hornblende*.

LAMPROPHYRE

Igneous rock intermediate in composition between the *syenite* and *gabbro* clans. A greyish-black porphyritic dyke. Contains *feldspar, biotite, hornblende, pyroxene* and *calcite*. The ferromagnesian phenocrysts are often chloritised and weather easily.
Occurrence: France—the Vosges Mountains. Brazil—Minas Gerais.
Variety is *cuselite*.

LANDERITE or ROSOLITE

Variety of *grossular*.

LANDSCAPE MARBLE

Sedimentary rock. A popular term for a Jurassic *argillaceous limestone* (the Cotham Stone) of the Bristol area of England.

LÅNGBANITE

$(Mn^{2+},Sb^{3+})_4(Mn^{4+},Fe^{3+},Mg)_3SiO_{12}$ Manganese iron antimony oxide and silicate. Hexagonal. Long prismatic crystals. Black. Metallic lustre. Reddish-brown streak. Conchoidal fracture. No cleavage. Hardness 6·5. SG 4·9. Soluble with difficulty in hydrochloric acid with the production of chlorine.
Occurrence: Sweden—Långban, Sjo. Rare mineral associated with *rhodonite*.

LANGBEINITE

$4[K_2Mg_2(SO_4)_3]$ Potassium magnesium sulphate. Cubic. Crystals rare, normally nodular masses or disseminated grains. Colourless. Vitreous lustre. Colourless streak. Conchoidal fracture. No cleavage. Hardness 3·5–4·0. SG 2·83. Slowly soluble in water.
Occurrence: USA—New Mexico (Carls-bad on the Pecos River, in Eddy County). Austria—Hall, Hallstatt. Germany. Oceanic *evaporite* mineral.

LANGITE

An emerald-green variety of *chalcanthite* found in copper gossans.

LANSFORDITE

$4[MgCO_3 5H_2O]$ Hydrated magnesium carbonate. Monoclinic. Elongated prismatic crystals; also stalactitic. Colourless or white. Vitreous lustre. White streak. Indistinct fracture. Cleavage in two directions. Hardness 2·5. SG 1·7. Soluble in dilute hydrochloric acid.
Occurrence: USA—Pennsylvania (Lansford). Canada—British Columbia. Italy—Piedmont. In *serpentinites*. ·

LAPILLI

Small lithic or glassy fragments ejected from a volcano through explosive eruption.

LAPIS LAZULI

Synonym of *lazurite*.

LARNITE

$4[Ca_2SiO_4]$ Calcium orthosilicate. Monoclinic. Crystals unknown; occurs as grains. White. Reacts with hydrochloric acid without effervescence.
Occurrence: Scotland—Argyllshire (Ardnamurchan, Isle of Mull). Northern Ireland—County Antrim (Scawt Hill). Contact metamorphism of *limestone* (*chalk*) by *dolerite*.

LARSENITE

$PbZnSiO_4$ Lead zinc silicate. Orthorhombic. Prismatic crystals. Colourless. Adamantine lustre.
Occurrence: USA—New Jersey (Franklin).
Member of the *chrysotile* family.

LARVIKITE

Igneous rock of the *syenite* clan. A greyish-black plutonic rock. Contains soda-*orthoclase*, *anorthoclase*, titaniferous *pyroxene*, *biotite* and *hornblende* with minor *nepteline*. The coarse-grained *feldspar* rhombs give a fine lustre. Widely used as an ornamental stone.
Occurrence: Norway—Larvik.

LASSALLITE

$Mg_3Al_4Si_{12}O_{33}.8H_2O$ A fibrous member of the *palygorskite* family.

LASSENITE

Unaltered form of *volcanic glass*.

LATERITE

Sedimentary rock. A white to cream to red, amorphous or crystalline residual deposit. Consists of hydrated aluminium iron oxides. May be featureless or pisolithic.
Occurrence: Western Australia. South America. Central Africa. In tropical areas.

LATERITITE

Sedimentary rock. Redeposited eroded *laterite*, normally mixed with *quartz* or *gneiss* fragments. Formed by mechanical deposition.
Occurrence: The Tropics. Associated with *laterites*.

LATEROID

Metasomatic replacement product. Contains hydrated iron or magnesium oxides.
Occurrence: The Tropics. Replaces quartzites, *argillaceous schists* and *phyllites* in areas of deep chemical weathering.

LATITE

Igneous rock intermediate in composition between the *syenite* and *diorite* clans. A fine-grained porphyritic extrusive rock. Contains phenocrysts of *plagioclase*, *hornblende* and *biotite* in a groundmass of *orthoclase*, *quartz* and *pyroxene*.
Occurrence: USA—Sierra Nevada Mountains. Italy—Monte Lattami.

LATIUMITE

Rare calcium potassium aluminium sulphatic silicate.
Occurrence: Italy—Latium (Albano). In ejected blocks of volcanic material.

LAUGENITE

Igneous rock of the *diorite* clan. A dark, coarse-grained plutonic rock. Contains *oligoclase* and *augite*.
Occurrence: Norway—Laugendal.

LAUMONTITE or HYPOSTILBITE

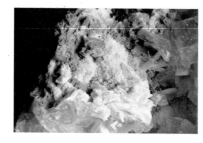

$4[CaAl_2Si_4O_{12}.4H_2O]$ Hydrated calcium aluminium silicate. Monoclinic. Prismatic crystals; often in columnar or radiating masses. Normally white; occasionally red or yellow. Vitreous lustre. Colourless streak. Uneven fracture. Cleavage in three directions. Hardness 3·5–4·0. SG 2·25–2·36. Gelatinises with hydrochloric acid.
Occurrence: Scotland—Dunbartonshire and Stirlingshire (Kilpatrick Hills). USA—New York (Tilley Foster Mine at Brewster). France—Brittany (lead mines). In basaltic lava cavities and *syenites*. Variety is *leonhardite*.

LAURDALITE

Igneous rock of the *syenite* clan. A very coarse-grained pseudoporphyritic plutonic rock. Contains *nepheline*, *anorthoclase*, soda-*orthoclase*, *diallage* and *diopside*. The nepheline crystals grow up to 100mm in diameter and are surrounded by rhombic *feldspars*.
Occurrence: Norway—near Laurdal. Large pluton.

LAVA

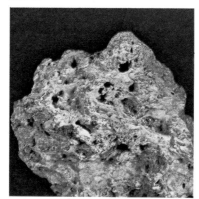

Igneous extrusion from a vent or a fissure. Usually basic, as acidic lavas have a high viscosity.

LAVENITE

$(Mn,Fe)(Si,Zr)O_3$ Manganese iron silicate with some zirconium substitution for silicon in the lattice. Monoclinic. Prismatic or tabular crystals; often as grains. Colourless or light yellow. Vitreous lustre. Perfect cleavage in one direction. Hardness 6·0. SG 3·51–3·55.
Occurrence: Southern Norway. Brazil—near Sao Paulo. Sierra Leone.

LAVIALITE

Low grade regional metamorphic rock. Contains relict phenocrysts of *labradorite* in a recrystallised groundmass of green *hornblende*. Veined with alteration passages of *quartz*, *microcline*, *biotite* and *hornblende*. Result of metamorphism of *basalt* or basaltic *tuff*.
Occurrence: Finland.

LAVROVITE or CHROME-AUGITE

Poor variety of *pyroxene*. A green, vanadium enriched aluminium.

LAWSONITE

$4[CaAl_2Si_2O_6(OH)_4]$ Hydrated calcium aluminium silicate. Orthorhombic. Prismatic crystals. Colourless; occasionally greyish-blue. Vitreous lustre. Prominent cleavage in three directions. Hardness 8·0. SG 3·0.
Occurrence: USA—California (Tiburan Peninsular). In crystalline *schists*. Metamorphic mineral.

LAZULITE

$2[(Mg,Fe)Al_2(PO_4)_2(OH)_2]$ Iron magnesium phosphate. Monoclinic. Acute pyramidal crystals; often massive and granular. Azure-blue. Vitreous lustre. White streak. Uneven fracture. Indistinct cleavage. Hardness 5·0–6·0. SG 3·0. Unaffected by acids.
Occurrence: USA—California (Death Valley), Georgia (Graves Mountain), New Hampshire (North Groton). Germany—

Salzburg. Switzerland—Zermatt. Brazil—Minas Gerais.
A high temperature vein mineral.

LAZURAPATITE

Sky-blue variety of *apatite*.

LAZURITE, LAPIS LAZULI or ULTRA MARINE

$Na_{4-5}Al_3Si_7O_{12}S$ Sodium aluminium silicate with sulphur. Cubic. Crystals rare; generally granular masses or disseminated grains. Azure-blue. Vitreous lustre. Blue streak. Uneven fracture. Indistinct cleavage. Hardness 5·0–5·5. SG 2·38–2·45. Decomposes in hydrochloric acid to give gelatinous silica and hydrogen sulphide gas.
Occurrence: USA—Colorado. Italy—Monte Somma. Chile—Ovalle. USSR—Siberia (near Lake Baikal).
Contact metamorphism of *limestone*.

LEADHILLITE

$8[Pb_4(SO_4)(CO_3)_2(OH)_2]$ Basic lead sulphate and carbonate. Monoclinic. Tabular crystals. White. Pearly to adamantine lustre. Colourless streak. Conchoidal fracture. Perfect cleavage in one direction. Hardness 2·5. SG 6·25–6·44. Reacts with nitric acid to produce precipitate of white lead sulphate.
Occurrence: Scotland — Lanarkshire (Leadhills). USA—Arizona (Mammoth Mine near Tiger), Missouri (Beer Cellar Mine near Granby).
Mineral of the secondary weathered zone of lead deposits.

LEAD VITRIOL

Synonym of *anglesite*.

LECHATELIERITE

Naturally occurring fused silica *glass* found in quartzose enclosures in volcanic rocks.

LEDMORITE

Igneous rock of the *syenite* clan. A melanocratic holocrystalline-granular plutonic rock. Contains *orthoclase*, *andradite*, *pyroxene*, *biotite* and *apatite*.
Occurrence: Scotland—Sutherland (the Ledmore River near Assynt).

LEEUWFONTEINITE

Igneous rock of the *syenite* clan. A coarse-grained plutonic rock. Contains *barkevikite*, *pyroxene*, *anorthoclase* and *plagioclase*.
Occurrence: South Africa—Leeuwfontein.

LEIDLEITE

Igneous rock of the *diorite* clan. A porphyritic hypabyssal *pitchstone*. Contains phenocrysts of *plagioclase* in a groundmass of *feldspars*, *augite* and *glass*.
Occurrence: Scotland—Argyllshire (Isle of Mull). Forms sills.

LENS

Geologically a biconvex crystal or mineral body; lenticular—so shaped.

LENTICULAR

Having the shape of a biconvex lens.

LEONHARDITE

Partly dehydrated variety of *laumontite*.

LEOPARDITE

Igneous rock of the *granite* clan. A streaky fine-grained porphyritic hypabyssal rock. Contains phenocrysts of *quartz* in a groundmass of *quartz, orthoclase, albite* and *mica*. Often stained by hydroxides of iron and aluminium.
Occurrence: South Africa—Stormberg Mountains.

LEPIDOCROCITE

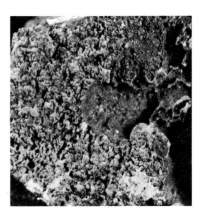

$4[FeO(OH)]$ Hydrous iron oxide. Orthorhombic. Fibrous or platy crystals. Reddish-brown. Dull lustre. Orange streak. Hardness 5·0. SG 4·1. Soluble in hydrochloric acid.
Occurrence: Worldwide.
A secondary weathering product of *iron* ore oxidation.

LEPIDOLITE

$4[K_2Li_3Al_4Si_7O_{21}(OH,F)_3]$ Hydrous lithium potassium aluminium fluosilicate. Monoclinic. Crystals rare, normally medium grained aggregates. Lilac. Pearly lustre. White streak. No fracture. Perfect basal cleavage. Hardness 2·5–4·0. SG 2·8–2·9. Reacts slightly to hydrochloric acid.
Occurrence: USA—California (San Diego County), Connecticut (Portland), Maine (Auburn). Sweden—Varütrask. In *pegmatite* veins.
A *mica*.

LEPIDOMELANE

Iron rich variety of *biotite*.
Varieties are *alurgite*, *annite* (*of Dana*) and *hydroxyl-annite*.

Illustration over page

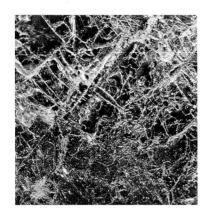

Above: lepidomelane

LEPTITE

Scandinavian synonym of *granulite*.

LEPTOCHLORITES

Group name used to describe iron rich *chlorites*.

LEPTYNITE

Medium grade regional metamorphic rock. A light coloured, coarse-grained granular rock. Contains *quartz, feldspar, mica* and *chlorite*. Result of metamorphism of a *quartz porphyry* or a *rhyolite*.
Occurrence: Scandinavia.

LEPTYNOLITE

Schistose type of *hornfels*.

LESTIWARITE

Igneous rock of the *syenite* clan. A snowy-white sacchiroidal *aplite*. Contains *microperthite, aegirite* and *arfvedsonite*.
Occurrence: Norway—Kvelle, Lysebofjord. USSR—Kola Peninsular.

LEUCHTENBERGITE

$2[(Mg,Fe,Al)_6(Si,Al)_4O_{10}(OH)_8]$ Hydrous magnesium iron aluminium aluminosilicate. Hexagonal. Tabular crystals. White. Pearly lustre. White streak. Hardness $2 \cdot 0$–$2 \cdot 5$. SG $2 \cdot 65$–$2 \cdot 78$.
Occurrence: USA—New York (Tilley Foster Mine near Brewster), Pennsylvania (Chester County). With *serpentine* and *talc*.
Rare variety of *clinochlore*.

LEUCITE

$16[KAlSi_2O_6]$ Potassium aluminium silicate. Tetragonal. Pseudocubic crystals. Normally grey, white or colourless. Dull lustre. No streak. Conchoidal fracture. Indistinct cleavage. Hardness $5 \cdot 5$–$6 \cdot 0$. SG $2 \cdot 4$–$2 \cdot 5$. A *feldspathoid*.

Occurrence: USA—Arkansas (Magnet Cove in the Ozark Mountains), Wyoming (Leucite Hills). Austria. Italy—Mount Vesuvius. Brazil.

LEUCITITE

Igneous rock of the *gabbro* clan. A fine-grained porphyritic extrusive rock. Composed of phenocrysts of *leucite* in a groundmass of *augite, olivine* and *nepheline*.
Occurrence: USA—Wyoming (Leucite Hills).

LEUCITOPHYRE

Igneous rock of the *gabbro* clan. A porphyritic extrusive rock. Contains phenocrysts of *leucite* in a groundmass of *leucite, nepheline* and *aegirine*.
Occurrence: Italy—Monte Somma.

LEUCOCRATIC

Light in colour. Such igneous rocks have a high proportion of *feldspar, quartz, topaz* or *mica*.

LEUCOPHYRE

Igneous rock of the *gabbro* clan. A porphyritic plutonic rock. Contains *feldspar* phenocrysts (often partly altered to *saussurite* and *kaolinite*) in a groundmass of *pyroxene, ilmenite* and *quartz*.
Occurrence: Germany—Fichtelgebirge Mountains. In altered *diabase*.

LEUCOTEPHRITE

Term describing *leucite* rich, *nepheline* free, *tephrites*.

LEUCOXENE

Alteration product of *sphene, ilmenite, perovskite* or *rutile*. Dull, yellow or brown, fine-grained material rich in titanium.
Occurrence: Worldwide.

LEVYNE

Synonym of *levynite*.

LEVYNITE or LEVYNE

$CaAl_2Si_3O_{10} \cdot 5H_2O$ Hydrated calcium aluminium silicate. Orthorhombic. Crystals rare, normally massive. White or greyish-green. Vitreous lustre. Subconchoidal fracture. Indistinct cleavage. Hardness $4 \cdot 0$–$4 \cdot 5$. SG $c2 \cdot 1$. Gelatinises with hydrochloric acid.
Occurrence: Scotland—Glasgow (Hartfield Moss). Northern Ireland—County Antrim. USA—Colorado (Table Mountain). In cavities in *basalts*.

LHERZITE

Igneous rock of the *ultrabasic* clan. A coarse-grained hypabyssal rock. Contains *hornblende, biotite, ilmenite* and *garnet*.
Occurrence: Spain—Pyrenees (Lherz).

LHERZOLITE

Igneous rock of the *ultrabasic* clan. A green, coarse-grained granular hypabyssal rock. Contains *olivine* (70–80 per cent), *enstatite, diopside* and *diallage*.
Occurrence: New Zealand—Olivine Range (Cascade Valley). Spain—Pyrenees (Lherz).

LIEBENERITE PORPHYRY

Altered *nepheline porphyry*. Contains *nepheline* phenocrysts replaced by a scaly *sericite* aggregate.
Occurrence: Austria—the Tyrol.

LIEBETHENITE

$4[Cu_3PO_4(OH)_3]$ Basic copper phosphate.

Orthorhombic. Prismatic crystals; also globular and reniform masses. Dark green. Resinous lustre. Olive-green streak. Uneven fracture. Indistinct cleavage. Hardness 4·0. SG 3·6–3·8. Soluble in nitric acid.

Occurrence: Britain—Cornwall (Redruth). Germany—the Rhine (Ehl). Chile—Coquimbo.

LIEVRITE

Synonym of *ilvaite*.

LIGNITE or BROWN COAL

Low rank *coal*. Contains carbon, volatiles and water. Woody texture with plant remains in a dark brown matrix.

Occurrence: Worldwide.
Variety is *jet*.

LIMAITE

Tin rich variety of *gahnite* (*of Von Moll*).

LIMESTONE

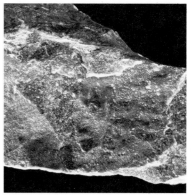

Below: limonite. *Left, above and over page:* limestone

Type of sedimentary rock which contains a high percentage of calcium carbonate as either *calcite* or *aragonite*. May be formed chemically, biologically or mechanically.
Occurrence: Worldwide.

LIMONITE

$FeO(OH).nH_2O$ Hydrous ferric oxide. Amorphous. Botryoidal, fibrous, concretionary, massive or earthy. Brown, black or ochre-yellow. Silky or dull lustre. Brown or yellow streak. Conchoidal to earthy fracture. No cleavage. Hardness 5·0–5·5. SG 3·6–4·0.
Occurrence: Worldwide.
Secondary weathering product of *iron* ores. Varieties are *bog iron-ore* and *pea iron-ore*.

LIMURITE

High grade contact metamorphic rock. A coarse granular rock. Contains *axinite*, *diopside*, *actinolite*, *zoisite*, *albite* and *quartz*. Produced at *granite*-calcareous rock contacts.

LINARITE

$2[PbCu(SO_4)(OH)_2]$ Hydrated lead copper sulphate. Monoclinic. Tabular crystals; also crusts or aggregates. Azure-blue. Vitreous lustre. Pale blue streak. Conchoidal fracture. Perfect cleavage in one direction. Hardness 2·5. SG 5·33–5·35. Soluble in dilute nitric acid with the precipitation of lead sulphate.
Occurrence: England—Cumberland (Keswick). Scotland—Lanarkshire (Leadhills). USA—Arizona (Mammoth Mine near Tiger), Montana (Butte). Spain—Linares. Secondary mineral of oxidation zone of lead and copper deposits.

LINDÖITE

Igneous rock of the *granite* clan. A medium to coarse-grained hypabyssal rock. Contains *microcline-microperthite*, *arfvedsonite*, *quartz* and *zircon*.
Occurrence: Norway—Lindö.

LINNAEITE

$8[Co_3S_4]$ Cobalt sulphide. Cubic. Normally granular masses. Pale steel-grey, tarnishing to copper-red. Metallic lustre. Blackish-grey streak. Uneven fracture. Imperfect cleavage. Hardness 5·5. SG 4·8–5·0. Soluble in nitric acid with the separation of sulphur.
Occurrence: USA—Maryland (Mineral Hill), Missouri (Mine le Motte). Sweden—Riddarhyttan. In veins in *gneiss*.

LINTONITE

Green variety of *thomsonite*.

LIPARITE

Synonym of *rhyolite*.

LIQUID

State of matter, intermediate between solid and gas. No long range order so no crystalline structure or properties, even on submicroscopic scale.

LIROCONITE

$4[Cu_2Al(AsO_4)(OH)_4 . 4H_2O]$ Basic hydrated copper arsenate. Monoclinic. Acicular crystals. Sky-blue to green. Vitreous lustre. Sky-blue to green streak. Uneven fracture. Indistinct cleavage. Hardness 2·0–2·5. SG 2·9–3·0. Soluble in nitric acid.
Occurrence: England—Cornwall (Redruth, St Day). USA—California (Cerro Gardo Mine in Ingo County). Germany—Saxony (Saida).
Secondary mineral of oxidation zone of *copper* deposits.

LISTWANITE

Low grade regional metamorphic rock. A yellowish-green schistose rock. Contains *quartz, dolomite, magnesite, talc* and *limonite*.
Occurrence: USSR—Ural Mountains (Beressowsk). Regional metamorphism of calcareous sediments.

LITCHFIELDITE

Igneous rock of the *syenite* clan. Snow-white, coarse-grained granular plutonic rock. Contains *sodaclase, nepheline, lepidomelane, cancrinite, sodalite* and *biotite*.
Occurrence: USA—Maine (Litchfield in Kennebec County). Canada—Ontario (Glamorgan, Monmouth).
Variety is *canadite*.

LITHARGE

2[PbO] Lead oxide. Tetragonal. Massive. Red. Greasy to dull lustre. Hardness 2·0. SG *c*9·2. Soluble in hydrochloric acid.
Occurrence: USA—California (Cucamarga Peak in San Bernardino County), Idaho (the Mineral Hill district of Blaine County).
Alteration product of *massicot*.

LITHIC TUFF

Type of volcanic *tuff*. Dominant constituent of later formed rock fragments.

LITHIFICATION

Completion of processes of diagenesis in which the original loose sediment becomes a compact rock.

LITHIOPHILITE

4[LiMn(PO₄)] Lithium manganese phosphate. Orthorhombic. Crystals rare, normally massive. Salmon coloured. Vitreous lustre. Greyish-white streak. Uneven fracture. Perfect cleavage in one direction. Hardness 4·5. SG 3·5–3·58. Soluble in hydrochloric acid.
Occurrence: USA—California (Pola in San Diego County), New Hampshire

(North Groton), South Dakota (Keystone). Canada—Manitoba (Ponte du Bois), North West Territories (Beaulieu, Yellowknife).

LITHIOPHORITE

Lithium rich variety of *wad*. Amorphous. Fine scales. Bluish-black. Metallic lustre. Blackish-grey streak. Hardness 3·0. SG 3·14–3·36.
Occurrence: Germany—Saxony (Schneeberg district).

LITHOID, -AL

Stone-like.

LITHOMARGE

Indurated variety of *kaolinite*.

LITHOPHYSAE

Hollow spherulites.

LITMUS

The best known indicator for acids and alkalis, with which it becomes red and blue respectively.

LIZARDITE

Mg(Si₂O₅)(OH)₄ Hydrous magnesium silicate. Monoclinic. Normally fibrous.

Greenish-white. Silky lustre.
Occurrence: England—Cornwall (Lizard Peninsular).
Variety of *serpentine*.

LODE

Vein of metallic ore.

LODESTONE

Name given to any magnetic variety of *magnetite*.

LODRANITE

Type of *meteorite*.

LOELLINGITE or LÖLLINGITE

2[FeAs₂] Iron diarsenide. Orthorhombic. Prismatic crystals; also massive. Silver-white to steel-grey. Metallic lustre. Greyish-black streak. Uneven fracture. Good cleavage in two directions. Hardness 5·0. SG 7·40–7·58.
Occurrence: USA—Colorado (San Juan), Maine (Auburn), New Jersey (Orange County). Canada—Ontario (Cobalt). Norway—Leyesund district. Spain—Andalusia.
Mesothermal vein mineral.

LOESS

Argillaceous sedimentary rock. A soft

crumbly aeolian deposit containing small sharply angular detrital grains of *quartz, feldspar, calcite* and *mica*.
Occurrence: Widespread in North America, Central Europe and China.

LÖLLINGITE

Synonym of *loellingite*.

LOMONOSOVITE

$Na_5Ti_2Si_2PO_{13}$ Sodium titanium phosphorosilicate. Crystal system uncertain. Platy masses. Dark brown. Uneven fracture. Poor cleavage. Hardness 3·0–4·0. SG 3·13.
Occurrence: USSR—Kola Peninsula.

LOPARITE

Alkali bearing variety of *perovskite* from the Kola Peninsular, USSR.

LOVOZERITE

$(Na,K)_2(Mn,Ca)ZrSi_6O_{16}.3H_2O$ Hydrated sodium manganese zirconium silicate. Trigonal. Irregular grains. Black or pink. Resinous lustre. Brown streak. Hardness 5·0. SG 2·38. Insoluble in acids.
Occurrence: USSR—Kola Peninsula (Lovozero).

LOW GRADE METAMORPHISM

Level of metamorphism (regional or contact) where some new minerals are developed and existing textures are reinforced or partly altered. e.g. *schist* or *slate*.

LOXOCLASE

Sodium rich variety of *orthoclase*.

LUBLINITE

$CaCO_3$ Variety of *calcite*. Encrustations on *chalk marl*. Matted aggregate of capillary or acicular crystals.
Occurrence: Poland—Lublin.

LUDWIGITE

$(Mg,Fe)_2FeBO_5$ Magnesium iron borate. Crystals rare, normally fibrous masses. Dark green. Silky lustre. Blackish-green streak. Uneven fracture. No cleavage.

Hardness 5·0. SG *c*3·6. Slightly soluble in hydrochloric acid.
Occurrence: USA—Montana (Philipsburg in Granite County), Nevada (Pioche in Lincoln County), Utah (Martin Lake). Norway—Norberg.

LUGARITE

Igneous rock intermediate in composition between the *syenite* and *gabbro* clans. A coarse-grained hypabyssal rock. Contains *titanaugite, labradorite, barkevikite* and *nepheline*.
Occurrence: Scotland—Ayrshire (Lugar). Forms sills.

LUJAVRITE

Igneous rock of the *syenite* clan. A coarse-grained plutonic rock. Contains *microcline-microperthite, oligoclase, nepheline, aegirine* and *arfvedsonite*.
Occurrence: USSR—Lapland (Lujavr).

LUSAKITE

Cobalt rich variety of *staurolite*.

LUSCLADITE

Igneous rock intermediate in composition between the *syenite* and *gabbro* clans. A coarse-grained plutonic rock. Contains *plagioclase, orthoclase, olivine, biotite* and *nepheline*.
Occurrence: France—Auvergne (Mont-Doré).

LUSITANITE

Igneous rock of the *syenite* clan. A medium-grained plutonic rock. Contains *riebeckite, aegirite, orthoclase, microcline-microperthite* and *sodaclase*.
Occurrence: Portugal—Lusitania.

LUSSATINE

Synonym of *cristobalite*.

LUSTRE

Descriptive terms used to classify the surface appearance and light reflective properties of an unweathered face of a mineral. Types (self-explanatory except where defined below) are:
Adamantine—diamond-like, e.g. *cinnabar*.
Vitreous—*glass-like*, e.g. *quartz*.
Resinous, e.g. *amber*.
Waxy, e.g. *brucite*.
Greasy, e.g. *saponite*.
Silky, e.g. *malachite*.
Satiny, e.g. *satin spar*.
Pearly, e.g. *ankerite*.
Pitchy, e.g. *pitchstone*.
Dull—no lustre, e.g. *bauxite*.

Earthy, e.g. *red ochre*.
Metallic, e.g. *pyrite*.
Supplementary descriptive terms are:
Splendent—mirror-like.
Brilliant—very bright.
Shining—bright.
Glistening—fairly dull.
The use of the prefix sub- indicates that the lustre is not very pronounced.

LUTÉCINE or LUTÉCITE

Fibrous form of silica.

LUTÉCITE

Synonym of *lutécine*.

LUXULLIANITE

Igneous rock of the *granite* clan. A porphyritic plutonic rock. Contains phenocrysts of *orthoclase* in a groundmass of *quartz, tourmaline, orthoclase, mica* and *cassiterite*.
Occurrence: England—Cornwall (Luxulyan).

LUZONITE

Reddish-grey variety of *famatinite*.
Occurrence: Philippine Islands—Island of Luzon (Mancayan).

LYDIAN STONE

Synonym of *basanite*.

LYDITE

Sedimentary siliceous deposit. Very fine-grained *quartz* and *chalcedony*. Coloured greyih-black by carbonaceous matter or greenish-brown by *chlorite* and iron hydroxides.
Occurrence: Worldwide. In Palaeozoic and Pre-Cambrian rocks representing silicified *shales, limestones* or *tuffs*.

MACEDONITE

Igneous rock of the *gabbro* clan. A bluish-black, very fine-grained extrusive rock. Contains *anorthoclase, plagioclase, biotite* and *olivine*.
Occurrence: Australia—Victoria (Mount Macedon).

MACERALS

The individual constituents of *coal*.

MACKENSITE

$Fe_6Si_3AlO_{15}(OH_3)$ Basic iron aluminosilicate. Monoclinic. Normally massive. Greenish-black. Black streak.
Occurrence: Czechoslovakia—Northern Moravia. In *diabase*.
Variety of the *thuringite* series of *chlorites*.

MACKINAWITE

2[FeS] Iron sulphide. Tetragonal. Crystals rare, normally irregular grains. Grey. Black streak. Hardness 4·0. SG $c3·1$. Insoluble in hydrochloric acid.
Occurrence: USA—Washington (Mackinawite Mine in Snohomish County). Finland—Varislakhti.

MACLE

Synonym of *chiastolite*.

MACRODOME

A large dome.

MADEIRITE

Igneous rock of the *ultrabasic* clan. A green and black porphyritic plutonic rock. Contains phenocrysts of *augite* and *olivine* in a very fine groundmass of *plagioclase*, *calcite* and *magnetite*.
Occurrence: Madeira Islands—Ribeira de Massapez.

MADUPITE

Feldspathoid igneous rock. A finegrained porphyritic extrusive rock. Contains phenocrysts of *diopside* and *phlogopite* in a groundmass of *leucite* composition *glass*.
Occurrence: USA—Wyoming (Leucite Hills).

MAENAÏTE

Igneous rock of the *diorite* clan. A fine-grained porphyritic hypabyssal rock. Contains phenocrysts of *plagioclase* in a groundmass of *oligoclase*, *orthoclase*, *pyroxene* and *hornblende*.
Occurrence: Norway—Kristiana, Maena.

MAFIC

Silicate minerals based on iron (Fe) and/or magnesium (Mg); also called ferromagnesian.

MAFRAÏTE

Igneous rock intermediate in composition between the *syenite* and *gabbro* clans. A medium to coarse-grained hypabyssal rock. Contains *labradorite, sanidine, hornblende* and *pyroxene*.
Occurrence: Portugal—near Cintra (Tifâo de Mafra).

MAGHEMITE

Fe_2O_3 Iron oxide. Cubic. Cubic crystals. Brown. Brown streak. Alteration product of *magnetite* and *lepidocrocite*.
Occurrence: Worldwide. In gossans. Formed by oxidation at low temperatures.

MAGMA

Igneous rock-forming fluid, perhaps carrying solid material in suspension, and with dissolved gases, formed in the crust or upper mantle of the Earth.

MAGNESIAN LIMESTONE

Sedimentary rock composed of granular *dolomite*. Normally associated with *evaporites*.

MAGNESIOARFVEDSONITE

Magnesium rich variety of *arfvedsonite*.

MAGNESIOCHROMITE, CHROME-SPINEL or MITCHELLITE

Magnesium rich variety of *chromite*.

MAGNESIOFERRITE

$MgFeO_3$ Magnesium iron oxide. Cubic. Octahedral crystals. Iron-black. Metallic lustre. Black streak. Hardness 6·0–6·5. SG 4·57–4·65. Soluble with difficulty in hydrochloric acid.
Occurrence: Italy—Mount Vesuvius. Formed at the mouths of fumeroles.

MAGNESIOHASTINGSITE

Magnesium rich variety of *hastingsite*.

MAGNESIOKATOPHORITE

Magnesium rich end member of the *catophorite* series.

MAGNESIOMAGNETITE

Magnesium rich variety of *magnetite*.

MAGNESIORIEBECKITE

Magnesium rich variety of *riebeckite*.

MAGNESIOWÜSTITE or FERRO-PERICLASE

Variety of *periclase* with the formula (Mg.Fe)O.

MAGNESITE

2[$MgCO_3$] Magnesium carbonate. Hexagonal. Crystals rare, usually massive, earthy or fibrous. Colourless, often white or greyish-white. Vitreous lustre. White streak. Conchoidal fracture. Perfect cleavage in one direction. Hardness 3·5–4·0. SG 3·0–3·5. Soluble with effervescence in hydrochloric acid.
Occurrence: USA—California (coast range), Nevada (Muddy Valley district of Clark County). Canada—Yukon (Black River). Austria—Tyrol (Hall). In magnesium rich rocks.
Variety is *breunnerite*. Also synonym of *sepiolite*.

MAGNESITE ROCK

Carbonate sediment. Composed almost entirely of *magnesite*.
Occurrence: USA—California (Bissell). In ancient salt lakes.

MAGNETIC

Magnetic rocks exhibit some degree of magnetisation; freely suspended samples have a tendency to align themselves in a roughly North-South direction. The effect is a result of the magnetic 'dipole' nature of their atoms, molecules or crystallites, which are aligned by the Earth's magnetic field during solidification.

MAGNETITE

8[Fe_3O_4] Iron oxide. Cubic. Octahedral crystals, commonly massive or granular. Black. Metallic lustre. Black streak.

Above and below: magnetite

Uneven fracture. No cleavage. Hardness 6·0. SG 5·2. Naturally strongly magnetic. **Occurrence:** Worldwide. In plutonic and metamorphic rocks and in *placer deposits*. Varieties are *hydromagnetite* (hydrous) and *magnesiomagnetite*.

MAGNOPHORITE

$NaKCaMg_5Si_8O_{23}OH$ Basic sodium potassium calcium and magnesium silicate. Monoclinic. Elongated crystals. Reddish-brown. Rare type of *amphibole*. **Occurrence:** USA—Wyoming (Leucite Hills). Western Australia—Wolgidee Hills.

MALACHITE

$2[Cu_2CO_3(OH)_2]$ Hydrated copper car-

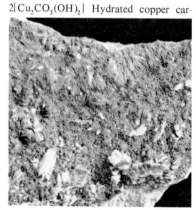

Below and right: malachite

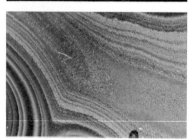

bonate. Monoclinic. Crystals rare, normally massive or in crusts. Various shades of green. Silky lustre. Pale green streak. Irregular fracture. Basal cleavage visible in crystals. Hardness 3·5–4·0. SG 3·9–4·0. Dissolves easily in hydrochloric acid with effervescence.
Occurrence: Worldwide. In secondary weathering zone of copper deposits.

MALACOLITE

Synonym of *diopside*.

MALACON

Altered *zircon*. Tetragonal. Crusts. Brown. Vitreous lustre. Reddish-brown streak. Hardness 6·5. SG *c*3·9.

Occurrence: Finland—Björkboda (Rosedal). Norway—Hitterö.

MALCHITE

Igneous rock of the *diorite* clan. A fine-grained porphyritic hypabyssal rock. Contains phenocrysts of *hornblende, labradorite* and *biotite* in a groundmass of *hornblende, andesine* and *quartz*.
Occurrence: Germany—Alsbach (Melibocus).

MALIGNITE

Igneous rock of the *syenite* clan. A mottled light grey, medium to coarse-grained plutonic rock. Contains *orthoclase, nepheline, apatite, aegirine-augite* and *biotite*.

Occurrence: Canada—Ontario (Maligna River, Poobah Lake).

MALLEABLE

Able to be hammered into a thin sheet.

MALMSTONE

Sedimentary siliceous deposit. Contains sponge spicules in a matrix of globular *opal* and *chalcedony*.
Occurrence: England—Kent and Sussex, Greensands.

MAMMILLATED

Aggregates with round surfaces, not unlike breasts.

MANASSEITE

$Mg_6Al_2(OH)_{16}.4H_2O$ Basic hydrated aluminium magnesium carbonate. Hexagonal. Normally forms scales. White. Waxy lustre. White streak. Hardness 2·0. SG $c2·1$. Easily soluble with effervescence in dilute hydrochloric acid.
Occurrence: USA—New Jersey (Vernon in Sussex County), New York (Rossie in St Lawrence County). Norway—Nordmark, Snarum.

MANGANANDALUSITE or VIRIDINE

Manganese rich variety of *andalusite*.

MANGANAPATITE

Manganese rich variety of *apatite*.

MANGANAXINITE or TINZENITE

Manganese rich variety of *axinite*.

MANGANESE NODULES

Ocean floor deposit of nodules of *manganite* and *goethite*. Accumulate around a nucleus of bone. Contain traces of many metals.
Occurrence: In all the large oceans.

MANGANITE

$8[MnO(OH)]$ Basic manganese oxide.

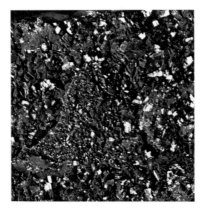

Monoclinic. Prismatic crystals. Grey to black. Submetallic lustre. Dark reddish-brown streak. Uneven fracture. Perfect cleavage in three directions. Hardness 4·0. SG 4·2–4·4. Soluble in hydrochloric acid with the production of chlorine.
Occurrence: Britain—Cornwall (Botallack Mine near St Just in Penwith), Cumberland (Egremont). USA—Michigan (Negauriee), Virginia (Powell Fort near Woodstock). Canada—Nova Scotia (Cheverie in Hants County). Germany—Harz Mountains (Ilfield). Low temperature vein mineral.

MANGANKNEBELITE

Manganese rich variety of *knebelite*.

MANGANOPHYLLITE

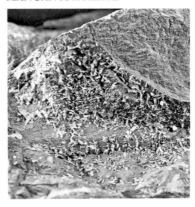

Manganese rich *biotite*. Monoclinic. Tabular crystals; also aggregates of thin plates. Bronze to copper-red. Pearly lustre. Pale red streak. Perfect basal cleavage. Hardness 2·5–3·0. SG 2·7–3·1. Soluble in hydrochloric acid, leaving residue of silica scales.
Occurrence: Sweden—Nordmark (Jakobsberg), Wermland (Harstig Mine in Pajsberg).

MANGANOSIDERITE

Iron rich variety of *rhodochrosite*, intermediate between *rhodochrosite* and *siderite*.

MANGANOSITE

$4[MnO]$ Manganese oxide. Cubic. Crystals rare, normally irregular masses or disseminated grains. Emerald-green, turning black on exposure. Vitreous lustre. Brown streak. Indistinct cleavage. Hardness 5·5. SG $c5·35$. Dissolves with difficulty in hydrochloric acid.
Occurrence: USA—New Jersey (Franklin). Sweden—Långban.

MANGERITE

Igneous rock of the *syenite* clan. A grey, medium-grained granular plutonic rock. Contains *microperthite*, *pyroxene*, *apatite*, *biotite* and *plagioclase*.
Occurrence: Norway—Manger parish (Kalsaas).

MANJAK

Type of *bitumen*. Black. Brilliant lustre. Conchoidal fracture. Hardness 2·0. SG $c1·06$.
Occurrence: West Indies—Barbados.

MANSJÖITE

Fluorine rich variety of *diopside*.

MARBLE

Limestone recrystallised by metamorphism to produce a granular mosaic of *calcite* or *dolomite* crystals.
Occurrence: Worldwide.

MARCASITE or WHITE PYRITES

$2[FeS_2]$ Iron sulphide. Orthorhombic. Tabular crystals; also concretionary masses. Pale bronze-yellow. Metallic lustre. Greyish-black streak. Uneven frac-

ture. Indistinct cleavage. Hardness 6·0–6·5. SG $c4·9$. Soluble in nitric acid.
Occurrence: Worldwide. Concretions in sedimentary rocks and replacement mineral in *limestones*.
See *pyrite*.

MARCELINE (of Beudant)

Synonym of *braunite* (*of Haidinger*).

MAREKANITE

Alteration product of *rhyolite perlite glass*. Glassy balls with concave indentations
Occurrence: USSR—Siberia (Okhotsh).

MAREUGITE

Igneous rock of the *gabbro* clan. A medium-grained granular plutonic rock. Contains *bytownite, häuyne, hornblende* and *augite*.
Occurrence: France—Auvergne (Mareuge).

MARGARITE

$CaAl_4Si_2O_{10}(OH)_2$ Hydrous calcium aluminium silicate. Monoclinic. Crystals rare, normally aggregates. White, violet or grey. Pearly lustre. Perfect basal cleavage. Hardness 3·5–4·5. SG $c4·0$. Slowly decomposes in boiling hydrochloric acid.
Occurrence: USA—Massachusetts (Chester), North Carolina (Madison County).

Greece—Island of Naxos.
A *mica* derived from altered *corundum*.

MARIALITE

$2[Na_4Al_3Si_9O_{24}Cl]$ Sodium aluminium silicate and chloride. Tetragonal. Prismatic crystals; also forms aggregates. Colourless or white. Vitreous lustre. Subconchoidal fracture. Indistinct cleavage. Hardness 5·5–6·0. SG $c2·6$.
Occurrence: USA—New York (Rossie in St Lawrence County). Canada—Ontario (Bedford). Norway—Arendal.
Also synonym of *haüyne*.

MARIPOSITE or CHROMEPHENGITE

Apple-green, chromium rich variety of *phengite*.

MARIUPOLITE

Igneous rock of the *diorite* clan. A coarse-

grained porphyritic plutonic rock. Contains phenocrysts of *nepheline* in a groundmass of *sodaclase, nepheline, aegirite* and *lepidomelane*.
Occurrence: USSR—Sea of Azov (Mariupol).

MARKFIELDITE

Igneous rock of the *granite* clan. A fine-grained porphyritic hypabyssal rock. Contains phenocrysts of *plagioclase* in a groundmass of *quartz, orthoclase* and *muscovite*.
Occurrence: England — Leicestershire (Markfield in Charnwood Forest).

MARL

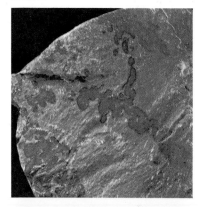

Sedimentary *argillaceous rock*. Contains a variable amount of calcareous material. Covers a range of rocks from calcareous *clays* to muddy *limestones*.
Occurrence: Worldwide.

MARLOESITE

Igneous rock of the *diorite* clan. A fine-grained porphyritic extrusive rock. Contains phenocrysts of *albite* and *olivine* in a groundmass of *augite, plagioclase* and *magnetite*.
Occurrence: Wales—Pembrokeshire (Skomer Island).

MARMATITE, CHRISTOPHITE or NEWBOLDITE

Iron rich variety of *blende*.

MARMOLITE or COOPERITE (of Adam)

$Mg_3Si_2O_5(OH)_4$ Hydrous magnesium silicate. Monoclinic. Thin foliated laminae. Greenish-white or bluish-white. Pearly lustre. Hardness 3·0. SG *c*2·41. Decomposes in hydrochloric acid to leave silica residue. Variety of *chrysotile*.
Occurrence: USA—New Jersey (Hoboken).

MAROSITE

Igneous rock of the *diorite* clan. A medium to coarse-grained plutonic rock. Contains *biotite, augite, sanidine, plagioclase, nepheline, sodalite* and *apatite*.
Occurrence: East Indies—Celebes (Pic de Maros).

MARTITE

Fe_2O_3 Iron oxide. Hexagonal. Octahedral crystals. Iron-black. Dull to submetallic lustre. Reddish or purplish-brown streak. Conchoidal fracture. Hardness 6·0–7·0. SG *c*4·8.
Occurrence: USA—Utah (Twin Peaks in Millard County). Canada—Nova Scotia (Digby Creek in Digby County). Brazil—Minas Gerais. Mexico—Durango (Cerro de Mercado).
Type of *hematite* pseudomorphed after *magnetite*.

MASANITE

Igneous rock of the *granite* clan. A fine-grained porphyritic hypabyssal rock. Contains phenocrysts of *plagioclase* and *quartz* in a groundmass of *quartz, orthoclase* and *muscovite*.
Occurrence: South Korea—Ma-san-Po. Variety is *masanophyre*.

MASANOPHYRE

Variety of *masanite*. Contains phenocrysts of *oligoclase* mantled by *orthoclase* in a groundmass of *quartz, hornblende* and *sphene*.
Occurrence: South Korea—Ma-san-Po.

MASCAGNITE

$4[(NH_4)_2SO_4]$ Ammonium sulphate. Orthorhombic. Crystals rare, normally crusts or stalactitic. Colourless or grey. Vitreous lustre. Uneven fracture. Good cleavage in one direction. Hardness 2·0–2·5. SG *c*1·77. Sharp and bitter taste. Soluble in water.
Occurrence: England—Staffordshire (Bradley). Scotland—Midlothian (Arniston). USA—California (The Geysers in Sonoma County). Italy—Mount Etna, Mount Vesuvius.

MASSICOT

$4[PbO]$ Lead oxide. Orthorhombic. Earthy or scaly masses. Yellow, sometimes with a reddish tint. Dull or greasy lustre. Light yellow streak. Cleavage visible in two directions. Hardness 2·0. SG *c*9·6. Decomposes in sulphuric acid with the precipitation of lead sulphate.
Occurrence: USA—California (Cucamaya Peak in San Bernardino County), Nevada (Redemption Mine in Esmeralda County), Virgina (Austins Mine in Wythe County). France—Vosges (La Croix-aux-Mines). Germany—Saxony (Freiberg). Italy—Sardinia (Oreddo Valley). Oxidation product of lead minerals.

MATILDITE

$8[AgBiS_2]$ Silver bismuth sulphide. Orthorhombic. Crystals rare, normally granular masses. Iron-black or grey. Metallic lustre. Light grey streak. Uneven fracture. No cleavage. Hardness 2·5. SG *c*6·9. Soluble in nitric acid with the precipitation of sulphur.
Occurrence: USA—Colorado (Lake City in Hinsdale County), Idaho (Mayflower Mine in Boise County). Canada—Ontario (O'Brian Mine near Cobalt). Germany—Black Forest (Schapbach). Spain—near Madrid (Bustarviejo).

MATRIX

The natural material in which a rock or mineral is embedded.

MAUCHERITE

$16[Ni_{11}As_8]$ Nickel arsenide. Tetragonal. Tabular crystals; also in granular or fibrous masses. Platinum-grey, tarnishing to coppery-red. Metallic lustre. Blackish-grey streak. Uneven fracture. No cleavage. Hardness 5·0. SG *c*8·0. Slowly decomposes in hydrochloric acid.
Occurrence: Canada—Ontario (Moose Horn Mine in Timeskaming County, Sudbury). Germany—Saxony (Eisleben). Spain—Malaga (Los Jarales).

MEASURES

A mining term formerly applied to sequences of sedimentary rocks in boreholes or mining shafts. Now obsolete, but preserved in the term 'coal measures'.

MEERSCHAUM

Synonym of *sepiolite*.

MEIONITE or TETRAKALSILITE

$2[Ca_4Al_6Si_6O_{24}(SO_4,CO_3,Cl_2)]$ Calcium aluminium silicate with carbonate. Tetragonal. Prismatic crystals. Colourless to white. Vitreous lustre. Conchoidal fracture. Good cleavage in two directions. Hardness 5·5–6·0. SG 2·70–2·74. Dissolves in hydrochloric acid.
Occurrence: USA—New York (Pierrepont in St Lawrence County). Canada—Ontario (Renfrew). Italy—Monte Somma.

MELACONITE

Massive black variety of *tenorite*.

MELANITE

Synonym of *andradite*.

MELANOCRATIC

Such igneous rocks mainly comprise mafic minerals with a resulting dark colour.

MELANOPHLOGITE

Pseudomorph of *quartz* after *cristobalite*. Light brown, with a vitreous lustre, spherical aggregates with an opaline crust and a fibrous interior.

Above: melaconite. *Below:* melanterite

Occurrence: Italy—Sicily (Caltanisetta, Girgenti). In *sulphur* deposits.

MELANTERITE or COPPERAS

$8[FeSO_4.7H_2O]$ Hydrous iron sulphate. Monoclinic. Prismatic crystals; also massive, fibrous, stalactitic or concretionary forms. Green, turning yellow on exposure. Vitreous lustre. Colourless streak. Conchoidal fracture. Perfect cleavage in one direction. Hardness $2·0$. SG $c 1·9$. Soluble in water.
Occurrence: Scotland—near Paisley (Hurlet). USA—California (Alma Mine in Alameda County), Utah (Lucky Boy Mine in Salt Lake County). Germany—Bavaria (Bodenmais). Sweden—Falun.
Oxidation product of *pyrite*.

MELILITE or HUMBOLDTILITE

$2[Ca_2MgSi_2O_7]$ Calcium magnesium silicate. Tetragonal. Tabular crystals. White or pale yellow, occasionally brown. Vitreous lustre. Conchoidal fracture. Good cleavage in one direction. Hardness 5·0. SG 2·9–3·1. Gelatinises with hydrochloric acid.
Occurrence: Italy—Monte Somma, near Rome (Capo di Bove). In lava flows.

MELILITITE

Feldspathoid igneous rock. A fine-grained extrusive rock. Contains *melilite, augite* and *magnetite*.
Occurrence: East Africa—Nyando River.

MELIPHANE

See *meliphanite*.

MELIPHANITE or MELIPHANE

$8[(Ca,Na)_2)Be(Si,Al)_2(O,F)_7]$ Beryllium calcium and sodium fluosilicate. Tetragonal. Platy masses. Yellow or red. Vitreous lustre. Uneven fracture. Good cleavage in one direction. Hardness 5·0–5·5. SGc3·0.
Occurrence: Norway—Fredriksvärn, Langodden, Stoksund. In plutonic igneous rocks.

MELNIKOVITE

Finely divided black variety of *pyrite*.

MELTEIGITE

Feldspathoid igneous rock. A medium-grained granular plutonic rock. Contains *nepheline, pyroxene, biotite* and *garnet*.
Occurrence: USA—Colorado (North Beaver Creek). Finland—Niskavaara. Norway—Fen district (Melteig).

MENACCANITE

Synonym of *ilmenite*.

MENILITE

Variety of *opal*. Occurs as dull greyish-brown concretions in *shales*.

MERCURY

Hg Liquid. Tin-white. Metallic lustre.
Occurrence: USA—California (New Almaden in Santa Clara County), Texas (Terlingua). Italy—near Gorizia (Idria). Spain—Almaden. In volcanic districts in association with *cinnabar*. Rare.

MERWINITE

$Ca_3MgSi_2O_8$ Calcium magnesium silicate. Monoclinic. Granular aggregate. White. Vitreous lustre. Perfect cleavage in one direction. Hardness 6·0. SG 3·14–3·3. Readily soluble in hydrochloric acid.
Occurrence: USA—California (Crestmore), Montana (Little Belt Mountains).

MESOCRATIC

Igneous rocks intermediate between leucocratic and melanocratic types, thus of intermediate colour.

MESOLITE

$8[Na_2Ca_2(Al_2Si_3O_{10}) \cdot 8H_2O]$ Hydrous sodium calcium aluminium silicate. Monoclinic. Acicular crystals; also fibrous tufts or nodules. White or grey. Vitreous lustre.

Perfect prismatic cleavage. Hardness 5·0. SG 2·2–2·4. Gelatinises with hydrochloric acid.
Occurrence: Scotland—Midlothian (Eigg near Edinburgh). Northern Ireland—County Antrim (Agnews Hill near Larne, Giant's Causeway). USA—Colorado (Table Mountain near Golden), Pennsylvania (Fritz Island in the Schuylkill River). In cavities in volcanic rocks.
A *zeolite*. Variety is *antrimolite*.

MESOSIDERITE

Type of *meteorite*.

MESOSTASIS

Partly amorphous, partly crystalline interstitial material which may occur in igneous rocks, the end product of magma consolidation. Forms a *glass* or a thin coating on earlier minerals.

MESOTHERMAL

Deposits formed hydrothermally at temperatures roughly between 200 degrees C and 300 degrees C.

MESOZOIC

See *geological time*.

METABASTITE

Metamorphic rocks originally of gabbroic composition.

METABOLITE

Altered *volcanic glass*.

METABRUCITE

Variety of *periclase* pseudomorphed after *brucite* (of Beudant).

METAL

Element with 'metallic' lustre, high electrical and thermal conductivity, ductile, malleable, of high density. Atomic structure and chemical behaviour are also relevant.

METALLOID

Element showing some, but not all, properties of a metal; somewhat metallic.

METAMICT

Mineral with external crystalline appearance but abnormal X-ray diffraction properties due to severe crystal lattice imperfections caused by bombardment from radioactive substances.

METAMORPHIC ROCK

Name given to any igneous or sedimentary rock which has been subjected to elevated temperatures and/or pressures. Constituent minerals may have been recrystallised and reorientated and new minerals grown without the whole rock melting. May be a result of contact, dynamic or regional metamorphism.

METAMORPHISM

Range of processes involving combinations of high temperature, high pressure and chemical action by which rocks in the crust change in nature. Some overlap with diagenesis, but weathering is specifically excluded.
Main types are:
(a) Contact or 'thermal' metamorphism, effect mainly of heat, usually associated with igneous intrusions.
(b) Regional metamorphism, effect of heat and high pressure over a large area, as always found with orogenesis.
(c) Dynamic or 'dislocation' metamorphism, associated with strong but very local forces.

METANATROLITE

High temperature dehydrated monoclinic variety of *natrolite*.

METASCOLECITE

High temperature polymorph of *scolecite*.

METASEDIMENT

Metamorphosed sedimentary rock.

METASOMATIC

Associated with metamorphism in which external material is introduced into the rock, perhaps with recrystallisation and/or with displacement of a pre-existing component.

METASTABLE

Existing in an unstable environment.

METATHOMSONITE

High temperature polymorph of *thomsonite*.

METEORITE

A body of sub-planetary dimensions which has fallen from space onto the earth. Divided into three main types: *stones* (composed dominantly of silicates), *stonyirons* (a mixture of silicates and nickel/iron minerals) and *irons* (dominantly nickel/-iron minerals). Stones are further divided into: *chondrites* (in which spherical aggregates of *olivine* and *pyroxene* occur) and *achondrites* (which are homogeneous and have the composition of a *gabbro*). *Stoneirons: pallasites* (a mixture of nickel/iron and *olivine*), *siderophyre* (nickel/iron, *bronzite* and *tridymite*), *lodranite* (*olivine* and *bronzite* in nickel/iron) and *mesosiderite* (nickel/iron and *gabbro*). *Irons: hexahedrite* (100 per cent *kamacite*), *octahedrite* (coarse intergrowth of *kamacite* and *taenite*) and *ataxite* (fine intergrowth of *kamacite* and *taenite*).
Occurrence: Worldwide.

MIARGYRITE

8[AgSbS₂] Silver antimony sulphide. Monoclinic. Tabular crystals; also massive. Iron-black to steel-grey. Metallic lustre. Cherry-red streak. Uneven fracture. Indistinct cleavage. Hardness 2·5. SG *c*5·25. Decomposes in nitric acid with the precipitation of sulphur.
Occurrence: USA—California (Randsbury district), Idaho (Flint in Owyhee County). Germany—Saxony (Braunsdorf near Freiberg).
Hydrothermal vein deposit.

MIASKITE

Igneous rock of the *syenite* clan. A mottled, medium to coarse-grained plutonic rock. Contains *microperthite*, *nepheline*, *mica* and *oligoclase*.
Occurrence: USSR—Ural Mountains (Miask).

MICACEOUS

Composed of *micas* or sub-*micas*. *Mica*-like in cleavage.

MICACEOUS HEMATITE

Variety of *hematite* with a micaceous and foliated habit.

MICA, CAT GOLD, CAT SILVER or ISINGLAS

Group of sheet silicate minerals. Silicates of aluminium and potassium with magnesium, iron, lithium or sodium. All members crystallise in the monoclinic

system and exhibit a pearly lustre and perfect basal cleavage. The average hardness is 2·5 and the SG between 2·7 and 3·1. The main members are *biotite*, *brittle micas*, *lepidolite*, *muscovite*, *paragonite*, *phlogopite* and *zinnwaldite*.

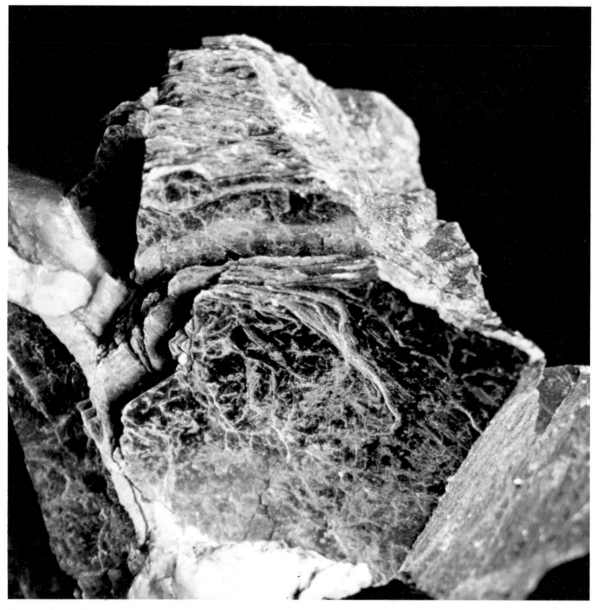

Above: sphere mica

MICA SCHIST

Low to medium grade regional metamorphic rock. Schistose and lineated texture. Contains essentially *micas*, *quartz* and *feldspars*; may also contain *garnet, staurolite* or other metamorphic minerals. Segregation banding developed between the *micas* and the quartzo-*feldspathic* minerals.
Occurrence: Worldwide in regional metamorphic areas.

MICRITE

Semi-opaque microcrystalline *calcite* which often forms the cement in *limestones*.
Variety is *cornstone*.

MICROADAMELLITE

Igneous rock of the *granite* clan. A medium-grained hypabyssal rock. Contains *biotite, hornblende* and *quartz*, with some *orthoclase* and *plagioclase*.
Occurrence: Worldwide.

MICROCLINE or AMAZON JADE

$4[KAlSi_3O_8]$ Potassium aluminium silicate. Triclinic. Crystals common; also granular masses. White to pale yellow, occasionally red or green. Vitreous lustre. Uneven fracture. Cleavage visible in two directions. Hardness $6 \cdot 0$–$6 \cdot 5$. SG $c2 \cdot 55$.
Occurrence: Worldwide. A *feldspar* found

in *granites* and *granite pegmatites*. Green gem variety is *amazonite*. Other varieties are *anorthoclase* and *rubidium microcline*.

MICROCLINE-MICROPERTHITE

Microscopically fine lamellar intergrowth of *microcline* and *albite*.

MICROCRYSTALLINE

Very fine-grained texture; individual crystals can be seen only with a microscope.

MICROGRANITE

Igneous rock of the *granite* clan. A medium-grained hypabyssal rock. Contains *quartz*, alkali-*feldspars*, *biotite*, *muscovite* and *plagioclase*.
Occurrence: Worldwide.

MICROGRANODIORITE

Igneous rock of the *granite* clan. A medium-grained hypabyssal rock. Contains *quartz*, *plagioclase*, *orthoclase*, *biotite*, *hornblende* and *pyroxene*.
Occurrence: Worldwide.

MICROLITE

A very small crystal.

MICROLITE or MIKROLITH

8[(Na,Ca)$_2$Ta$_2$O$_6$(O,OH,F)] Complex calcium sodium tantalum oxide with hydroxyl and fluorine. Cubic. Octahedral crystals; also massive. Pale yellow to brown. Resinous lustre. Pale yellowish-brown streak. Conchoidal fracture. Indistinct cleavage. Hardness 5·5. SG *c*5·5. Slowly decomposes in sulphuric acid.
Occurrence: USA—California (Hemet in Riverside County), Connecticut (Branchville), Massachusetts (Chesterfield). Finland—Skogbole. Norway—near Tele-Mark (Söve). In albitised *pegmatites*.

MICROPEGMATITE

Very fine-grained intergrowth of *quartz* and *orthoclase*.

MICROPERTHITE

Microscopically fine lamellar intergrowth of *orthoclase* and *albite*.
Varieties are *cryptoperthite*, *orthoclase microperthite*.

MICROSOMMITE

Chlorine rich variety of *cancrinite*.

MIEMITE

A yellowish-brown variety of *dolomite* found at Miemo in Tuscany.

MIGMATITE

Zone of mixed rock between a *granite* and the country rock. *Magma* and country rock intermingled randomly. Streaky appearance. Represents magmatic intrusion—*injection gneisses*—or deep seated regional metamorphism.
Occurrence: Worldwide. Associated with *granites* in regional metamorphic areas.

MIHARAITE

Igneous rock of the *gabbro* clan. A fine-grained porphyritic extrusive rock. Contains phenocrysts of *bytownite* and *hypersthene* in a groundmass of *plagioclase*, *augite* and *magnetite*.
Occurrence: Japan—Oshima (volcanic districts).

MIKROLITH

Synonym of *microlite*.

MILARITE

K$_2$Ca$_4$Be$_4$Al$_2$Si$_{24}$O$_{60}$.H$_2$O Hydrated Potassium calcium beryllium and aluminium silicate. Hexagonal. Prismatic crystals. Colourless to pale green. Vitreous lustre. Conchoidal fracture. Indistinct cleavage. Hardness 5·5–6·0. SG 2·55–2·59. Decomposes in hydrochloric acid.
Occurrence: Switzerland—Grisons (Val Giuf), Tavetschthal (Strim Glacier). In *granite*.

MILKY QUARTZ

A milky-white variety of *quartz*.

MILLERITE

3[NiS] Nickel sulphide. Hexagonal. Capillary crystals. Brass-yellow, tarnishing grey. Metallic lustre. Greenish-black streak. Uneven fracture. Perfect cleavage in two directions. Hardness 3·0–3·5. SG *c*5·5.
Occurrence: Wales—Glamorgan (Merthyr Tydfil). USA—New York (Sterling Mine near Antwerp), Pennsylvania (Gap Mine in Lancaster County). Canada—Ontario (Sudbury). Germany—Saxony (Freiberg, Johanngeorgenstadt). Low temperature vein mineral.

Illustration also over page

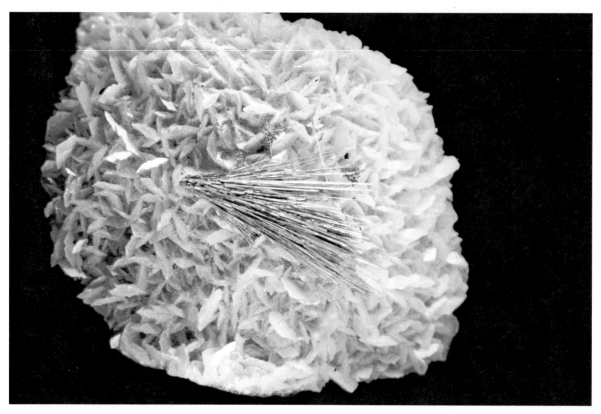

Above: millerite with linnaeite on dolomite

Below: mimetite

MILTONITE

Synonym of *bassanite*.

MIMETITE

$2[Pb_5(AsO_4,PO_4{}_3Cl]$ Lead chloride-phos-

MINERALISATION

Formation of mineral dposits; introduction tion of mineral veins or disseminated inclusions into an existing rock.

MINETTE

Igneous rock of the *syenite* clan. A fine-grained porphyritic hypabyssal rock. Contains phenocrysts of *biotite* in a groundmass of *orthoclase*, *hornblende* and *pyroxene*.
Occurrence: Germany—Freiberg (St Michaelis).

MINIUM or RED LEAD

$4[Pb_3O_4]$ Lead oxide. Earthy masses. Scarlet-red or brownish-red. Greasy or dull lustre. Orangish-yellow streak. Hardness 2·5. SG c9·0. Soluble in hydrochloric acid producing chlorine.
Occurrence: England—Yorkshire (Weardale). Scotland—Lanarkshire (Leadhills). USA—California (Azusa in Los Angeles County), Virginia (Austins Mine in Wythe County), Utah (Godiva Mine in the Tintic district). Germany—Baden (Badenweiler). Sweden—Långban. Secondary alteration product of *galena* or *cerussite*.

MINNESOTAITE

Iron rich variety of *talc*.

MIRABILITE

$Na_2SO_4.10H_2O$ Hydrated sodium sulphate. Monoclinic. Prismatic or tabular crystals, often massive. Colourless to white. Vitreous lustre. White streak. Conchoidal fracture. Perfect cleavage in one direction. Hardness 1·5–2·0. SG c1·5. Easily soluble in water.
Occurrence: Worldwide in *evaporite* deposits.

MISKEYITE

Synonym of *pseudophite*.

MISPICKEL

Synonym of *arsenopyrite*.

MISSOURITE

Feldspathoid igneous rock. A mottled grey, coarse-grained granular plutonic rock. Contains *leucite*, *olivine*, *augite* and *biotite*.
Occurrence: USA—Montana (Highwood Mountains).

MITCHELLITE

Synonym of *magnesiochromite*.

MIZZONITE

Synonym of *dipyre*.

MOCHA STONE

Synonym of *moss agate*.

MODE

The percentage (by weight) of the individual minerals which make up a rock.

MOLASSE

Sedimentary *rudaceous* deposit. Gravels deposited torrentially in terrestrial environment. Produced by intense erosion of rising mountain chains.
Occurrence: Europe—large areas to the North of the Alps and the Carpathian Mountains.

MOLDAVITE

Type of *tektite* found in Bohemia, Czechoslovakia.

MOLYBDENITE

$2[MoS_2]$ Molybdenum sulphide. Hexagonal. Tabular crystals; also foliated masses. Lead-grey. Metallic lustre. Bluish-grey streak on paper. No fracture. Perfect cleavage in one direction. Hardness 1·0–1·5. SG 4·62–4·73. Decomposes in nitric

---

phate and arsenate. Hexagonal. Acicular crystals; also massive. Colourless or pale yellow to yellowish-brown. Resinous lustre. White streak. Uneven fracture. Indistinct cleavage. Hardness 3·5–4·0. SG 7·0.
Occurrence: England—Cumberland (Dry Gill and Roughton Gill near Caldbeck). Scotland—Lanarkshire (Leadhills). USA—Nevada (Eureka district), Pennsylvania (Phoenixville). Germany—Baden (Badenweiler), Saxony (Johanngeorgenstadt). Secondary weathered zone of lead deposits.
Variety is *campylite*.

MIMOSITE

Igneous rock of the *gabbro* clan. A very dark, fine-grained extrusive rock. Contains *augite*, *ilmenite*, *olivine*, *magnetite* and *plagioclase*.
Occurrence: USA—New Jersey (Watchung Mountain in Orange County). Germany—Nassau.

MINERAL

Substance of specific chemical nature found naturally in geological structures. Sometimes restricted to solids, sometimes to inorganic origin, sometimes to materials of economic importance.

acid leaving residue of molybdenum oxide.
Occurrence: Worldwide in *pegmatites*, *aplites* and *granites*.

MOLYBDITE

4[MoO_3] Molybdenum oxide. Orthorhombic. Crusts. Straw-yellow or yellowish-white. Silky or earthy lustre. No fracture. Good cleavage in one direction. Hardness 1·0–2·0. SG c4·5.
Occurrence: USA—Georgia (Chester in Delaware County), New Hampshire (Westmoreland).
Rare.

MONAZITE

4[(Ce,La,Y,Th)(PO_4)] Thorium rare earths phosphate. Monoclinic. Prismatic crystals. Yellow or white or reddish-brown to brown. Waxy or resinous lustre. White streak. Conchoidal to uneven fracture. Good cleavage in two directions. Hardness

5·0–5·5. SG 4·6–5·4. Slowly decomposes in hydrochloric acid.
Occurrence: Worldwide. Accessory mineral in *granites* and *gneisses*.

MONCHIQUITE

Igneous rock of the *ultrabasic* clan. A fine-grained porphyritic hypabyssal rock. Contains phenocrysts of *olivine* and *augite* in a groundmass of *biotite*, *barkevikite* and *analcite*.
Occurrence: Portugal—Serra de Monchique.
Variety is *guimarrite*.

MONDHALDEITE

Igneous rock of the *diorite* clan. A very fine-grained porphyritic hypabyssal rock. Contains phenocrysts of *hornblende*, *augite*, *bytownite* and *leucite* in a glassy groundmass.
Occurrence: Germany—Baden (Mondhalde).

MONMOUTHITE

Igneous rock of the *syenite* clan. A coarse-grained granular plutonic rock. Contains *nepheline*, *sodacase*, *hastingsite*, *apatite* and *calcite*.
Occurrence: Canada—Ontario (Monmouth Township in Monmouth County).

MONOCLINIC

See *crystal*.

MONTASITE

Synonym of *amosite*.

MONTEBRASITE

2[(Li,Na)Al(PO_4)(OH,F)] Basic lithium sodium aluminium fluophosphate. Triclinic. Prismatic cystals; also massive. White or creamy-white. Vitreous lustre. Uneven fracture. Perfect cleavage in one direction. Hardness 5·5–6·0. SG c2·98. Soluble with difficulty in hydrochloric acid.

Occurrence: Worldwide. In *granite pegmatites* associated with *amblygonite*.

MONTICELLITE

4[$MgCaSiO_4$] Calcium magnesium silicate. Orthorhombic. Prismatic crystals; also disseminated grains. Colourless or grey. Vitreous lustre. Colourless streak. Uneven fracture. Good cleavage in one direction. Hardness 5·0–5·5. SG 3·03–3·25. Soluble in hydrochloric acid.
Occurrence: USA—Arkansas (Magnet Cove in the Ozark Mountains). Italy—Monte Somma. As grains in *calcite*.
Variety is *iron monticellite*.

MONTMORILLONITE

$Al_4Si_8O_{20}(OH)_4.nH_2O$ Basic hydrated aluminium silicate. Amorphous. Massive. Greyish-white, occasionally red or blue. Very soft.
Occurrence: Worldwide.
Alteration product of aluminium silicates.
Varieties are *beidellite*, *hectorite* and *saponite*.
Also synonym of *smectite*.

MONTREALITE

Igneous rock of the *ultrabasic* clan. A very dark, coarse-grained plutonic rock. Contains *pyroxene*, *hornblende* and *olivine*.

Occurrence: Canada—Quebec (Mount Royal near Montreal).
A type of *peridotite*.

MONTROSEITE

$4[(V,Fe)O.OH]$ Type of *diaspore*. Orthorhombic. Bladed crystals; also disseminated grains. Black. Submetallic lustre. Good cleavage in two directions. SG $c4\cdot0$.
Occurrence: USA—Colorado (Paradox Valley in Montrose County).

MONZONITE

Name given to plutonic igneous rocks containing *plagioclase* and *orthoclase* in roughly equal amounts. Forms the division between the *diorite* and the *syenite* clans.

MOONSTONE

Gem variety of *orthoclase*.

MORDENITE or PTILOLITE

$(Ca,Na_2,K_2)_4Al_8Si_{40}O_{96} \cdot 28H_2O$ Hydrated calcium sodium potassium aluminosilicate. Monoclinic. Tabular crystals; also massive. White, occasionally yellow or pink. Vitreous lustre. Uneven fracture. Perfect cleavage in one direction. Hardness $3\cdot0$–$4\cdot0$. SG $c2\cdot1$. Decomposes in hydrochloric acid.
Occurrence: USA—Wyoming (Hoodoo Mountain near the Yellowstone River).

Canada—Nova Scotia (Morden in Kings County).
Also synonym of *arduinite*.

MORENOSITE or NICKEL VITRIOL

$4[NiSO_4.7H_2O]$ Hydrated nickel sulphate. Orthorhombic. Crystals rare, normally forms crusts. Greenish-white to apple-green. Vitreous lustre. White streak. Conchoidal fracture. Good cleavage in one direction. Hardness $2\cdot0$–$2\cdot5$. SG $c1\cdot95$. Easily soluble in water.
Occurrence: USA—California (Julian in San Diego County), Pennsylvania (Gap Mine in Lancaster County). Canada—Ontario (Sudbury). Italy—Lombardy. Secondary mineral produced by the oxidation of nickel bearing sulphides.

MORGANITE

Synonym of *vorobyevite*.

MORION

Synonym of *'smoky' quartz*.

MOROXITE

Greenish-blue variety of *apatite*.

MOSS AGATE or MOCHA STONE

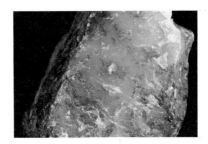

Variegated variety of *chalcedony*. Usually grey, bluish or milky stone. Contains branching inclusions of manganese oxide (black) or iron oxide (brownish or yellowish-red). Semi-precious stone.
Occurrence: Worldwide. Forms nodules.

MOSSITE

Cadmium rich variety of *tapiolite*.

MOTTRAMITE

$4[Pb(Cu,Zn)VO_4.OH]$ Basic lead copper zinc vanadate. Orthorhombic. Prismatic crystals; also massive. Brownish-red to blackish-brown. Greasy lustre. Orange to brownish-red streak. Conchoidal fracture. No cleavage. Hardness $3\cdot0$–$3\cdot5$. SG $c5\cdot9$. Easily soluble in hydrochloric acid.
Occurrence: England—Cheshire (Mottram St Andrews), Shropshire (Pin Hill near Shrewsbury). USA—Arizona

(Tombstone in Bisbee County), Montana (Silver Star district). Italy—Sardinia (Bena de Podnu near Sassaro). Secondary mineral of the oxidation zone of ore deposits.

MOUNTAINITE

$(Ca,Na,K_2)_{10}Si_{32}O_{80}.24H_2O$ Hydrated calcium sodium potassium silicate. Monoclinic. Fibrous. White.
Occurrence: South Africa—Kimberley (Bultfontein Mine).
A type of *zeolite*.

MUD

A mixture of *clay*-like material and water.

MUDSTONE or PERLITE

Argillaceous sedimentary rock. A type of *clay*, massive and structureless, which breaks irregularly.
Occurrence: Worldwide. A marine or a freshwater deposit.

MUGEARITE

Igneous rock of the *syenite* clan. A very fine-grained extrusive rock. Contains *oligoclase*, *orthoclase*, *olivine* and *augite*.
Occurrence: Scotland—Isle of Skye (Mugeary, Roineval).

MULLITE

$Al_6Si_2O_{13}$ Aluminium silicate. Orthorhombic. Prismatic crystals. Colourless or white. Perfect cleavage in one direction. Hardness 6·0–7·0. SG 3·15–3·26.
Occurrence: Scotland—Isle of Mull. In baked *shales*.

MUNIONGITE

Igneous rock of the *syenite* clan. A fine to medium-grained hypabyssal rock. Contains soda-*orthoclase*, *nepheline*, *aegirite* and *cancrinite*.
Occurrence: Australia—New South Wales (Muniong).

MURASAKITE

Low grade regional metamorphic rock. A fine-grained *schist*. Contains *piedmontite* and *quartz*.
Occurrence: Japan—Murasako.

MURCHISONITE

Red variety of *orthoclase*.

MURDOCHITE

$4[Cu_6PbO_8]$ Copper lead oxide. Cubic. Octahedral crystals. Black.
Occurrence: USA—Arizona (Mammoth Mine near Tiger).

MURMANITE

Sodium titanium silicate. Monoclinic. Scaly masses. Violet. Metallic lustre. Cherry-red Streak. Perfect basal cleavage. Hardness 2·0–3·0. Sg 2·84.
Occurrence: USSR—Kola Peninsula (Alluaiva).

MUSCOVITE

$4[KAl_3Si_3O_{10}(OH)_2]$ Hydrous potassium aluminium silicate. Monoclinic. Tabular crystals; also aggregates. Colourless or white. Pearly lustre. Colourless streak. No fracture. Perfect basal cleavage. Hardness 2·0–2·5. SG 2·76–3·0.
Occurrence: Worldwide. In *granites* and *pegmatites*.
A *mica*. Varieties are *adamsite*, *damourite*, *didymite*, *fuchsite*, *gilbertite*, *phengite*, *rose muscovite* and *sericite*.

MYLONITE

Dynamic metamorphic rock. Fine-grained flinty texture. Banded or streaked appearance. Formed by the shearing and granulation of rocks in severe dynamic metamorphism.
Occurrence: Widespread in areas of extreme regional metamorphism and thrusting.
Variety is *flinty crush-rock*.

MYLONITE GNEISS

Synonym of *augen-schist*.

MYRICKITE

Synonym of *chalcedony*.

MYRMEKITE

Coral like intergrowth of *quartz*, *orthoclase* and *microperthite*.
Occurrence: Austria—Fensteralp. In *granite gneiss*.

NACRITE

Synonym of *kaolinite* and *damourite*.

NAGATELITE

Formula uncertain. Hydrous cerium earths aluminium iron and calcium phosphosilicate. Monoclinic. Prismatic crystals. Black. Resinous lustre. Pale brown streak. Hardness 5·5. SG 3·91. Decomposes in hydrochloric acid to give a white precipitate.
Occurrence: Japan—Ishikawa (Nagatejima).
Pegmatite mineral.

NAGYAGITE

$8[Pb_5Au(Te,Sb)_4S_{5-8}]$ Lead gold tellurium and antimony sulphide. Monoclinic. Tabular crystals; also granular masses. Blackish-grey. Metallic lustre. Blackish-grey streak. Perfect cleavage in one direction. Hardness 1·0–1·5. SG c7·4. Soluble in nitric acid with a residue of gold.

Occurrence: USA—Colorado (Gold Hill in Boulder County), North Carolina (Kings Mountain Mine in Gaston County). Canada—Ontario (Huronian Mine). Western Australia—Kalgoorlie.

NAPOLEONITE

Synonym of *orthoclase* and *corsite*.

NASTURAN

Massive variety of *uraninite*.

NATIVE

Occurring in the free, uncombined state.

NATROALUNITE

$NaAl_3(SO_4)_2(OH)_6$ Basic sodium aluminium sulphate. Hexagonal. Tabular crystals; also granular masses. White. Vitreous lustre. White streak. Conchoidal fracture. Good cleavage in one direction. Hardness 3·5–4·0. SG 2·6–2·9. Slowly soluble in dilute sulphuric acid.
Occurrence: USA—Arizona (Sugarloaf Butte), California (Funeral Range in Death Valley), Colorado (Knickerbocker Hill in Custer County). Western Australia—Kalgoorlie. Chile—Salamanca.

NATROJAROSITE

$NaFe_3(SO_4)_2(OH)_6$ Basic sodium iron sulphate. Hexagonal. Earthy crusts. Yellow to brown. Vitreous lustre. Conchoidal fracture. Perfect cleavage in one direction. Hardness 3·0. SG c3·2. Slowly soluble in hydrochloric acid.
Occurrence: USA—Nevada (Soda Springs Valley in Esmeralda County), South Dakota (Buxton Mine in Lawrence County). Western Australia—Philips River Goldfield. Denmark—Zealand (Rifsnaes).
Oxidation zone of *pyrite* deposits.

NATROLITE

$Na_2Al_2Si_3O_{10} \cdot 2H_2O$ Hydrous sodium aluminium silicate. Orthorhombic. Prismatic crystals; also massive. Colourless or white. Vitreous lustre. Uneven fracture. Perfect cleavage in one direction. Hardness 5·0–5·5. SG 2·20–2·25. Gelatinises with hydrochloric acid.

Occurrence: Scotland—Fifeshire (Glen Farg). Northern Ireland—County Antrim (Glenarm, Magee Island, Portrush). USA—New Jersey (Bergen Hill), New York (New York Island). Canada—Nova Scotia (Gates Mountain). In cavities in *basalts* and related igneous rocks.
A *zeolite*. Varieties are *kondrikovite* and *metanatrolite*.

NATRON

$Na_2CO_3 \cdot lOH_2O$ Hydrated sodium carbonate. Monoclinic. Granular or columnar crusts. Colourless to white. Vitreous lustre. Conchoidal fracture. Good cleavage in one direction. Hardness 1·0–1·5. SG c1·5. Easily soluble in water. Effervescent in hydrochloric acid.
Occurrence: USA—California (Owens Lake), Nevada (Ragtown). Canada—British Columbia (Clinton in the Lillvet district). Italy—Mount Etna. In soda lakes.

NAUJAITE

Igneous rock of the *syenite* clan. A mottled, very coarse-grained granular plutonic rock. Contains alkali-*feldspar*, *nepheline*, *aegirite*, *arfvedsonite* and *eudialyte*.
Occurrence: Greenland—Julianehaab (Tunugdliarfik Fiord).

NAUMANNITE

Ag_2Se Silver selenide. Cubic. Cubes; also granular masses. Iron-black. Metallic lustre. Iron-black streak. Perfect cleavage in one direction. Hardness 2·5. SG c7·0.
Occurrence: USA—Idaho (De Lamar Mine at Silver City in Owyhee County). Germany—Harz Mountains (Tilkerode).

NAVITE

Igneous rock of the *gabbro* clan. A medium-grained porphyritic hypabyssal rock. Contains phenocrysts of *labradorite*, *olivine* and *augite* in a groundmass of *feldspar* and *augite*.
Occurrence: Germany—Nave Valley (Nave).

NELSONITE

Igneous rock of the *ultrabasic* clan. A medium-grained granular hypabyssal rock. Contains *ilmenite* and *apatite*.
Occurrence: USA—Virginia (Rose's Mill in Nelson County).

NEMALITE

Fibrous variety of *brucite* (*of Beudant*).

NEPHELINE

$8[NaAlSiO_4]$ Sodium aluminium silicate. Hexagonal. Prismatic crystals; also col-

umnar masses. Colourless, white or yellow. Vitreous or greasy lustre. Subconchoidal fracture. Good cleavage in one direction. Hardness 5·5–6·0. SG 2·55–2·65. Gelatinises with hydrochloric acid.
Occurrence: Worldwide. In *silica* poor soda rich igneous rocks.
A *feldspathoid*. Variety is *eleolite*.

NEPHELINE BASALT

Silica deficient variety of *basalt*. Contains *feldspathoids* in place of *feldspars*.
Occurrence: Worldwide.

NEPHELINE SYENITE

Silica deficient soda rich variety of *syenite*. Contains *feldspathoids*, alkali-*feldspars*, soda-*pyroxenes* and soda-*amphiboles*.
Occurrence: Worldwide.

NEPHELINITE

Igneous rock of the *gabbro* clan. A fine-grained, silica deficient extrusive rock. Contains *nepheline* and *augite*.
Occurrence: Worldwide.

NEPHRITE

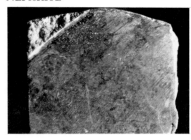

Spanish variety of *actinolite*. Also synonym of *saponite*.

NEPTUNITE

$8[(Na,K)_2(Fe,Mn)TiSi_4O_{12}]$ Sodium potassium iron manganese and titanium silicate. Monoclinic. Octahedral crystals. Black. Vitreous lustre. Perfect cleavage in one direction. SG c 3·23.

Occurrence: Greenland—near Igaliko (Narsasik).

NESQUEHONITE

$4[MgCO_3.3H_2O]$ Hydrated magnesium carbonate. Orthorhombic. Prismatic crystals; also radiating acicular masses. Colourless or white. Vitreous lustre. Splintery fracture. Perfect cleavage in one direction. Hardness 2·5. SG c 1·85. Easily soluble in hydrochloric acid with effervescence.
Occurrence: USA—Pennsylvania (Nesquehoning near Lansford in Carbon County). France—near Isère (La Mure).
D GEHYDRATION product of *lansfordite*.

NEVADITE

Igneous rock of the *granite* clan. A fine-grained porphyritic extrusive rock. Contains phenocrysts of *quartz, sanidine, biotite* and *hornblende* in a very fine groundmass.
Occurrence: USA—Colorado (Chalk Mountain).

NEWBOLDITE

Synonym of *marmatite*.

NEWLANDITE

Igneous rock of the *ultrabasic* clan. A medium-grained granular hypabyssal rock. Contains *garnet, enstatite* and *diopside*.
Occurrence: South Africa—Kimberley. In *diamond* pipes of *kimberlite*.

NICCOLITE or KUPFERNICKEL

$2[NiAs]$ Nickel arsenide. Hexagonal. Crystals rare, normally massive. Pale copper-red, tarnishing to greyish-black. Metallic lustre. Pale brownish-black streak. Brittle fracture. No cleavage. Hardness 5·0–5·5. SG c 7·8. Soluble in sulphuric acid.
Occurrence: USA—Colorado (Silver Cliff in Custer County), New Jersey (Franklin). Canada—Ontario (Sudbury). Germany—Saxony (Freiberg). In *norites*.
Important ore of nickel.

NICKEL BLOOM

Synonym of *annabergite*.

NICKEL VITRIOL

Synonym of *morenosite*.

NIGERITE

$(Zn,Fe,Mg)(Sn,Zn)_2(Al,Fe)_{12}O_{22}(OH)_2$ Basic zinc iron magnesium tin and aluminium oxide. Hexagonal. Platy crystals. Dark brown. Vitreous lustre. Brown streak. Brittle fracture. Poor basal cleavage.
Occurrence: Nigeria—Kabba Province (Egbe). In *pegmatite*.

NINGYOITE

$CaU(PO_4)_2.1-2H_2O$ Hydrated calcium uranium phosphate. Orthorhombic. Acicular crystals. Brownish-green to brown.
Occurrence: Japan—Tottori Prefecture (Ningyo-Toge Mine). In cavities in *pyrite* vein.

NITRATINE or SODA NITRE

2[$NaNO_3$] Sodium nitrate. Hexagonal. Crystals rare, normally granular masses or crusts. Colourless or white. Vitreous lustre. Colourless streak. Conchoidal fracture. Perfect cleavage in one direction. Hardness 1·5–2·0. SG 2·24–2·29. Easily soluble in water.
Occurrence: Worldwide as surface deposit in arid regions.

NITRE

4[KNO_3] Potassium nitrate. Orthorhombic. Crystals rare, normally thin earthy crusts. Colourless or white. Vitreous lustre. Colourless or white streak. Subcon-

Below: nodule (thunder egg)

choidal fracture. Perfect cleavage in one direction. Hardness 2·0. SG $c2·2$. Easily soluble in water.
Occurrence: Worldwide as a surface deposit in arid regions.

NITRIC ACID

HNO_3, a powerful testing agent.

NITROMAGNESITE

$Mg(NO_3)_2.6H_2O$ Hydrated magnesium nitrate. Monoclinic. Thin earthy crusts. Colourless or white. Vitreous lustre. Colourless or white streak. Uneven fracture. Perfect cleavage in one direction. SG 1·46. Easily soluble in water.
Occurrence: USA—Kentucky (Madison County). France—Jura Mountains (Salina). Efflorescence in *limestone* caves.

NODULAR

Regular, rounded, generally applied to certain concretions called nodules.

NODULES

Nodular concretions.

NONESITE

Igneous rock of the *gabbro* clan. A fine-grained porphyritic extrusive rock. Contains *labradorite*, *augite* and *enstatite*

phenocrysts in a groundmass of *plagioclase* and *augite*.
Occurrence: Austria—Tyrol (Cles in Nonsberg).

NON-FISSILE

A substance which cannot be split in a reproduceable fashion.

NONTRONITE or CHLOROPAL

A canary-yellow *clay mineral*.

NORBERGITE

4[$Mg(OH,F)_2.Mg_2SiO_4$] Hydrous magnesium oxide fluoride and silicate. Orthorhombic. Granular masses. Pinkish-white. Dull lustre. White streak. Hardness 5·5. SG 3·2. Soluble in warm hydrochloric acid.
Occurrence: USA—New Jersey (Franklin). Sweden—Norberg.

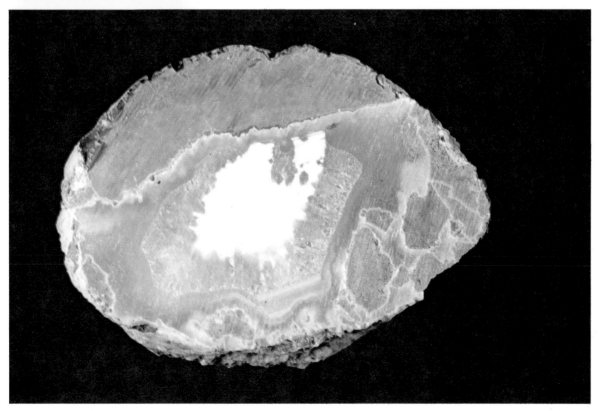

NORDMARKITE

Igneous rock of the *syenite* clan. A reddish, medium-grained granular plutonic rock. Contains *microperthite, oligoclase, biotite, hornblende* and *quartz*.
Occurrence: Norway—near Christiania (Tonsenås), near East Åker (Grorund), near Holmestrand (Hillisstadvand).

NORITE, HYPERSTHENFELS or HYPERSTHENITE

Igneous rock of the *gabbro* clan. A coarse-grained granular plutonic rock. Contains *labradorite* and orthorhombic *pyroxenes*.
Occurrence: USA—California (Plumas County), New York (Centreville, Cortlandt Township). Norway—Risör, Romsaas.

NORTHFIELDITE

Igneous rock of the *granite* clan. A white coarse-grained granular plutonic rock. Contains *quartz* and *muscovite*.
Occurrence: USA — Massachusetts (Northfield).
Also synonym of *esmeraldrite*.

NOSEAN or NOSELITE

$Na_8Al_6Si_6O_{24}SO_4$ Sodium aluminium silicate and sulphate. Cubic. Dodecahedral crystals; also granular masses. Grey, blue or brown. Subvitreous lustre. Variable

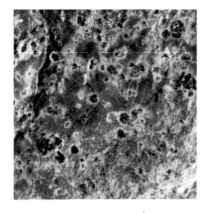

streak. Uneven fracture. Poor cleavage. Hardness 5·5. SG 2·25–2·4. Gelatinises with hydrochloric acid.
Occurrence: Worldwide in silica deficient alkali rich volcanic rocks.

NOSELITE
Synonym of *nosean*.

NOUMEITE
Dark green variety of *garnierite*.

NOVACEKITE

Magnesium enriched form of *autunite*.

Below: novacékite. *Right:* obsidian

NOVACULITE

Sedimentary *chert*. A white, very fine, even grained organic deposit. Contains *quartz* and other forms of silica.
Occurrence: USA—Arkansas (the palaeozoic rocks of the Ouachita Mountains).

OBSIDIAN

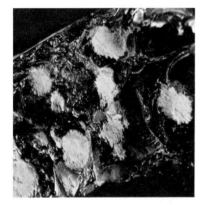

An igneous rock wholly composed of black, natural *glass*. Gemstone variety is *apache tears*.

OBSIDIANITE

Type of *tektite*.

OCTAHEDRAL

See *crystal*.

OCTAHEDRITE

Type of *meteorite*. Also synonym of *anatase*.

ODINITE or GABBROPHYRE

Igneous rock of the *gabbro* clan. A medium-grained porphyritic hypabyssal rock. Contains phenocrysts of *labradorite*, *augite* and *hornblende* in a groundmass of *feldspar* laths and *hornblende* needles.
Occurrence: Germany—Odenwald (Frankenstein). Rumania—Banat.

ODONTOLITE or BONE TURQUOISE

Fossil bone or tooth coloured by *vivianite* and partly replaced by *apatite* or *calcite*. Amorphous. Pale blue. Hardness 5·0. SG 1·8–3·2. Effervesces in hydrochloric acid.
Occurrence: France—Gers Department (Simmare). Bones of *Mastodon* and *Dinotherium*.

OIL SHALE

Fine black or dark brown *shale* containing kerogen.
Occurrence: Scotland—West Lothian (Scottish Oil Shales). USA—Colorado and Wyoming (Green River Shales). Canada—New Brunswick (Albert Shales).

OISANITE

Synonym of *anatase* and *axinite*.

OKKOLITE

Gem variety of *epidote*.

OLD RED SANDSTONE

See *geological time*.

OLIGOCENE

See *geological time*.

OLIGOCLASE, AVENTURINE or AVENTURINE FELDSPAR

A *plagioclase*. Triclinic. Crystals uncommon, normally massive. White, green or red. Vitreous lustre. Conchoidal or uneven fracture. Perfect cleavage in one direction. Hardness 6·0–7·0. SG 2·65–2·67.

Unaffected by acids.
Occurrence: Worldwide in igneous rocks. Variety is *sunstone*.

OLIGOCLASITE

Igneous rock of the *gabbro* clan. A coarse-grained granular plutonic rock. Contains *oligoclase* and minor *microcline*, *zircon*, *rutile*, *muscovite* and *hornblende*.
Occurrence: Norway—Ostvaagö Island (Preston).

OLIGONITE

Synonym of *oligon spar*.

OLIGON SPAR or OLIGONITE

Magnesium rich variety of *chalybite*.

OLIVENITE

$4[Cu_2(AsO_4)(OH)]$ Basic copper arsenate. Orthorhombic. Prismatic or acicular crystals; also fibrous or granular masses. Olive-green to brown. Adamantine or vitreous lustre. Olive-green to brown streak. Conchoidal fracture. Indistinct

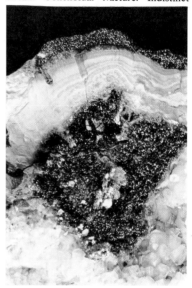

cleavage. Hardness 3·0. SG c4·4. Soluble in hydrochloric acid.
Occurrence: England—Cumberland (Alston Moor), Devonshire (Tavistock). USA—Nevada (Mojuba Hill in Pershing County), Utah (American Eagle Mine near Mammoth).
Oxidation zone of copper deposits.

OLIVINE

Mineral family with the general formula ·R_2SiO_4 where R equals magnesium, iron or manganese with or without calcium. Important members are *forsterite, fayalite, tephroite, monticellite, glaucochroite*. Solid solution series exist between *forsterite* and *fayalite, fayalite* and *tephroite*.
Varieties are *chrysolite* (*of Wallerius*) and *ferrite* (*of Heddle*).

OLIVINE ROCK

Synonym of *dunite*.

OLLENITE

High grade regional metamorphic rock. Schistose texture. Contains *hornblende, epidote, sphene, rutile* and *garnet*.
Occurrence: Italy—Piedmont (Col d'Ollen).

OMPHACITE

Variety of *augite*. A pale green granular mass. Occurs in *eclogites*.

ONKILONITE

Igneous rock of the *gabbro* clan. A steel-grey, fine-grained porphyritic extrusive rock. Contains phenocrysts of *olivine* in a groundmass of *augite, nepheline* and *leucite*.
Occurrence: Arctic Ocean—the Henriette Islands (Wilkitski Island).

ONYX

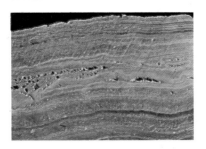

Evenly banded variety of *agate* used for jewellery.

ONYX MARBLE

Synonym of *alabaster*.

OOLITH

Rounded grains up to 2mm in diameter. Composed of calcium carbonate or *silica, hematite, chamosite* or *limonite*. Normally show radial or concentric structures around a nucleus of *quartz* or shell fragment.
Occurrence: Worldwide. Essential constituent of sedimentary oolitic rocks.

OOLITIC

Including or comprising spherical or nearly spherical particles (up to a few millimetres in diameter), with a resulting fish-roe appearance.
The 'ooliths' are usually formed by growth (often of *chalk* or *limestone*) round an existing nucleus.

OOLITIC IRONSTONE

Marine sedimentary rock. *Ironstone* formed of *ooliths* composed of *chamosite, limonite* or *hematite* in a fine-grained matrix of *calcite, chamosite, chalybite, limonite* or *hematite*.
Occurrence: Worldwide. Important ore of *iron*.

OOLITIC LIMESTONE

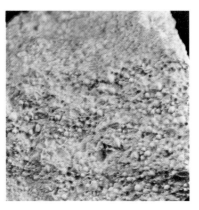

Limestone composed of *ooliths* of *aragonite* or *calcite* in a carbonate matrix.
Occurrence: Worldwide.

OOZE

Very fine-grained deep sea sedimentary deposit. Composed of shell debris from foraminiferans, diatoms, radiolarians or pteropods. Slow accumulation of organic material on the ocean floor with very little *argillaceous* material.
Occurrence: Ocean basins.

OPAL

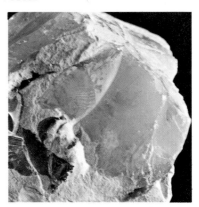

$SiO_2 . nH_2O$ Hydrous silica. Amorphous (crystalline aggregate of submicroscopic *cristobalite*). Crusts, concretions, veins or massive. Colourless or white, often tinted. Vitreous, resinous or pearly lustre. White streak. Hardness 5·5. SG 1·8–2·25.
Occurrence: Worldwide.
Varieties are *fire opal, hyalite, hydrophane,*

Right: opal

menilite, opal agate, opal jasper, wax opal and wood opal.

OPAL AGATE

Variety of *opal* displaying an *agate*-like banding of different colours.

OPAL JASPER

Yellow variety of *opal* with inclusions of iron oxide.

OPHICALCITE

Type of *marble* containing *serpentine*.

Above: ophicalcite

OPHIOLITE

A massive and mottled variety of *serpentine*.

OPHITIC

With a structure of laths of one mineral fully or partially enclosed by large crystals of another.

ORANGITE

Yellowish-orange variety of *thorite*.

ORBICULAR

With a structure of concentric spheres or near-spheres, each layer having a different composition; sometimes with a nucleus.

ORBICULAR GABBRO

Gabbro with orbicular structure due to concentric growth of *plagioclase* and *amphibole* layers around a core of normal gabbroic texture.
Occurrence: USA—California (San Diego County). Canada—Ontario (Kenara). Italy—Corsica (*corsite*). Norway— Romsaas.

ORBICULAR GRANITE

Granite with orbicular structure. Composed of orbs of concentric *quartz* and *mica* in a feldspathic matrix.
Occurrence: Finland—Rapakivi.

ORBICULAR JASPER

Variety of *jasper*. Composed of white, red or black orbs in a white, red or yellow matrix.
Occurrence: USA—California (Morgan Hill in Santa Clara County).

ORDANCHITE

Igneous rock of the *diorite* clan. A fine-grained porphyritic extrusive rock. Contains phenocrysts of *labradorite* and *andesine* in a groundmass of *hauynite*, *augite* and *hornblende*.
Occurrence: France—Auvergne (Laqueville).

ORDOVICIAN

See *geological time.*

ORE

Rock including ore minerals, those suitable for the extraction of required substances. Also includes country rock and gangue minerals.

ORENDITE

Igneous rock of the *syenite* clan. A reddish-grey, fine-grained porphyritic vesicular extrusive rock. Contains phenocrysts of *phlogopite* in a groundmass of *leucite* and *sanidine.*
Occurrence: USA—Wyoming (Leucite Hills).

ORIENTAL ALABASTER

Old usage for stalagmitic variety of *calcite.*

ORIENTAL AMETHYST

A purple gemstone variety of *corundum.*

ORIENTAL EMERALD

A green gemstone variety of *corundum.*

ORIENTAL TOPAZ

A yellow gemstone variety of *corundum.*

OROGENESIS

Mountain-building.

ORÖITE

Igneous rock of the *diorite* clan. A medium-grained granular plutonic rock. Contains *hornblende, plagioclase, microcline, diopside, biotite* and *quartz.*
Occurrence: Sweden—Ornö Hufvud.

ORPIMENT

$4[As_2S_3]$ Arsenic sulphide. Monoclinic. Small prismatic crystals; also columnar or fibrous masses. Lemon-yellow. Pearly lustre. Pale lemon-yellow streak. Irregular fracture. Perfect cleavage in one direction. Hardness $1·5–2·0$. SG $c3·5$. Soluble in sulphuric acid.
Occurrence: USA—Nevada (Manhatten in Nye County), Utah (Mercur in Tooele County). France—Maritime Alps (Luceram). Germany—Saxony (St Andresberg).
Mineral of low temperature veins and hot springs.

ORTHITE

Synonym of *allanite.*

ORTHOCLASE or NAPOLEONITE

4[KAlSi$_3$O$_8$] Potassium aluminium silicate. Monoclinic. Prismatic or tabular crystals, often granular masses. White, pink, yellow or brown. Vitreous lustre. Conchoidal fracture. Good cleavage in two directions. Hardness 6·0. SG 2·6. Insoluble in acids.
Occurrence: Worldwide in igneous rocks and as a detrital mineral in sediments. A *feldspar*.
Varieties are *adulasia*, *erythrite* (*of Thomson*), *moonstone* and *Murchisonite*.

ORTHOCLASE MICROPERTHITE

Variety of *microperthite* with a high *orthoclase* content.

ORTHOCLASITE or ORTHOSITE

Igneous rock of the *syenite* clan. A medium to fine-grained granular hypabyssal rock. Contains *orthoclase* and minor ferromagnesian minerals.
Occurrence: USA—Alaska (Kasaan).

ORTHOGNEISS

Gneiss derived from igneous rocks by regional metamorphism.

ORTHORHOMBIC

See *crystal*.

ORTHOSITE

Synonym of *orthoclasite*.

OSMIRIDIUM

Synonym of *iridosmine*.

OSSANNITE

Synonym of *riebeckite*.

OSSIPITE or OSSYPITE

Igneous rock of the *gabbro* clan. A coarse-grained granular plutonic rock. Contains *labradorite, olivine, pyroxene* and *magnetite*.
Occurrence: USA—New Hampshire (Waterville).

OSSYPITE

Synonym of *ossipite*.

OSTEOLITE

Massive impure variety of *apatite*.

OSUMILITE

(K, Na, Ca)(Mg, Fe)$_2$(Al, Fe, Fe)$_3$(Si. Al)$_{12}$ O$_{30}$.H$_2$O Hydrated potassium sodium calcium magnesium iron and aluminium aluminosilicate. Hexagonal. Prismatic or tabular crystals. Black. Vitreous lustre. Black streak. Uneven fracture. No cleavage.
Occurrence: Germany—Laacher See. Japan—Kyusyu (Sakurazima Volcano near Sakkabira). In cavities in extrusive igneous rocks.

OTTAJANITE

Igneous rock of the *diorite* clan. A fine-grained porphyritic extrusive rock. Contains phenocrysts of *augite* and *leucite* in a groundmass of calcic *plagioclase, leucite, augite, sanidine* and *nepheline*.
Occurrence: Italy—Monte Somma, Mount Vesuvius.

OTTRELITE

Manganese rich variety of *chloritoid*. Also synonym of *diallage*.

OTTRELITESCHIST

Type of *schist* characterised by abundant crystals of *ottrelite*.

OUACHITITE

Igneous rock of the *ultrabasic* clan. A very coarse-grained porphyritic hypabyssal rock. Contains phenocrysts of *biotite* and *augite* in a groundmass of *analcime glass* and *augite*.
Occurrence: USA—Arkansas (Ouachita Mountains).

OUENITE

Igneous rock of the *gabbro* clan. A fine-grained granular hypabyssal rock. Contains *anorthite, diopside, olivine* and *bronzite*.
Occurrence: New Caledonia—Island of Ouen.
Forms dykes in *peridotite*.

OXIDATION

Chemical combination of a substance with oxygen, giving various oxides. (More generally, any reaction in which atoms of the element lose electrons.)

OXYHORNBLENDE

Synonym of basaltic *hornblende*.

OZARKITE

Massive variety of *thomsonite*.

OZOKERITE

Beeswax-like variety of *bitumen*.
Occurrence: USA—Utah.

PADPARADSCHAH

Orange gem variety of *corundum*.

PAHOEHOE or DERMOLITH

Type of *basalt* lava flow with a smooth billowy surface.
Occurrence: USA—Hawaii.

PAISANITE

Igneous rock of the *granite* clan. A fine-grained porphyritic hypabyssal rock. Contains phenocrysts of *quartz* and *riebeckite* in a groundmass of *quartz* and *microperthite*.
Occurrence: USA—Massachusetts (Magnolia in Essex County), Texas (Paisano Pass).
Variety is *dahamite*.

PALAEOZOIC

See *geological time*.

PALAGONITE

Yellow or orange devitrified hydrated basaltic *glass*.
Occurrence: Worldwide. Fills cavities in *basalt* or forms *palagonite tuff*.

PALAGONITE TUFF

Igneous rock of the *gabbro* clan. A brown, fine-grained *sandstone* like *tuff*. Contains *palagonite*, *augite* and *olivine* fragments and *feldspar* microliths.
Occurrence: Sicily. The Canary Islands. The islands of the Pacific.

PALLASITES

Type of *meteorite*.

Below: paligorskite

PALYGORSKITE or ATTAPULGITE

A rare group of *clay minerals*.
Variety is *pilolite*.

PAN

A basin or depression in the ground containing minerals, normally deposited by a process of evaporation.

PANDERMITE

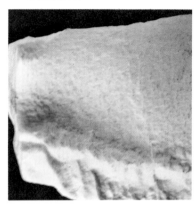

A hard, white, earthy variety of *colemanite*.

PANTELLERITE

Igneous rock of the *granite* clan. A fine-grained porphyritic extrusive rock. Contains phenocrysts of *aegirine-augite*, *diopside*, *anorthoclase* and *cossyrite* in a groundmass of *quartz*, *feldspar* and *aegirine*.
Occurrence: Italy—South of Sicily (Island of Pantelleria).
Leucocratic variety is *comendite*.

PARACELSIAN

Orthorhombic phase of *celsian*.

Above: paracelsian

PARAGNEISS

Gneiss derived from clastic sedimentary rocks by regional metamorphism.

PARAGONITE

4[NaAl$_3$Si$_3$O$_{10}$(OH)$_2$] Basic sodium aluminium silicate. Monoclinic. Normally massive. Colourless or pale yellow. Pearly lustre. No fracture. Perfect basal cleavage. Hardness 2·5–3·0. SG c2·85.
Occurrence: Germany—Saxony (Ochsenkopf near Schwarzenberg). Switzerland—Tessin Canton (Monte Campiane).

PARAMORPHISM

The conversion of a mineral from one distinct form to another as a result of a change in appearance only.

PARAWOLLASTONITE

Monoclinic phase of *wollastonite*.

PARGASITE

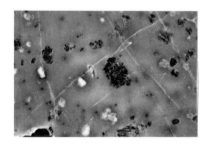

NaCa$_2$Mg$_4$(Al, Fe)(Si$_6$Al$_2$O$_{22}$)(OH, F)$_2$ Basic sodium calcium magnesium aluminium and iron fluo-aluminosilicate. Monoclinic. Tabular crystals. Colourless or light brown. Vitreous lustre. Uneven fracture. Good cleavage in one direction. Hardness 5·0–6·0. SG 3·05. Insoluble in hydrochloric acid. Variety of *ederite*.
Occurrence: Widespread in igneous and metamorphic rocks.
A *hornblende*.

PARKERITE

Ni$_3$(Bi,Pb)$_2$S$_2$ Nickel bismuth lead sulphide. Monoclinic. Platy crystals; also grains. Bronze. Metallic lustre. Black streak. Perfect cleavage in one direction. Hardness 3·0. SG c8·74.
Occurrence: Canada—Ontario (Sudbury). South Africa—Griqualand (Insizwa Range).

PARSETTENSITE

Hydrous manganese silicate. Monoclinic. Normally massive. Copper-red. Metallic lustre. Perfect basal cleavage. SG c2·6. Soluble in hot hydrochloric acid with the separation of gelatinous silica.
Occurrence: Switzerland—Val d'Err (Alp Parsettans).

PATRONITE

VS$_4$ Vanadium sulphide. Dark green.
Occurrence: Peru—near Cerro de Pasco (Minas Ragra). Extracted from vanadium bearing layer interbedded with *shales*.

PAULINGITE

Potassium calcium aluminosilicate. Cubic. Perfect rhombic crystals. Colourless. Vitreous lustre. Colourless streak. Uneven fracture. No cleavage. Hardness 5·0.
Occurrence: USA—Washington (Rock Island Dam at Wenatchee). In vesicular *basalt*.
A *zeolite*.

PEACH

Name for massive *ripidolite* in Cornwall, England.

PEACOCK ORE

Descriptive name used for tarnished *chalcopyrite*. Also synonym of *bornite*.

PEA IRON-ORE

A pisolitic variety of *limonite*.

PEARCITE

An arsenical variety of *polybasite*.

PEARL SPAR

A white, grey, pale yellowish or brownish variety of *dolomite* with a pearly lustre. Often associated with *galena* and *blende*.

PEARLSTONE

Synonym of *perlite*.

PEAT

Dark brown or black material produced by

the partial decomposition and disintegration of plant life in marshes.
Occurrence: Worldwide in humid climates. Used as fuel.

PEBBLE

A fragment of rock with a particle size between 4mm and 65mm.

PEBBLE BED

Sandstone bed containing pebbles.
Occurrence: England—Devonshire (Budleigh Salterton Pebble Bed), Midlands (Bunter Pebble Beds).

PEBBLE PHOSPHATES

Sedimentary phosphate beds formed by the erosion of pre-existing phosphate deposits by fluvial or marine action. Eroded fragments normally well rounded and redeposited as terraces.
Occurrence: England—Cambridgeshire (Cambridge Greensand). Scoland—Fife (Anstruther). USA—Wyoming (Bighorn Valley).

PECTOLITE

2[NaCa$_2$Si$_3$O$_8$OH] Basic sodium calcium silicate. Monoclinic. Acicular crystals. Whitish-grey. Silky lustre. White streak. Uneven fracture. Perfect cleavage in two directions. Hardness 5·0. SG 2·68–2·78.

Decomposes in hydrochloric acid with the separation of silica.
Occurrence: Scotland—Ayrshire (Knockdolian Hill). USA—New Jersey (Bergen Hill). Greenland—Disco Island. In basic lavas.
Variety is *schizolite*.

PEGMATITE

Igneous rock of the *granite* clan. A coarse-grained rock. Contains *orthoclase, quartz, microcline microperthite* and many possible ferromagnesian constituents.
Occurrence: Worldwide. Forms dykes, veins and irregular pockets. End product of *granite* magma differentiation.

PELITE

Synonym of *mudstone*.

PENCATITE

Synonym of *predazzite*.

PENNANTITE

$Mn_9Al_6Si_5O_{20}(OH)_{16}$ Basic manganese aluminium silicate. Monoclinic. Normally scales. Orange. Good basal cleavage.
Occurrence: Wales—Caernarvonshire (Benallt Mine).
A *chlorite*.

PENNINE or PENNINITE

$2[(Mg, Fe, Al)_6(Si, Al)_4O_{10}(OH)_8]$ Hydrated magnesium aluminium iron aluminosilicate. Monoclinic. Thick tabular crystals; also massive. Shades of green.

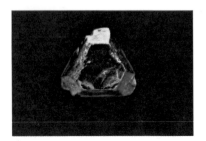

Pearly lustre. No fracture. Perfect cleavage in one direction. Hardness 2·0–2·5. SG 2·6–2·85. Decomposes in sulphuric acid.
Occurrence: USA—Pennsylvania (Lancaster County). Switzerland—Zermatt. In *serpentines* and metamorphic rocks.
A *chlorite*. Varieties are *kämmeresite* and *pseudophite*.

PENNINITE

Synonym of *pennine*.

PENTLANDITE

$4[(Fe,Ni)_9S_8]$ Iron nickel sulphide. Cubic. Granular masses. Bronze-yellow. Metallic lustre. Bronze-brown streak. Conchoidal fracture. No cleavage. Hardness 3·5–4·0. SG 4·6–5·0.
Occurrence: USA—Alaska (Key West Mine in Clark County). Canada—British Columbia (Yale Mine on Emory Creek), Ontario (Sudbury). South Africa—Transvaal (Rustenberg). In basic and *ultrabasic* rocks.
Important ore of nickel.

PERICLASE

$4[MgO]$ Magnesium oxide. Cubic. Crystals rare, normally irregular grains. Colourless or greyish-white. Vitreous lustre. White streak. Irregular fracture. Perfect cleavage in one direction. Hardness 5·5. SG c3·56. Easily soluble in dilute hydrochloric acid.
Occurrence: USA—California (Crestmore), New Mexico (Organ Mountains). Italy—Monte Somma, Sardinia (Teulado). Sweden—Nordmark. High temperature

metamorphic mineral found in *marbles*. Varieties are *magnesiowüstite* and *metabrucite*.

PERICLINE

Opaque white variety of *albite* found in chloritic *schists* in the Alps.

PERIDOT

Yellowish-green gem variety of *tourmaline*. Also synonym of *chrysolite* (*of Wallerius*).

PERIDOTITE

Igneous rock of the *ultrabasic* clan. A fine to coarse-grained granular plutonic rock. Contains *olivine, amphibole, biotite* and *pyroxene*.
Occurrence: Worldwide.

PERISTERITE

Iridescent variety of *albite*.

PERKNITE

Igneous rock of the *ultrabasic* clan. A fine to coarse-grained granular plutonic rock. Contains monoclinic *pyroxenes* and *amphiboles*.
Occurrence: Worldwide.

PERLITE or PEARLSTONE

Igneous rock of the *granite* clan. A colourless, or shades of grey and black, extrusive rock. Glassy with concentric partings.
Occurrence: Italy—Eiyanean Hills (Monte Menone), Sardinia—Monte Arci at Punta Brenta).

PERMEABLE

Capable of being penetrated by water or other liquid.

PERMIAN

See *geological time.*

PERMO-TRIAS

The Permian and Triassic Systems.

PEROVSKITE

8[$CaTiO_3$] Calcium titanium oxide. Cubic or monoclinic. Striated cubes or granular masses. Black, greyish-black, brownish-black, reddish-black or yellow. Metallic adamantine lustre. Colourless or grey streak. Uneven fracture. Indistinct cleavage. Hardness 5·5. SG $c4·0$. Decomposes in hot sulphuric acid.
Occurrence: USA—Arkansas (Magnet Cove in the Ozark Mountains), New York (Syracuse). Germany—Baden (Kaiserstuhl). Switzerland—Zermatt.
Varieties are *dysanalyte, knopite* and *loparite.*

PERTHITE

Coarse intergrowth of *orthoclase* and *albite.*

PETALITE

4[$LiAl(Si_2O_5(_2)$] Lithium aluminium sili-

cate. Monoclinic. Crystals rare, normally foliated masses. Colourless, white of grey. Vitreous lustre. Colourless streak. Conchoidal fracture. Perfect cleavage in one direction. Hardness 6·0–6·5. SG 2·39–2·46. No reaction with acids.
Occurrence: USA—Main (Peru), Massachusetts (Bolton). Canada—near Toronto (York). Sweden—Utö.
Variety is *castor.*

PETRIFIED WOOD

Cryptocrystalline *quartz* pseudomorphed after wood. Replacement of cell structures

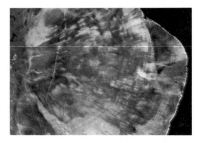

preserves original wood texture in *agate* or *opal* (*wood opal*).
Occurrence: World-wide.

PETZITE

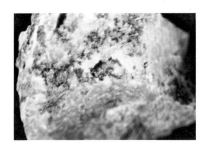

(Ag,Au)$_2$Te Silver gold telluride. Cubic. Normally granular masses. Steel-grey to iron-black. Metallic lustre. Subconchoidal fracture. Good cleavage in one direction. Hardness 2·5–3·0. SG 8·7–9·02. Decomposes in nitric acid with a residue of *gold*.
Occurrence: USA—California (Stanislans Mine in Calavera County), Colorado (Goldhill in Boulder County). Canada—Ontario (Hollinger Mine at Timmins). Western Australia—Kalgoorlie.

PHACOLITE

Colourless variety of *chabazite*.

PHARMACOLITE

4[CaHAsO$_4$.2H$_2$O] Hydrated calcium acid arsenate. Monoclinic. Crystals rare, normally silky fibres or acicular clusters. White or grey. Vitreous lustre. Uneven fracture. Perfect cleavage in one direction.

Hardness 2·0–2·5. SG 2·53–2·73. Readily soluble in acids.
Occurrence: USA—California (OK Mine in San Gabriel Canyon in Los Angeles County), Nevada (White Caps Mine at Manhatten). France—Alsace (Sainte-Marie-aux-Mines). Germany—Black Forest (Wittichen).

PHARMACOSIDERITE

(H,K)Fe$_4$(AsO$_4$)$_3$(OH)$_4$. nH$_2$O Hydrated basic iron arsenate. Cubic. Tetrahedral crystals. Variable. Adamantine lustre. Green or brown streak. Uneven fracture. Poor cleavage. Hardness 2·5. SG 2·9–3·0. Soluble in hydrochloric acid.
Occurrence: England—Cornwall (Wheal Gorland, Wheal Jane and Wheal Unity). USA—Utah (Tintic district). Germany—Saxony (Schneeberg).

PHENACITE or PHENAKITE

6[Be$_2$SiO$_4$] Beryllium silicate. Orthorhombic. Lenticular crystals. Colourless, yellow or pale red. Vitreous lustre. Conchoidal fracture. Good cleavage in one direction. Hardness 7·5–8·0. SG 2·97–3·0.
Occurrence: USA—Colorado (Topaz Butte near Flaissant), New Hampshire (Bald Face Mountain near North Chatham). Switzerland—Valais.

PHENAKITE

Synonym of *phenacite*.

PHENOCRYST

Larger crystal as found in a mass of smaller crystals in igneous rocks.

PHILLIPSITE, BONITE or SPANGSITE

(Ca,Na,K)$_3$(Al$_3$Si$_5$O$_{16}$).6H$_2$O Hydrated calcium sodium potassium aluminosilicate. Monoclinic. Normally radiating tufts or spheres of fine crystals. White. Vitreous lustre. Colourless streak. Uneven fracture. Good cleavage in two directions.

Above: phenocrysts (feldspar in granite)
Below: phillipsite

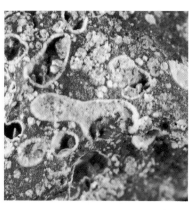

Hardness 4·0–4·5. SG c2·2. Gelatinises with hydrochloric acid.
Occurrence: Northern Ireland—County Antrim (Giant's Causeway). Italy—Monte Somma. Sicily (Aci Castello). A *zeolite*.

Below: phlogopite

PHILLIPSITE (of Beudant)

Synonym of *bornite*.

PHLOGOPITE

$4[KMg_3AlSi_3O_{10}(OH)_2]$ Hydrated potassium magnesium aluminium silicate. Monoclinic. Tabular crystals. Yellowish-brown to brownish-red. Pearly lustre. No fracture. Perfect basal cleavage. Hardness 2·5–3·0. SG 2·78–2·85. Decomposes in sulphuric acid with silica residue.
Occurrence: Worldwide in *serpentines* and *crystalline limestones*.
A *mica*.

PHONOLITE

Igneous rock of the *syenite* clan. A fine-grained porphyritic extrusive rock. Contains phenocrysts of *sanidine* in a groundmass of *sanidine*, *nepheline*, *nosean* and *aegirine*.
Occurrence: South Africa—Pretoria (Leeuwkraal). Czechoslovakia—Bohemia (Bilin). Italy—Rome (Monte di Cuma).

PHOSGENITE

$4[Pb_2CO_3Cl_2]$ Lead chlorocarbonate. Tetragonal. Prismatic or tabular crystals; also granular masses. Yellowish-white to yellowish-brown, pale brown or smoky brown. Adamantine lustre. White streak. Conchoidal fracture. Good cleavage in two directions. Hardness 2·0–3·0. SG $c6·1$. Soluble with effervescence in dilute nitric acid.
Occurrence: England—Derbyshire (Matlock). USA—Colorado (Terrible Mine in Custer County), Massachusetts (Southampton Mine in Hampshire County). Australia—New South Wales (Broken Hill).
Secondary alteration mineral of *galena*.

PHOSPHATIC NODULES

Sedimentary product of marine diagenesis. A grey, brown or black, earthy, organic phosphate deposited around a nucleus, normally a fossil.
Occurrence: South East England—the Gault. Scotland—Fifeshire (Anstruther near Fife).

PHOSPHOCHALCITE

Synonym of *pseudomalachite*.

PHOSPHORITE

Massive variety of *apatite*.

PHOSPHOROCHALCITE

Synonym of *pseudomalachite*.

PHOSPHURANYLITE

Hydrated uranium calcium phosphate. Tetragonal. Earthy crusts. Deep yellow to golden-yellow. Hardness 2·5. Easily soluble in acids.
Occurrence: USA—Maine (Newry in Oxford County), New Hampshire (Palermo near North Groton), North Carolina (Flat Rock in Mitchell County). Germany—Bavaria (Wölsendorf). Portugal—Sabugal (Rosmaneira).

PHOTICITE

Carbonate rich variety of *rhodonite*.

PHYLLITE

Medium grade regional metamorphic rock. Lustrous sheen and schistose texture. Contains *clay minerals* plus *chlorite* and *mica*.
Occurrence: Worldwide in areas of regionally metamorphosed *argillaceous rocks*.

PHYSALITE

Synonym of *pyrophysalite*.

PICOTITE or CHROME-SPINEL

Chromium rich variety of *spinel*.

PICRANALCIME

Magnesium bearing variety of *analcite*.

PICRITE

Igneous rock of the *ultrabasic* clan. A medium to fine-grained granular extrusive rock. Contains *olivine*, *augite*, *hornblende*, *biotite* and *magnetite*.
Occurrence: Germany—Nassau (Trinjenstein), Rhineland (Medenbach).

PICRITE (of Brongniart)

Synonym of *dolomite*.

PICROLITE

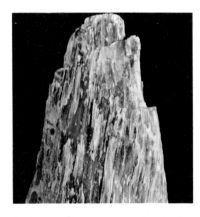

Dark green columnar variety of *antigorite*.

PICROTEPHROITE

Magnesium rich variety of *tephroite*.

PIEDMONTITE

Synonym of *piemontite*.

PIEMONTITE or PIEDMONTITE

$2[Ca_2(Al,Fe,Mn)_3Si_3O_{12}OH]$ Basic calcium aluminium iron and manganese silicate. Monoclinic. Prismatic crystals; also massive. Reddish-brown or reddish-black. Vitreous lustre. Red streak. Uneven fracture. Perfect cleavage in one direction.

Hardness 6·5. SG c3·4. Unaffected by acids.
Occurrence: France—Brittany (Ile de Croix). Italy—Piedmont (St Marcel).

PIENAARITE

Igneous rock of the *syenite* clan. A medium-grained granular plutonic rock. Contains *anorthoclase, titanite, aegirine-augite* and *nepheline*.
Occurrence: South Africa—Pretoria (Leeuwfontein).

PIETERSITE

Red gem variety of *crocidolite* with *limonite*.

PIGEONITE

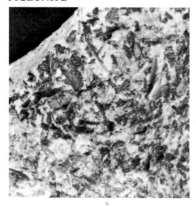

8[(Mg,Fe,Ca)SiO₃] Magnesium iron calcium silicate. Monoclinic. Tabular crystals; also grains. Colourless, brown or black. Good cleavage in one direction. Hardness 6·0. SG 3·3–3·46. Insoluble in hydrochloric acid.
Occurrence: Scotland—Isle of Mull. USA—Minnesota (Pigeon Point), New Jersey (Lambertville), Virginia (Goose Creek). Japan—Hakone. In lava flows and minor intrusions.

PILLOW LAVA

Basalt, andesite or *spilite* lava flows which display pillow structure. Rounded masses of lava with a smooth fine grained surface and a coarser-grained interior. The rounded masses interlock and overlie one another. Result of underwater deposition.

PILOLITE

Fibrous variety of *palygorskite*.

PINACOID, -AL

Major crystal face, e.g. basal, or pair of crystal faces each parallel to a major axis.

PINITE

Hydrous aluminium potassium silicate. Amorphous. Granular masses. Greyish-white or green. Waxy lustre. Hardness 2·5–3·5. SG 2·6–2·85.
Occurrence: Widespread alteration product of *cordierite, spodumene, scapolite* and *feldspar*. Usually admixed with *clay minerals*.

PINOLITE

Medium grade regional metamorphic rock. Contains *magnetite* aggregates in a schistose matrix.
Occurrence: Worldwide in area of regional metamorphism of iron enriched *argillaceous rocks*.

PIPE

Tube-like body of igneous rock.

PISOLITE

Name given to *ooliths* with a diameter greater than 2mm.

PISOLITIC IRONSTONE

Oolitic *ironstones* containing *ooliths* greater than 2mm in diameter.

PISOLITIC LIMESTONE

Oolitic *limestone* containing *ooliths* greater than 2mm in diameter.

Pisolite. *Below:* pisolitic limestone

PISTACITE

Synonym of *epidote*.

PISTOMESITE

Magnesium rich variety of *chalybite*.

PITCH

The angle between horizontal and the axis of a fold.

PITCHBLENDE

Massive variety of *uraninite*.

PITCHSTONE

Igneous rock of the *granite*, *diorite* or *syenite* clan. A black, grey, green, brown or red extrusive rock. Dense *glass* with a pitchy lustre and conchoidal fracture.
Occurrence: Scotland—Arran. Germany—Saxony (Meissen). Iceland—Berufjord.
Variety is *cantalite*.

PITCHY

Resinous, i.e. dull but definitely shiny.

PLACER DEPOSITS

Rudaceous or *arenaceous* sedimentary deposits produced by the erosion of rocks containing precious metals or gems. Heavy minerals concentrated by erosion and redeposited as placers. May be either fluvial or marine.
Occurrence: England—Cornwall (Cape Cornwall—*cassiterite*, the Lizard—*platinum*). Scotland—Caithness (*gold*), Sutherland (*gold*). USA—Alaska (*gold*), California (*gold*). South Africa—Orange Free State (*diamonds*), Vaal River (*diamonds*).

PLAGIAPLITE

Igneous rock of the *syenite* clan. A medium-grained granular hypabyssal rock. Contains *oligoclase*, *hornblende*, *quartz*, *biotite* and *muscovite*.
Occurrence: USSR—Ural Mountains (Koswa Range).

PLAGIOCLASE FELDSPARS

Soda lime *feldspars* forming a solid solution series from $NaAlSi_3O_8$ to $CaAl_2Si_2O_8$. Individual minerals are *albite*, *oligoclase*, *andesine*, *labradorite*, *bytownite* and *anorthite*.

PLANE, CRYSTAL

See *crystal*.

PLANT BEDS

Fine-grained sedimentary rocks containing the fossil remains of plants.
Occurrence: Widespread, particularly in Tertiary and younger deposits.

PLASMA

Opaque-green microgranular or microfibrous variety of *heliotrope*.

PLASTIC

Able to undergo deformation without fracture and without returning to original form.

PLATE

See *crystal*.

PLATELETS

See *crystal*.

PLATINIFEROUS

Platinum-bearing.

PLATINUM

4[Pt] Cubic grains. Whitish steel-grey to dark-grey. Metallic lustre. Hackly fracture. No cleavage. Hardness 4·0–4·5. SG 14·0–19·0. Soluble in hot sulphuric acid.

Occurrence: Republic of Ireland—County Wicklow. USA—California (Oroville in Butte County), North Carolina (Rutherford County).

PLATTNERITE

2[PbO_2] Lead dioxide. Tetragonal. Prismatic crystals; also massive. Black or brownish-black. Metallic adamantine lustre. Chestnut-brown streak. Conchoidal fracture. No cleavage. Hardness 5·5. SG 9·42. Easily soluble in hydrochloric acid with the evolution of chlorine.
Occurrence: Scotland—Dumfriesshire (Wanlockhead). USA—Idaho (Morning Mine at Mullon in Shoshone County).

PLATY

See *crystal*.

PLAUENITE

Igneous rock of the *syenite* clan. A coarse-grained granular plutonic rock. Contains *orthoclase*, *oligoclase*, *hornblende*, *sphene* and *apatite*.
Occurrence: Germany—Dresden (Plauen).

PLAYA CONGLOMERATE

Type of *conglomerate* deposited torrentially in desert basins during flooding of arid areas.

PLAZOLITE

Hydrated calcium aluminium silicate and carbonate. Cubic. Cubic crystals; also massive. Colourless. Vitreous lustre. Conchoidal fracture. No cleavage. Hardness 6·5. SG c3·1. Easily soluble in hydrochloric acid.

Occurrence: USA—California (Crestmore). In metamorphosed *limestone*.

PLEONASTE

Synonym of *ceylonite*.

PLUMASITE

Igneous rock of the *diorite* clan. A medium-grained granular hypabyssal rock. Contains *oligoclase* and *corundum*.
Occurrence: USA—California (Spanish Peak in Plumas County).

PLUMBOGUMMITE or BISCHOFITE (of Fischer)

$PbAl_3(PO_4)_2(OH)_5.H_2O$ Basic hydrated lead aluminium phosphate. Hexagonal. Crystals rare, normally botryoidal or reniform masses. Greyish-white, yellowish-grey or yellow. Dull to resinous lustre. Colourless or white streak. Uneven fracture. No cleavage. Hardness 4·5–5·0. SG $c4·0$. Soluble in hot hydrochloric acid.
Occurrence: England—Cumberland (Dry Gill and Roughton Gill near Caldbeck). USA—Missouri (Mine la Motte). France—Brittany (Huelgoat). Secondary mineral in lead deposits.

PLUMBOJAROSITE

$PbFe_6(SO_4)_4(OH)_{12}$ Basic lead iron sulphate. Hexagonal. Earthy crusts. Golden-brown to dark brown. Dull lustre. Pale brown streak. No fracture. Poor cleavage. Hardness $c1·0$. SG $c3·6$. Very slowly soluble in hydrochloric acid.
Occurrence: USA—Nevada (Boss Mine in Clark County), New Mexico (San Jose Mine in Grant County). Turkey—Anatolia (Bolkardaz). Secondary mineral of lead deposits.

PLUMOSE

Feather-like.

PLUTONIC

Igneous intrusions or other bodies solidifying deep in the crust.

PNEUMATOLYSIS

Changes brought about by the action of hot fluids associated with igneous activity.

POLIANITE

Synonym of *pyrolusite*.

POLLENITE

Igneous rock of the *diorite* clan. A greenish-grey, fine-grained porphyritic extrusive rock. Contains phenocrysts of *sanidine* and *andesine* in a groundmass of *olivine*, *andesine*, *augite*, *hornblende*, *biotite* and *glass*.
Occurrence: Italy—Monte Somma.

POLLUCITE

$16[(C.S,Na)AlSi_2O_6.nH_2O]$ Hydrated aluminium sodium caesium silicate. Cubic. Striated cubes; also massive. Colourless. Vitreous lustre. Colourless streak. Conchoidal fracture. Poor cleavage. Hardness 6·5. SG $c2·9$. Slowly decomposes in hydrochloric acid with the separation of silica.
Occurrence: Italy—Island of Elba. In cavities in *granite*.
A *zeolite*.

POLYADELPHITE

Massive brownish-yellow variety of *andradite*.

POLYBASITE

$16[Ag_{16}Sb_2S_{11}]$ Silver antimony sulphide. Monoclinic. Tabular crystals; also massive. Iron-black. Metallic lustre. Black streak. Uneven fracture. Poor cleavage. Hardness 2·0–3·0. SG $c6·1$. Decomposes in nitric acid.
Occurrence: USA—Colorado (Mollie Gibson Mine in Pitkin County), Montana (Drumlummon Mine in Marysville), Nevada (Tanopoh in Esmeralda County). Canada—Ontario (Cobalt).
Variety is *pearcite*.

POLYHALITE

$K_2MgCa_2(SO_4)_4 2H_2O$ Hydrated potas-sium magnesium calcium sulphate. Triclinic. Crystals rare, normally fibrous masses. Colourless, white or grey. Vitreous lustre. Perfect cleavage in one direction. Hardness 3·5. SG 2·78. Decomposes in water with the separation of *gypsum*.
Occurrence: Worldwide. Oceanic salt deposit.

POLYMORPHOUS

Existing in more than one form although chemically identical. Dimorphism is term used where two such forms obtain.

POLZENITE

Feldspathoid igneous rock. A medium-grained porphyritic hypabyssal rock. Contains phenocrysts of *anomite* in a groundmass of *olivine*, *melilite*, *lazurite* and *phlogopite*.
Occurrence: Czechoslovakia—Bohemia (Modibar).

PONITE

Iron rich variety of *rhodochrosite*.

PONZITE

Igneous rock of the *syenite* clan. A fine-grained porphyritic extrusive rock. Contains phenocrysts of *diopside* and *aegirine-augite* in a groundmass of *orthoclase*, *augite* and *magnetite*.
Occurrence: Pacific—Paya Islands.

PORCELLANITE

High grade contact metamorphic rock. Light-coloured, compact and porcelain-like.
Occurrence: Widespread. Contact metamorphism of *marls* and *shales*.

POROUS

Containing voids between the constituent particles through which liquids may pass or in which they may be contained. Depending on the shape of the particles (but not their size) and on the degree of

Above: porcellanite

cementation a rock's porosity may reach up to 50 per cent, i.e. 50 per cent of its volume is void.

PORPHYRITE

Hypabyssal igneous rock containing *plagioclase* as the dominant mineral.

PORPHYRITIC

Igneous rock texture: large crystals which have formed first are later surrounded by a fine grained crystalline mass of minerals as the whole magma consolidates.

PORPHYRY

Hypabyssal igneous rock containing *orthoclase* as the dominant mineral.

POTASH ALUM

Synonym of *alum*.

POTASH ANALCIME

Potassium rich variety of *analcite*.

POTSTONE

A massive greyish-green, dark-green, iron-grey or brownish-black variety of *talc*.

POWELLITE

4[CaMoO$_4$] Calcium molybdate. Tetragonal. Pyramidal or tabular crystals; also foliated masses. Straw-yellow, brown, greenish-blue or black. Subadamantine lustre. Uneven fracture. Indistinct cleavage. Hardness 3·5–4·0. SG 4·23. Decomposes in hydrochloric acid.
Occurrence: USA—Texas (Barringer Hill, Llanos County), Utah (Seven Devils Mine in Adens County). Turkey—Vilayet Ankara (Kasa Keskin).
Secondary alteration mineral of *molybdenite*.

PRASE

Transluscent leek-green variety of silica with minute inclusions of *hornblende*.

PRASINITE

High grade regional metamorphic rock. Green. Schistose texture. Contains *albite, chlorite, epidote, glaucophane* and *hornblende.*
Occurrence: Widespread. Regional metamorphism of *spilites*.

PRE-CAMBRIAN

See *geological time.*

PRECIOUS GARNET

A deep-red transparent variety of *almandine*.

PRECIOUS STONE

A translucent variety of *serpentine*.

PRECIPITATION

(a) Deposition of insoluble particles formed in solution as result of chemical action. Useful observation in chemical testing; process leading to certain sedimentary rocks.
(b) Deposition of atmospheric water on Earth's surface in form of rain, sleet, snow, hail, dew, frost, freezing fog.

PREDAZZITE, FENCATITE or PENCATITE

Contact metamorphic rock. A dedolomitised crystalline *limestone*. Contains *calcite* and *brucite*.
Occurrence: Italy—Predazzo.

PREHNITE

2[Ca$_2$Al$_2$Si$_3$O$_{10}$(OH)$_2$] Hydrous calcium aluminium silicate. Orthorhombic. Rare tabular crystals, normally massive. White,

Above: prehnite. *Below:* priceite

grey or light green. Vitreous lustre. Colourless streak. Uneven fracture. Good cleavage in one direction. Hardness 6·0–6·5. SG 2·8–2·95. Decomposes slowly in hydrochloric acid.
Occurrence: Scotland—Dunbartonshire (Friskie Hall), Renfrewshire (Hartfield Moss). USA—Connecticut (Woodly), New Jersey (Bergen Hill).

PRICEITE

A soft, white, earthy variety of *colemanite*.

PRIMARY

Original, as applied to a rock which is later altered in some way.
See secondary.

PRISMATIC

See *crystal*.

PRISMATINE

Synonym of *kornerupine*.

PROCHLORITE

Synonym of *ripidolite*.

PROPYLITE

Igneous rock of the *diorite* clan. A green fine-grained porphyritic extrusive rock, hydrothermally altered to produce a complex suite of minerals. Contains phenocrysts of *feldspar, hornblende* and *biotite* partly altered to *chlorite, quartz, carbonates, zoisite* and *epidote*, in a groundmass of *epidote, chlorite, serpentine, calcite, pyrite* and *actinolite*.
Occurrence: USA—Nevada (Wahoe district and the Comstock Lode in the Sierra Nevada Mountains). Hungary—Kremnitz.

PROTOGINE

Igneous rock of the *granite* clan. A coarse-grained gneissose plutonic rock. Contains *orthoclase, microcline, oligoclase, quartz* and *mica*.
Occurrence: France—Monte Blanc. Switzerland—Aar Massif, Gotthard Massif.

PROTOMYLONITE

Low grade dynamic metamorphic rock. Transitional between *breccia* and *mylonite*. Lenticular particles coarsely brecciated with traces of original structures and bedding.
Occurrence: In areas of thrusting associated with regional metamorphism.

PROUSTITE

$2[Ag_3AsS_3]$ Silver arsenic sulphide. Hexagonal. Prismatic crystals; also compact masses. Scarlet-vermilion. Adamantine lustre. Vermilion streak. Conchoidal to uneven fracture. Good cleavage in one direction. Hardness 2·0–2·5. SG $c5·6$. Decomposes in nitric acid with the separation of sulphur.
Occurrence: USA—Colorado (Red River in San Juan County), Idaho (Poorman Mine in Owyhee County). France—Alsace (Sainte-Marie-aux-Mines). Germany—Saxony (Himmelfürst Mine near Freiberg).

PSEUDO-ALBITE

Synonym of *andesine*.

PSEUDOBRECCIA

Limestone with the textural appearance of a *breccia*. Produced by the accumulation of detrital algae or by the differential recrystallisation of an originally fine-grained *limestone*.
Occurrence: England—Yorkshire (Lower Carboniferous). Australia—New South Wales (Devonian Limestones).

PSEUDOCONGLOMERATE

Type of *conglomerate* formed in situ by mechanical fragmentation and rolling caused by tectonic activity.

PSEUDOCUBIC

Any crystal habit falsely resembling cubic system.

PSEUDOHEXAGONAL

Any crystal habit falsely resembling hexagonal system.

PSEUDOJADE

Any mineral falsely resembling *jade*.

PSEUDOLEUCITE

Leucite altered to *orthoclase* and *nepheline*.
Occurrence USA—Arkansas (Magnet Cove in the Ozark Mountains).

PSEUDOMALACHITE, PHOSPHO-CHALCITE or PHOSPHOROCHAL-CITE

$2[Cu_5(PO_4)_2(OH)_4H_2O]$ Basic hydrated copper phosphate. Monoclinic. Crystals rare, normally aggregates or massive. Dark emerald-green to blackish-green. Vitreous lustre. Dark green streak. Splintery fracture. Good cleavage in one direction. Hardness 4·5–5·0. SG 4·35. Soluble in hydrochloric acid.
Occurrence: England—Cornwall (Liskeard). USA—Pennsylvania (Wheatley Mine in Chester County). Western Australia—Collier Bay. Secondary mineral associated with *malachite*.

PSEUDOMORPHED

See pseudomorphic.

PSEUDOMORPHIC

Applied to a mineral appearing in the

crystal form of another, due to alteration, for example, where *serpentines* may retain the crystal shape of the original *olivines*.

PSEUDOPHITE or MISKEYITE

Green serpentine-like variety of *pennine*.

PSILOMELANE

$2[(Ba,Mn)Mn_4O_8(OH)_2]$ Basic barium manganese oxide. Orthorhombic. Massive. Iron-black. Submetallic lustre. Brownish-black or black streak. Hardness $5 \cdot 0 - 6 \cdot 0$. SG $c4 \cdot 7$. Soluble in hydrochloric acid with the evolution of chlorine.
Occurrence: Scotland—Orkney (Lead Geo). USA—Arizona (Tucson), Virginia (Austinville in Wythe County). Germany—Saxony (Schneeberg). Weathering product of manganese deposits.

PTEROPOD OOZE

Type of *ooze* rich in pteropod shells.

PTILOLITE

Synonym of *mordenite*.

PUDDINGSTONE

Type of *conglomerate* with well rounded pebbles in a siliceous matrix.

PUGLIANITE

Igneous rock of the *diorite* clan. A coarse-grained granular plutonic rock. Contains *augite, leucite, anorthite, sanidine, hornblende* and *biotite*.
Occurrence: Italy—Monte Somma. Fragments in lava flows.

PULASKITE

Igneous rock of the *syenite* clan. A blue, coarse-grained granular plutonic rock. Contains *microperthite, hornblende, diopside, nepheline* and *apatite*.
Occurrence: USA—Arkansas (Farche Mountain in Pulaski County). Family member is *bostonite*.

PULVERULENT

Powdery or readily powdered.

PUMICE

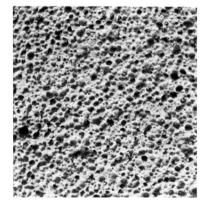

Igneous rock of the *granite* clan. A white or light grey extrusive 'rock froth'.
Occurrence: Worldwide. Forms very light and crumbly air-filled crusts on compact lavas.

PUMPELLYITE

$2[Ca_4(Al, Fe^{2+}, Fe^{3+}, Mg)_6Si_6O_{23}(OH)_3 \cdot 2H_2O$ Basic hydrated calcium aluminium iron and magnesium silicate. Monoclinic. Normally fibrous or platy masses. Bluish-green or brown. Brittle fracture. Perfect basal cleavage. Hardness $3 \cdot 2$. SG $5 \cdot 5$. Insoluble in hydrochloric acid.
Occurrence: USA—Michigan (Keweenaw Peninsular). New Zealand—Southlands (Howells Point).

PURPURITE

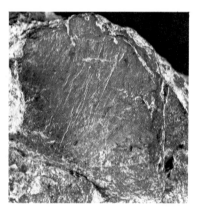

$(Mn,Fe)PO_4$ Iron manganese phosphate. Orthorhombic. Prismatic crystals; also massive. Reddish-purple. Satiny lustre. Reddish-purple streak. Uneven fracture. Good cleavage in one direction. Hardness $4 \cdot 0 - 4 \cdot 5$. SG $c3 \cdot 3$. Easily soluble in hydrochloric acid.
Occurrence: USA—California (Pala in San Diego County), North Carolina (Faires Mine in Gaston County), South Dakota (Custer in Pennington County). Western Australia—Wodgina. France—Chanteloube.

PYARGYRITE

Ag_3SbS_3 Silver antimony sulphide. Hexagonal. Prismatic crystals; also massive. Deep red. Adamantine lustre. Purplish-red streak. Conchoidal to uneven fracture. Cleavage visible in one direction. Hardness 2·5. SG c5·8. Decomposes in nitric acid with the separation of silver and antimony oxide.

Occurrence: USA—Colorado (Ruby district of Gunnison County), Idaho (Silver City in Owyhee County). Germany—Saxony (Himmelfürst Mine near Freiberg).

Important ore of *silver*.

PYCNITE

Compact columnar variety of *topaz*.

PYRALMANDITE or PYRANDITE

Garnet intermediate between *pyrope* and *almandine*.

PYRALSPITE GARNETS

Group name for *pyrope*, *almandine* and *spessartite garnets*.

PYRAMIDAL

See *crystal*.

PYRANDITE

Synonym of *pyralmandite*.

PYRARGYRITE

$2[Ag_3SbS_3]$ Silver antimony sulphide. Hexagonal. Prismatic, rhombohedral crystals; also massive. Black to cochineal-red. Metallic, adamantine lustre. Cochineal-red streak. Conchoidal fracture. Fairly good cleavage. Hardness 2·0–3·0. SG 5·7–5·9. When heated in open tube gives sulphurous fumes and antimony oxide sublimate. Decomposes in nitric acid.

Occurrence: USA—Nevada (Comstock Lode). Canada—Ontario (Cobalt).

Czechoslovakia—Pribram. Germany—Harz Mountains (Andreasberg). Bolivia—Potosi.

PYRENEITE

A black or greyish-black variety of *andradite*.

PYRIBOLE

Group name covering all *pyroxenes* and *amphiboles*.

PYRITE, FOOL'S GOLD or IRON PYRITES

$4[FeS_2]$ Iron sulphide. Cubic. Cubic crystals; also massive. Pale brassy-yellow. Metallic lustre. Greenish-black to brown-

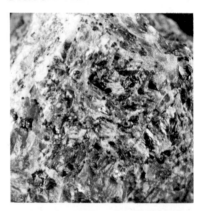

ish-black streak. Conchoidal to uneven fracture. Indistinct cleavage. Hardness 6·0–6·5. SG c5·0. Insoluble in hydrochloric acid.

Occurrence: Worldwide.

Family member is *bravoite*.

Variety is *melnikovite*.

PYROCHLORE

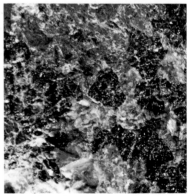

$8[(Ca, Na, Ce)(Nb, Ti, Ta)_2(O, OH, F)_7]$ Complex oxide containing essentially cal-

cium sodium niobium (columbium) and tantalum with hydroxyl and fluorine. Cubic. Octahedral crystals; also massive. Brown or black. Vitreous or resinous lustre. Light brown or yellowish-brown streak. Subconchoidal to uneven fracture. Indistinct cleavage. Hardness $5 \cdot 0$–$5 \cdot 5$. SG $4 \cdot 2$–$4 \cdot 5$. Slowly soluble in sulphuric acid.
Occurrence: USA—California (Pala in San Diego County), Maine (Newry). France—Cantal (Brocq en Menet). Germany—Baden (Kaiserstuhl). Norway—Laurvik.
Variety is *koppite*.

PYROCLASTIC

Formed by ejection into the air by volcanic activity. Many types, dependent on manner of, and state in, ejection and on subsequent cooling.

PYROCLASTIC DEPOSITS

Fragmentary volcanic deposits. Covers *volcanic conglomerates*, agglomerates, *tuffs* and ashes.

PYROLUSITE or POLIANITE

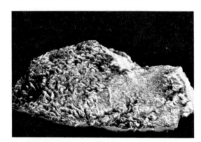

$2[MnO_2]$ Manganese dioxide. Tetragonal. Crystals rare, normally columnar or fibrous masses. Steel-grey to iron-grey. Metallic lustre. Black or bluish black streak. Uneven fracture. Perfect cleavage in one direction. Hardness $6 \cdot 0$–$6 \cdot 5$. SG $c5 \cdot 0$.
Occurrence: Worldwide. In oxidised zone of bog, lacustrine or shallow marine manganese deposits.
Common manganese mineral.

PYROMORPHITE

$2[Pb_5(PO_4)_3Cl]$ Lead chloride phosphate. Hexagonal. Prismatic crystals; also massive. Green, yellow or brown.

Resinous to subadamantine lustre. White streak. Uneven fracture. Indistinct cleavage. Hardness $3 \cdot 5$–$4 \cdot 0$. SG $c7 \cdot 0$. Soluble in nitric acid.
Occurrence: England—Cumberland (Roughton Gill near Caldbeck). Scotland—Lanarkshire (Leadhills). USA—Georgia (Canton Mine in Cherokee County), Pennsylvania (Wheatley Mine in Chester County). Canada—British Columbia (Mayie). France—Brittany (Huelgoat).

PYROPE, ARIZONA RUBY, CAPE RUBY or ELIE RUBY

$8[Mg_3Al_2Si_3O_{12}]$ Magnesium aluminium silicate. Cubic. Normally rounded or angular fragments. Deep crimson-red. Vitreous lustre. Dark red streak. Conchoidal fracture. No cleavage. Hardness $7 \cdot 5$. SG $3 \cdot 7$. Gelatinises in hydrochloric acid.
Occurrence: Scotland—Fife (Elie). Czechoslovakia—Bohemia (Budweis). Germany—Saxony (Greifendorf). South Africa—Kimberley. In *ultrabasic* rocks.
Variety is *rhodolite*.

PYROPHANITE

$2[MnTiO_3]$ Manganese titanium oxide. Hexagonal. Tabular crystals; also massive. Deep red. Metallic to submetallic lustre. Ochre-yellow streak. Conchoidal to subconchoidal fracture. Perfect cleavage in one direction. Hardness $5 \cdot 0$–$6 \cdot 0$. SG $4 \cdot 58$. Slowly soluble in hot hydrochloric acid.
Occurrence: Sweden—Vermland (Harstig Mine at Pajsberg). Brazil—Minas Gerais (Queluz).

PYROPHYLLITE

$4[Al_2Si_4O_{10}(OH)_2]$ Basic aluminium silicate. Monoclinic. Crystals rare; normally massive. White, apple-green, greyish to brownish-green, yellow or grey. Pearly or dull lustre. Variable streak. Uneven fracture. Perfect basal cleavage. Hardness $1 \cdot 0$–$2 \cdot 0$. SG $2 \cdot 8$–$2 \cdot 9$. Partly decomposes in sulphuric acid.
Occurrence: USA—Arkansas (Kellogg Lead Mine near Little Rock), North Carolina (Cottonstone Mountain in Mecklenburg County). USSR—Ural Mountains (Berezov).

PYROPYSALITE or PHYSALITE

Coarse opaque yellowish-white variety of *topaz*.

PYROXENES

Group of chemically and physically related

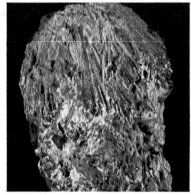

minerals. Iron magnesium calcium silicates with or without aluminium, sodium and lithium. Possess orthorhombic or monoclinic symmetry. Form a number of solid solution series. Members are *enstatite, hypersthene, diopside, hedenbergite, augite, pigeonite, acmite, aegirite, aegirine-augite, jadeite* and *spodumene*. Variety is *lavrovite*.

PYROXENITE

Igneous rock of the *ultrabasic* clan. A coarse-grained granular plutonic rock. Contains *pyroxenes, hornblende* and *mica*.
Occurrence: Northern Ireland—Newry Complex. Canada—Stillwater Complex. South Africa—Bushveld Complex.

PYROXMANGANITE or IRON RHODONITE

16[(Mn,Fe)SiO$_3$] Manganese iron silicate. Triclinic. Massive. Yellowish-brown, reddish-brown or dark brown. Vitreous to resinous lustre. Good cleavage in one direction. Hardness 5·5–6·0. SG 3·8. Insoluble in acids.
Occurrence: USA—Idaho (Boise in Ada County), South Carolina (Iva in Anderson County).

PYRRHOSIDERITE

Synonym of *goethite*.

PYRRHOTITE

2[FeS] Iron sulphide. Hexagonal. Tabular or platy crystals; also granular masses. Bronze-yellow to brown. Metallic lustre.

Dark greyish-black streak. Uneven to sub-conchoidal fracture. No cleavage. Hardness 3·5–4·5. SG 4·58–4·65. Decomposes in hydrochloric acid with the production of hydrogen sulphide.
Occurrence: USA—Connecticut (Standish Mine near Trumbull), Tennessee (Ducktown). Canada—Ontario (Sudbury). Germany—Saxony (Freiberg). Variety is *troilite*.

QUARTZ

3[SiO$_2$] Silicon dioxide. Hexagonal. Hexagonal prisms; also massive and crypto-crystalline forms. Colourless, white, violet,

Right: rutilated quartz

230

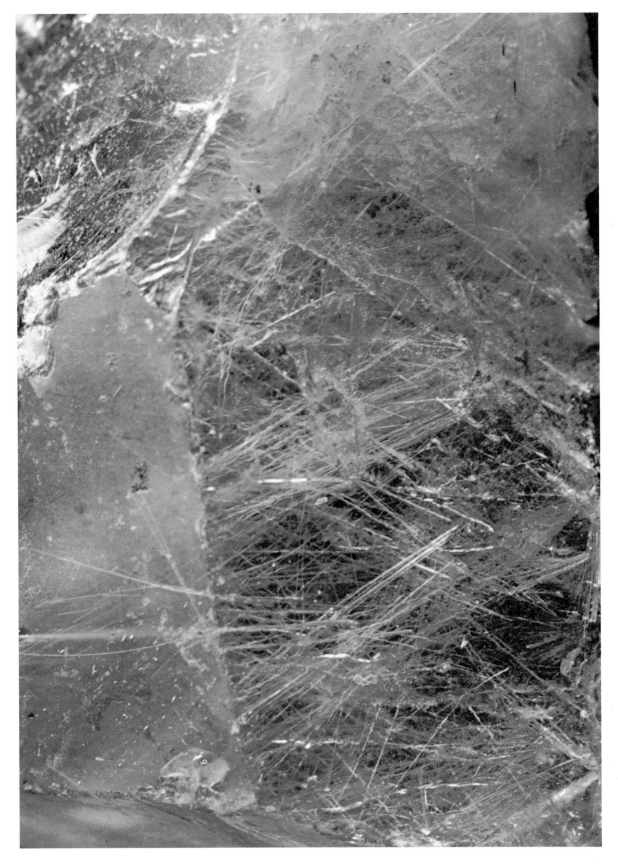

red, brown or black. Vitreous lustre. Conchoidal fracture. No cleavage. Hardness 7·0. SG 2·6. Insoluble in acids.

Occurrence: Worldwide in igneous, metamorphic and sedimentary rocks. Polymorphs: *cristobalite* and *tridymite*. Crystalline varieties: *amethyst, aventurine quartz, cairngorm stone, citrine, milky quartz, rose quartz, royite* and *smoky quartz*.

QUARTZ DIORITE

Synonym of *tonalite*.

QUARTZINE

Synonym of *chalcedony*.

QUARTZITE

Rock composed entirely of granular *quartz*. Produced by contact or regional metamorphism of a pure *sandstone*.
Occurrence: Worldwide.
Variety is *eosite*.

QUARTZ PORPHYRY

Igneous rock of the *granite* clan. A fine-grained porphyritic hypabyssal rock. Contains phenocrysts of *quartz, orthoclase* and *mica* in a groundmass of cryptocrystalline *glass*.
Occurrence: England—Cornwall (Prah Sands). Canada—British Columbia (North Bed). Germany—Saxony (Nossen).

QUARTZ SCHIST

Type of *schist* containing streaks and lenticles of *quartz*.

QUEENSTOWNITE or DARWIN GLASS

Small shattered and worn greenish-yellow transparent fragments of silica *glass*.
Occurrence: Australia—Tasmania (Mount Darwin).
Thought to be the result of *meteorite* impact fusing deserts sands.

QUENSELITE

4[$PbMnO_2.OH$] Basic lead manganese oxide. Monoclinic. Tabular crystals. Pitch-black. Metallic to adamantine lustre. Dark brown to grey streak. Perfect basal cleavage. Hardness 2·5. SG 6·84. Soluble in dilute hydrochloric acid with the evolution of chlorine.
Occurrence: Sweden—Långban.

QUERCYITE

Type of *phosphorite* from Quercy in France.

RADIATING

Radial, arranged 'spoke-like' from centre.

RADIOACTIVE

Radioactive substances have atoms whose nuclei (some or all) are unstable, with resulting tendency to break down and to emit some form of particulate or electromagnetic radiation. In natural rocks such activity normally slight and long-lived; may be detected and measured in the field or in the laboratory by devices such as the Geiger-Müller counter.

RADIOLARIAN OOZE

Type of *ooze* rich in skeletal remains of radiolarian Protozoa.

RADIOLARITES

A type of *chert* which contains plentiful remains of radiolarian Proroza.

RAMMELSBERGITE

2[$NiAs_2$] Nickel arsenide. Orthorhombic. Normally massive. Tin-white with a red tint. Metallic lustre. Greyish-black streak. Uneven fracture. No cleavage. Hardness 5·5–6·0. SG 7·1.
Occurrence: Canada—North West Territories (Eldorado Mine near the Great Bear Lake), Ontario (Cobalt). France—Alsace (Saint-Marie-aux-Mines). Ger-

many—Saxony (Schneeberg). Italy—Torino (Bruzolo).

RAMSDELLITE

4[MnO_2] Manganese dioxide. Orthorhombic. Tabular crystals; also massive. Iron-grey to black. Metallic lustre. Black streak. Good cleavage in two directions. Hardness 3·0. SG 4·7.
Occurrence: USA—New Mexico (Lake Valley in Sierra County). Canada—Nova Scotia (East River in Picton County). Turkey—Roumelia (Kodjas Karil Mine).

RANCIÉITE

(Ca,Mn)$Mn_4O.3H_2O$ Hydrated calcium manganese oxide. Amorphous. Massive. Silver-grey, brown or brownish-black. Metallic lustre. Purplish-brown streak. Hardness *c*1·0. SG 3·3.
Occurrence: USA—Arkansas (Batesville in Independence County), North Dakota (Dunseith in Rolette County). Cuba—Oriente Province.
Related to *wad*.

RANITE

Synonym of *hydronepheline*.

RANKINITE

4[$Ca_3Si_2O_7$] Calcium silicate. Monoclinic. Normally irregular grains. Colourless. No cleavage. Hardness 5·5. SG 2·96.
Occurrence: Scotland—Argyllshire (Ardnamurchan). Northern Ireland—County Antrim (Scawt Hill).

RAPAKIVI GRANITE

Igneous rock of the *granite* clan. A coarse-grained orbicular plutonic rock. Contains *quartz, orthoclase, oligoclase, hornblende* and *biotite*.
Occurrence: Finland—Rapakivi.

RARE EARTH

Member of group of elements (lanthanides) numbered 57 to 71, all having very similar properties.

RASHLEIGHITE

Light bluish-green variety of *turquoise* from St Austell in Cornwall, England.

RASORITE

Synonym of *kernite*.

REALGAR

4[As_4S_4] Arsenic sulphide. Monoclinic. Prismatic crystals; also granular masses. Red to orangish-yellow. Resinous to greasy lustre. Red to orangish-yellow

Above: rashleighite
Below: realgar

streak. Conchoidal fracture. Good cleavage in one direction. Hardness 1·5–2·0. SG 3·56. Decomposes in nitric acid.
Occurrence: USA—Nevada (Manhattan in Nye County), Utah (Mercur in Tooele County), Wyoming (Norris Geyser Basin in the Yellowstone National Park). Germany—Saxony (Schneeberg). Italy—Naples (Pozzuoli).

RED BEDS

Group name for red coloured clastic sediments. Normally *sandstones*, *shales* or marls. Coloured by *hematite* coating the constituent grains of the rock.
Occurrence: England—the Permo-Trias Red Beds. Scotland—the Devonian Old Red Sandstone Beds.
Indicates an arid terrestrial environment.

REDDLE

Synonym of *red ochre*.

RED LEAD

Synonym of *minium*.

RED OCHRE or REDDLE

Earthy red variety of *hematite*.

REFRACTION

Process whereby light is bent on passing from one substance into another, the bending being dependent on angle of entering, colour of the light, and nature of the substances.
'Double refraction' is occasionally observed, the incident light splitting into two components, one normal and following the laws of refraction, the other polarised and behaving differently.

REFRACTORY

Able to be raised to high temperatures without damage.

REGIONAL METAMORPHISM

Heat and high pressure process always associated with orogenesis and igneous intrusions by which rocks in the crust change in nature.

REGRESSION CONGLOMERATE

Type of *conglomerate* resulting from a relative fall in sea level and a rejuvenation of fluvial acticity.

REINITE

Wolframite pseudomorphs after *scheelite*.

RELATIVE DENSITY

See specific gravity.

RELICT STRUCTURE

Relics of the original rock texture or structure which may still be visible in a metamorphic rock and may have modified the superimposed texture or structure.

RENIFORM

Kidney-shaped.

RESINOUS

Dull but definitely shiny.

RETICULATED

Net-like.

RETINALITE

Massive yellowish-green waxy variety of *chrysotile*.

RHODESITE

$(Ca,Na_2,K_2)_8Si_{16}O_{40} \cdot 11H_2O$ Hydrated calcium sodium potassium silicate. Orthorhombic. Normally fibrous masses. White. Silky lustre. Cleavage visible in one direction. Hardness 4·0. SG 2·36.
Occurrence: South Africa—Kimberley (Bultfontein Mine).
A *zeolite*.

RHODIZITE

$NaKLi_4Be_3Al_4B_{10}O_{27}$ Complex beryllium aluminium alkali borate. Dodecahedral crystals. Colourless or white. Vitreous lustre. Conchoidal fracture. Indistinct cleavage. Hardness 8·0. SG 3·38. Insoluble in acids.
Occurrence: USSR—Ural Mountains (Mursinsk). Madagascar—Antandrokomby, Manjaka.

RHODOCHROSITE or DIALOGITE

$2[MnCO_3]$ Manganese carbonate. Hex-

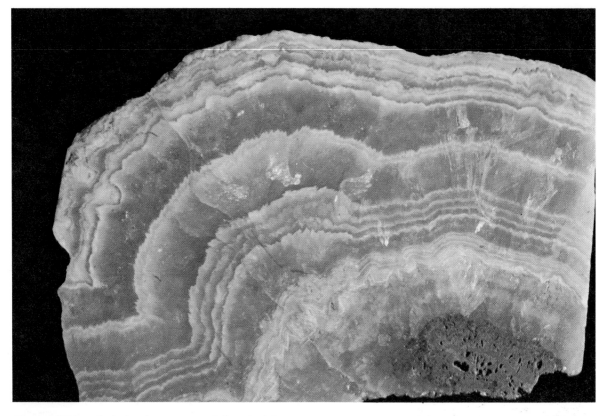

Above and below: rhodochrosite

agonal. Crystals rare, normally granular or columnar masses. Shades of pink and red. Vitreous lustre. White streak. Uneven to conchoidal fracture. Perfect cleavage in one direction. Hardness 3·5–4·0. SG 3·7. Soluble with effervescence in warm hydrochloric acid.
Occurrence: USA—Montana (Butte), Nevada (Austin), Utah (Park City). Germany—Saxony (Freiberg). Yugoslavia—Ljubija.
Varieties are *capillitite*, *kutnahorite*, *manganosiderite* and *ponite*.

RHODOLITE

Rose-red variety of *pyrope*.

Below: rhodolite

RHODONITE or CUMMINGTONITE (of Rammelsberg)

10[$MnSiO_3$] Manganese silicate. Triclinic. Tabular prismatic crystals; also massive.

Pink to grey. Vitreous lustre. White streak. Conchoidal to uneven fracture. Perfect cleavage in two directions. Hardness 5·5–6·0. SG 3·5–3·7. Slightly affected by hydrochloric acid.
Occurrence: USA—Massachusetts (Cummington), Rhode Island (Cumberland). Australia—New South Wales (Broken Hill). Sweden—Långban.
Varieties are *bustamite*, *fowkrite* and *photicite*.

RHODUSITE

Fibrous variety of *glaucophane*.

RHOMBOHEDRAL

See *crystal*.

RHOMB PORPHYRY

Igneous rock of the *syenite* clan. A fine-grained porphyritic hypabyssal or extrusive rock. Contains phenocrysts of soda-*microcline* and *anorthoclase* in a groundmass of *orthoclase*, *microperthite*, *apatite* and *magnetite*.
Occurrence: Norway—Larvik (Audvik), Oslo (Ringeriks).

RHOMB SPAR

An *iron*-bearing variety of *dolomite* which turns brown on exposure.

RHONITE

Calcium rich variety of *aenigmatite*.

RHYOLITE or LIPARITE

Igneous rock of the *granite* clan. A fine-grained porphyritic extrusive rock. Contains phenocrysts of *quartz* and *sanidine* in a groundmass of *feldspar*, *augite* and *glass*.
Occurrence: Widespread.
The extrusive equivalent of a *granite*.

RICHTERITE

$2[(Na,K)_2 (Mg,Mn,Ca)_6 Si_8O_{22} (OH)_2]$ Hydrous sodium potassium magnesium manganese and calcium silicate. Monoclinic. Elongated crystals. Brown, yellow or red. Variable streak. Perfect cleavage in one direction. Hardness 5·0–6·0. SG 3·09.
Occurrence: USA—Colorado (Iron Hill), Montana (Libby). Sweden—Långban.
An *amphibole*. Variety is *szechenyiite*.

RICOLITE

Trade name for green *serpentine* from New Mexico.

RIEBECKITE or OSSANNITE

$2[Na_2(Fe^{2+}Fe^{3+})Si_8O_{22}(OH)_2]$ Hydrous sodium iron silicate. Monoclinic. Prismatic crystals; also massive. Blue or bluish-black. Vitreous lustre. Perfect cleavage in two directions. Hardness 4·0. SG 3·43.
Occurrence: Wales—Caernarvonshire (Mynydd Mawr). USA—Colorado (St Peters Dome in El Paso County). Varieties are *crocidolite* (fibrous), *magnesio-riebeckite* and *tigers eye* (silicified).

RIPIDOLITE or PROCHLORITE

$2[(Mg,Fe,Al)_6(Si,Al)_4O_{10}(OH)_8]$ Hydrous magnesium iron aluminium aluminosilicate. Monoclinic. Tabular prismatic crystals; also massive. Shades of green. Pearly lustre. Colourless streak. No fracture. Good cleavage in one direction. Hardness 1·0–2·0. SG 2·78–2·96.
Occurrence: England—Cornwall (St Just). USA—North Carolina (Steeles Mine in Montgomery County), Virginia (Castle Mountain near Batesville). Norway—Arendal.
A *chlorite*. Variety is *aphrosiderite*.

ROCK

Body of mineral material with usually more than one constituent.

ROCKALLITE

Igneous rock of the *granite* clan. A coarse-grained granular plutonic rock. Contains *quartz*, *albite* and *aegirite*.
Occurrence: North Atlantic—Island of Rockall.

ROCK FLOUR

Rock ground to a clay-like particle size.

ROCK MEAL

A white earthy variety of *calcite* found in caverns.

ROCK SALT

Synonym of *halite*.

ROEBLINGITE

Synonym of *haüyne*.

ROEPPERITE

$4[(Fe,Mn,Zn)_2SiO_4]$ Iron manganese zinc silicate. Orthorhombic. Normally massive. Yellow, weathering to black. Vitreous lustre. Yellow to reddish-grey streak. Good cleavage in two directions. Hardness 5·5–6·0. SG c4·0. Gelatinises in hydrochloric acid.
Occurrence: USA—New Jersey (Franklin Furnace, Sterling Hill).

RÖMERITE

$Fe^{2+} Fe^{3+}_2(SO_4)_4 . 12H_2O$ Hydrated iron sulphate. Triclinic. Tabular crystals; also massive. Brown to yellow. Vitreous lustre. Uneven fracture. Perfect cleavage in one direction. Hardness 3·0–3·5. SG 2·17. Saline taste.
Occurrence: USA—Arizona (United Verde Mine in Jerome), California (Island Mountain in Trinity County). Germany—Harz Mountains (Rammelsberg).

ROSASITE

$4[(Cu,Zn)_2CO_3(OH)_2]$ Copper zinc car-

bonate-hydroxide. Monoclinic. Massive. Green or blue. Hardness 4·5. SG *c*4·0. Soluble in hydrochloric acid.
Occurrence: USA—Nevada (Jack Pot Claim near Wellington). Italy—Sardinia (Rosas Mine near Sulais).

ROSCOELITE

$K(V,Al)_3Si_3O_{10}(OH)_2$ Hydrous potassium aluminium vanadium silicate. Monoclinic. Normally scales. Clove-brown to greenish-brown. Pearly lustre. No fracture. Perfect basal cleavage. Hardness 1·0–2·0. SG 2·92–2·94.
Occurrence: USA—California (Granite Creek in El Dorado County), Colorado (Magnolia district).
A *mica*.

ROSELITE

$2[Ca_2(Co,Mg)(AsO_4)_2.2H_2O]$ Hydrated calcium cobalt magnesium arsenate. Monoclinic. Prismatic crystals; also spherical masses. Pink to dark red. Vitreous lustre. Red streak. Perfect cleavage in one direction. Hardness 3·5.

SG 3·5–3·74. Easily soluble in hydrochloric acid.
Occurrence: Germany—Bavaria (Schapback), Saxony (Schneeberg).

ROSE MUSCOVITE

Manganese rich variety of *muscovite*.

ROSENBUSCHITE

$4[(Ca, Na)_3(Zr, Ti)Si_2O_8F]$ Calcium sodium zirconium and titanium fluosilicate. Monoclinic. Crystals rare, normally radiating masses. Orangish-grey. Vitreous lustre. Uneven fracture. Perfect cleavage in one direction. Hardness 5·0–6·0. SG 3·3.
Occurrence: Norway—Langesund Fiord.

ROSE QUARTZ

A pale pink or rose-coloured variety of *quartz*.

ROSOLITE

Synonym of *landerite*.

ROSTERITE

Synonym of *vorobyevite*.

ROUGEMONTITE

Igneous rock of the *gabbro* clan. A medium to fine-grained granular plutonic rock. Contains *olivine*, *anorthite*, *augite*, *biotite* and *hornblende*.
Occurrence: Canada—Quebec (Rougemont).

ROUMANITE or RUMANITE

Brittle brownish-yellow massive variety of *amber* from Rumania.

ROUTIVARITE

Igneous rock of the *gabbro* clan. A fine-grained granular plutonic rock. Contains *plagioclase*, *garnet* and *orthopyroxene*.
Occurrence: Sweden—Lapland (Routivara).

ROUVILLITE

Igneous rock of the *gabbro* clan. A mottled, medium to coarse-grained plu-

tonic rock. Contains *labradorite*, *nepheline*, *hornblende* and *augite*.
Occurrence: Canada—Quebec (St Hilaire Mountain).

ROYITE

Variety of *quartz* with a bladed habit.

RUBELLITE or ELBAITE (of Vernadsky)

Red variety of *tourmaline*.

RUBICELLE

Yellow or orangish-red gem variety of *spinel*.

RUBIDIUM MICROCLINE

Rubidium rich variety of *microcline*.

RUBY

Red gem variety of *corundum*.

RUBY COPPER

A crystallised variety of *cuprite*.

RUBY SPINEL

Red gem variety of *spinel*.

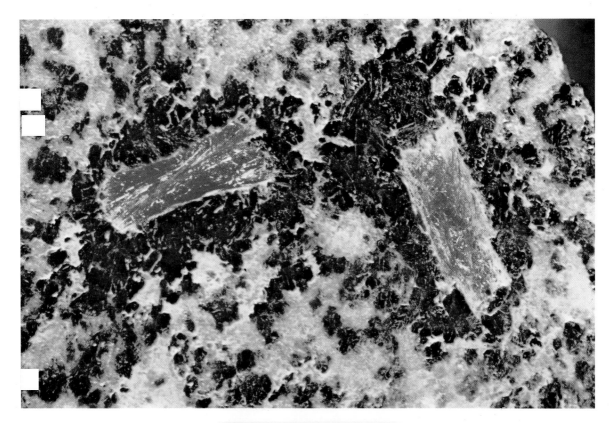

Above: ruby in zoisite
Below: ruby spinel

RUDACEOUS ROCKS

General term for clastic rocks with a particle size greater than 2mm. Includes all *conglomerates* and *breccias*.

RUMANITE

Synonym of *roumanite*.

RUTILE

2[TiO_2] Titanium dioxide. Tetragonal. Prismatic or acicular crystals; also granular masses. Reddish-brown, yellow, blue or black. Metallic adamantine lustre.

Pale brown to yellowish or black streak. Conchoidal to uneven fracture. Good cleavage in one direction. Hardness 6·0–6·5. SG 4·23. Insoluble in acids.
Occurrence: Worldwide in *gneisses*, *schists*, plutonic igneous rocks. and *pegmatities*.
Varieties are *flèches d'amour*, *ilmenorutile* and *strüverite*.

SAAMITE

Variety of *apatite* containing strontium and rare earths in place of calcium.

SACCHAROIDAL

'Sugary', i.e. consisting of closely massed small crystals.

SAGENITE

Complex intergrowth of acicular *rutile* and transparent *quartz*.

SAHLITE or SALITE

4[$Ca(Mg,Fe)Si_2O_6$] Calcium magnesium iron silicate. Monoclinic. Normally granular masses. Greyish-green to green or black. Dull to subvitreous lustre. Good cleavage in one direction. Hardness 5·0–6·0. SG 3·25–3·4. Variety of *diopside*
Occurrence: Scotland—Argyllshire (Kentallen), Shiant Isles (Gabh Eilean). Northern Ireland—County Antrim (Scawt Hill). USA—Pennsylvania (West Chester). Australia—New South Wales (Black Jack Sill near Gunnedah).
A *pyroxene*.

SAL AMMONIAC

4[NH_4Cl] Ammonium chloride. Cubic.

Trapezohedral crystals; also massive crusts and stalactitic. Colourless or white. Vitreous lustre. Conchoidal fracture. Cleavage visible in one direction. Hardness 1·5–2·0. SG 1·53. Soluble in water.
Occurrence: England—Northumberland (Newcastle). USA—Hawaii (Kilauea). Italy—Mount Vesuvius. In volcanic fumaroles.

SALINAS or SALT PANS

Areas of precipitation of *halite* from sea water. Coastal depressions in arid areas separated from the sea by permeable sands.
Occurrence: Widespread in tropical areas, particularly Eastern Central America and North East Africa.

SALINE

Containing salt.

SALITE

Synonym of *sahlite*.

SALT

(a) Chemical compound formed when a metal takes the place of some or all of the hydrogen in an acid.
(b) Common salt—NaCl.

SALT DEPOSITS

Synonym of *evaporite*.

SALT PANS

Synonym of *salinas*.

SAMARSKITE

(Y, Er, Ce, U, Ca, Fe, Pb, Th)(Nb, Ta, Ti, Sn)$_2$O$_6$ Uranium iron rare earths niobate (columbate) and tantalate. Orthorhombic. Prismatic crystals; also massive. Velvet-black with a brownish tint. Vitreous to resinous lustre. Dark reddish-brown or black streak. Conchoidal fracture. Cleavage visible in one direction. Hardness 5·0–6·0. SG 5·7. Decomposes in hot concentrated hydrochloric acid when finely powdered.
Occurrence: USA—California (Nuevo), Colorado (Ohio City in Gunnison County), North Carolina (Wisemans Mica Mine in Mitchell County). Norway—Moss. Brazil—Minas Gerais (Divino de Ubá). In *granite pegmatites*.

SAND

A fragment of rock with a grain size of between 0·05mm and 2mm.

SANDSTONES

Lithified *arenaceous rocks* composed

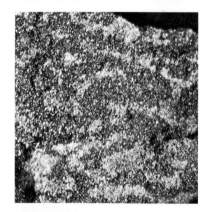

dominantly of *quartz* and *feldspar* fragments and cemented by silica, *clay min-*

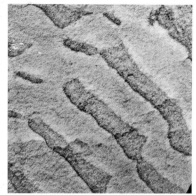

erals, *iron* minerals or *calcite*.
Occurrence: Worldwide.

SANIDINE

High temperature polymorph of *orthoclase*.

SANTORINITE

Igneous rock of the *diorite* clan. A fine-grained porphyritic extrusive rock. Contains phenocrysts of *labradorite* with *oligoclase* rims in a groundmass of *oligoclase* and *hypersthene*.
Occurrence: Italy—Santorini.

SANUKITE or KANKAN-ISHI

Igneous rock of the *diorite* clan. A very

fine-grained extrusive rock. Contains *bronzite* in a *glass* of *andesine* composition.
Occurrence: Japan—Shikolu Island (Sanuki).

SAPONITE, BOWLINGITE, NEPHRITE or SOAPSTONE

Hydrous magnesium aluminium silicate. Formula and crystal system uncertain. Massive. White with a yellowish, greenish, bluish or reddish tint. Greasy lustre. Soft *clay*, brittle when dry. Decomposes in sulphuric acid.
Occurrence: England—Cornwall (Lizard Point). USA—Connecticut (Roaring Brook near New Haven). Canada—Ontario (Pigeon Point).
Variety is *sauconite*.
Also white plastic variety of *montmorillonite* from Plombières in France.

SAPPHIRE

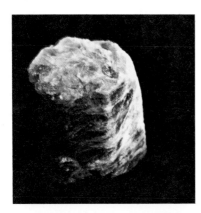

Blue gem variety of *corundum*.

SAPPHIRINE

$8[Mg_2Al_4SiO_{10}]$ Magnesium aluminium silicate. Monoclinic. Tabular crystals; also disseminated grains. Pale blue or green. Vitreous lustre. Subconchoidal fracture. Indistinct cleavage. Hardness 7·5. SG 3·42–3·48.
Occurrence: Greenland—Fiskenäs.
Also blue variety of *chalcedony*.

SAPROPELIC COALS

Group name for *coals* formed from homogeneous masses of spores, algae and plant remains. Characterised by being massive, unbanded and breaking with a conchoidal fracture. Includes *cannel coal* and *boghead coal*.

SARCOLITE

$12[NaCa_4Al_3Si_5O_{19}]$ Sodium calcium aluminium silicate. Tetragonal. Pseudocubic crystals. Flesh-red to rose-red. Vitreous lustre. Conchoidal fracture. No cleavage. Hardness 6·0. SG 2·54. Gelatinises with hydrochloric acid.
Occurrence: Italy—Monte Somma, Mount Vesuvius. In ejected blocks.

SARCOLITE (of Vauquelin)

Synonym of *gmelinite*.

SARD

Light to dark brown translucent semi-precious variety of *chalcedony*.

Below: sardonyx

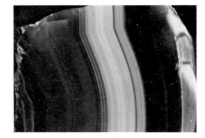

SARDONYX

Layered and banded variety of *chalcedony* consisting of alternating layers of *onyx* and *sard*.

SASSOLITE

$4[H_3BO_3]$ Boric acid. Triclinic. Tabular crystals; also scaly masses and stalactites. White to grey. Pearly lustre. White streak. Perfect basal cleavage. Hardness 1·0. SG 1·48. Soluble in water.
Occurrence: USA—California (Lake County). Germany—Aachen. Italy—Tuscany (Sasso).

SATIN SPAR

Silky fibrous variety of *aragonite* or *calcite*. Finely fibrous opalescent variety of *gypsum*.

SATURATED

(a) State of a solution which can normally take up no more solute. Supersaturation is a rare exception.
(b) State of a rock with perfect balance between silica content and silica demand of other components. Very rare. Rocks may be over-saturated (containing free silica) or under-saturated (containing 'unsaturated' minerals, which could not develop in situations of excess silica). Occasionally the degree of saturation of rocks may be expressed with reference to oxides other than silica.

SAUCONITE

Zinc rich variety of *saponite*.

SAUSSURITE

Alteration product of *plagioclase*. Consists of a mixture of *zoisite*, *epidote* and *plagioclase*.

SAXONITE

Igneous rock of the *ultrabasic* clan. A coarse-grained granular plutonic rock. Contains *olivine*, *enstatite*, *bronzite* and *hypersthene*.
Occurrence: USA—Minnesota (Minnesota River Valley). Germany—Harz Mountains (Radautal). Sweden—Mansjö (Toppgriwan).

SCAPOLITE GROUP

Group of tetragonal minerals forming a solid solution series between $3(NaAlSi_3O_8).NaCl$ and $3(CaAl_2Si_2O_8).CaCO_3$. Important members are *marialite*, *wernerite*, *dipyre* and *meionite*.

SCAWTITE

$Ca_7Si_6O_{16}CO_3(OH)_4$ Hydrous calcium carbonate and silicate. Monoclinic. Platy masses. Colourless. Vitreous lustre. Perfect cleavage in one direction. Hardness 4·5–5·0. SG 2·77. Decomposes in hydrochloric acid with effervescence.
Occurrence: Northern Ireland—County Antrim (Scawt Hill).

SCHEELITE

$4[CaWO_4]$ Calcium tungstate. Tetragonal. Octahedral crystals; also granular or columnar masses. Colourless or white, often tinted. Vitreous lustre. White streak. Uneven to subconchoidal fracture. Good cleavage in one direction. Hardness 4·5–5·0. SG 6·1. Decomposes in hydrochloric acid leaving a yellow residue of hydrous tungstic oxide.
Occurrence: Widespread. In high temperature hydrothermal veins, *pegmatites* and contact metamorphic zones between *limestones* and *granites*.

SCHEFFERITE

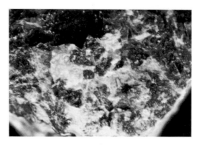

$4[(Ca,Mn)(Mg,Fe,Mn)Si_2O_6]$ Manganese iron magnesium and calcium silicate. Monoclinic. Tabular crystals; also massive. Yellowish-brown to reddish-brown or black. Subvitreous lustre. Good cleavage in two directions. Hardness 5·5–6·0. SG 3·25–3·4.
Occurrence: USA—New Jersey (Franklin Furnace). Sweden—Långban.
A *pyroxene*.

SCHILLER

Remarkable reflection of light from crystal or cleavage planes of certain minerals when viewed in specific directions. Unexpected colours sometimes also appear in schiller as result of interference effects.

SCHIST

Regional metamorphic rock. May contain *quartz*, *feldspar*, *mica*, *chlorite*, *talc* or *hornblende*. Schistose and lineated texture.
Occurrence: Worldwide in areas of regional metamorphism.

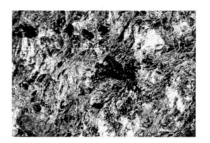

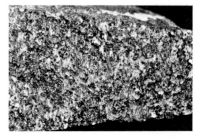

Above: schist

SCHISTOSE

Closely foliated texture produced by the parallel alignment of *micas* or similar minerals.

SCHIZOLITE

Manganese rich variety of *pectolite*.

SCHÖNFELSITE

Igneous rock of the *gabbro* clan. A fine-grained porphyritic extrusive rock. Contains phenocrysts of *olivine* and *augite* in a groundmass of *augite, bronzite* and *plagioclase*.
Occurrence: Germany—Saxony (Altschönfels).

SCHORL

Synonym of *tourmaline*.

SCHORLOMITE

$8[Ca_3(Fe,Ti)_2(Si,Ti)_3O_{12}]$ Calcium iron

titanium silicate. Cubic. Crystals rare, normally massive. Black, tarnishing to blue. Vitreous lustre. Greyish-black streak. Conchoidal fracture. Hardness 7·0–7·5. SG 3·81–3·88. Gelatinises with hydrochloric acid.
Occurrence: USA—Arkansas (Magnet Cove in the Ozark Mountains).
A *garnet*.

SCHORL ROCK

Cornish term for a plutonic igneous rock consisting entirely of *quartz* and acicular *tourmaline*.
Occurrence: England—Cornwall (Roche Rock near Roche).

SCHREIBERSITE

(Fe,Ni)$_2$P Iron nickel phosphide. Tetragonal. Crystals rare, normally rounded grains. Silver-white to tin-white, tarnishing yellow or brown. Metallic lustre. Perfect basal cleavage. Hardness 6·5–7·0. SG 7·0–7·3. Soluble with difficulty in hydrochloric acid.
Occurrence: Iron rich *meteorites*.

SCHRIESHEIMITE

Igneous rock of the *ultrabasic* clan. A medium-grained granular hypabyssal rock. Contains *hornblende, olivine, phlogopite* and *diopside*.
Occurrence: Germany—Baden (Schriesheim).

SCOLECITE

8[CaAl$_2$Si$_3$O$_{10}$3H$_2$O] Hydrated calcium aluminium silicate. Monoclinic. Prismatic crystals; also fibrous or radiating masses. White. Vitreous or silky lustre. Good cleavage in one direction. Hardness 5·0–5·5. SG 2·16–2·4. Gelatinises with hydrochloric acid.
Occurrence: Scotland—Argyllshire (Staffa), Isle of Mull. USA—Colorado (Table Mountain).
A *zeolite*.

SCORALITE

(Fe,Mg)Al$_2$(PO$_4$)$_2$(OH)$_2$ Basic iron magnesium aluminium phosphate. Monoclinic. Pyramidal crystals; also granular masses. Shades of blue. Vitreous lustre. White

Above: scolecite

streak. Uneven fracture. Indistinct cleavage. Hardness 5·5–6·0. SG c3·38. Slowly soluble in hot hydrochloric acid.
Occurrence: USA—South Dakota (Custer in Custer County). Brazil—Minas Gerais (Corrego Frio). In *pegmatites*.

SCORIA

Lava containing rough vesicles.

SCORODITE

8[FeAsO$_4$.2H$_2$O] Hydrated iron arsenate.

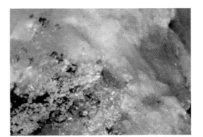

Orthorhombic. Pyramidal or tabular crystals; also massive. Leek-green to brown. Vitreous to subadamantine lustre. Subconchoidal fracture. Indistinct cleavage. Hardness 3·5–4·0. SG 3·28. Soluble in hydrochloric acid.
Occurrence: England—Cornwall (St Stephens), Devon (Virtuous Lady Mine near Tavistock). USA—Georgia (Canton Mine in Cherokee County), Idaho (Black Pine in Cassia County), New York (Carmel in Putnam County).

SCREE DEPOSITS

Sedimentary *rudaceous rocks*. Unconsolidated material eroded from a slope and deposited at its base.
Occurrence: Worldwide, particularly in upland areas.

SCYELITE

Igneous rock of the *ultrabasic* clan. A medium-grained granular plutonic rock. Contains *hornblende, olivine, biotite* and *serpentine*.
Occurrence: Scotland—Caithness (Achavarasdale Moor at Loch Scye).

SEAT EARTH

Type of *clay* underlying *coal* seams in deltaic rhythms of sedimentation.

SEBASTIANITE

Igneous rock of the *diorite* clan. A coarse-grained granular plutonic rock. Contains *anorthite, biotite, augite* and *apatite*.
Occurrence: Italy—Monte Somma. As ejected fragments in lava.

241

SECONDARY

Later, as opposed to original or primary.

SECTILE

Able to be cut without cracking or crumbling.

SEDIMENTARY ROCKS

One of the three major classes (with igneous and metamorphic) produced and laid down on the surface of the Earth. Includes products of erosion and weathering and remains of living organisms. Chemical action is often involved. When first deposited, the material is a sediment. Processes of diagenesis lead eventually to a consolidated rock.

SEEBACHITE or HERSCHELITE

Orthorhombic form of *chabazite*.

SELAGITE

Igneous rock of the *syenite* clan. A fine-grained porphyritic extrusive rock. Contains phenocrysts of *biotite* in a groundmass of *orthoclase, diopside, oligoclase* and *apatite*.
Occurrence: Italy—Tuscany (Monte Catini).

SELENITE

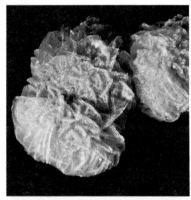

Crystalline variety of *gypsum*.

SENAITE

(Fe,Mn,Pb)TiO$_3$ Iron manganese lead titanate. Crystal system uncertain. Rough crystals and rounded fragments. Black. Submetallic lustre. Brownish-black streak. Conchoidal fracture. No cleavage. Hard-

Below: senarmontite.

ness 6·0—7·0. SG 5·3. Decomposes in boiling sulphuric acid.
Occurrence: Brazil—Minas Gerais (Dattas). In *diamond* bearing sands.

SENARMONTITE

16[Sb$_2$O$_3$] Antimony oxide. Cubic. Octahedral crystals; also granular masses. Colourless or greyish-white. Resinous lustre. White streak. Uneven fracture. Poor cleavage. Hardness 2·0—2·5. SG 5·5. Easily soluble in hydrochloric acid.
Occurrence: Canada—Quebec (South Ham in Wolf County). Germany—Westphalia (Arnsberg). Italy—Sardinia (Nieddoris).

SEPIOLITE or MEERSCHAUM

Mg$_3$Si$_4$O$_{11}$.5H$_2$O Hydrated magnesium silicate. Clay masses; also fibrous. Greyish-white or white. Hardness 2·0—2·5. SG 2·0. Fine earthy texture. Gelatinises with hydrochloric acid.
Occurrence: USA—North Carolina (Webster in Jackson County). Czechoslovakia—Moravia (Hrubschitz).

SEPTARIAN NODULES

Clay ironstone nodules. Cracked internally due to shrinkage while consolidating. Irregular cracks infilled latter by *calcite*.
Occurrence: Widespread in *shales* and *clays*.

SERANDITE

Na$_6$(Ca,Mn)$_{15}$Si$_{20}$O$_{58}$.2H$_2$O Hydrated sodium calcium manganese silicate. Monoclinic. Elongated crystals. Peach-blossom red. Silky lustre. Cleavage visible in two directions. Hardness 4·5—5·0. SG 3·2.
Occurrence: Africa—Guinea (Los Island).

SERENDIBITE

Ca$_2$Mg$_4$Al$_6$B$_2$Si$_4$O$_{26}$ Calcium magnesium aluminium and boron silicate. Triclinic. Massive. Greyish-blue to green. Vitreous lustre. Subconchoidal fracture. No cleavage. Hardness 6·5—7·0. SG 3·42.
Occurrence: USA—New York (Johnsberg

Right: septarian nodule

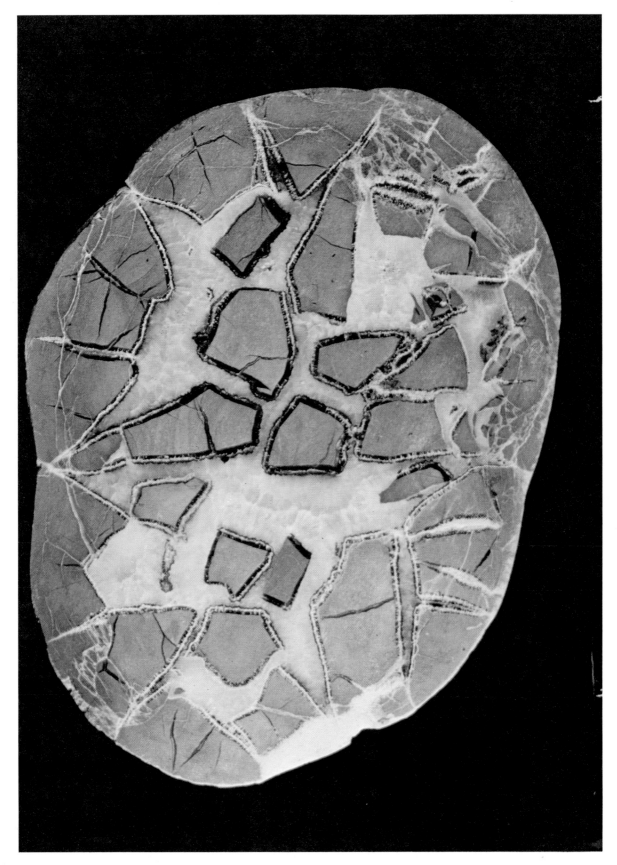

243

in Warren County). Sri Lanka—Kandy (Gragapitiya).

SERICITE

Scaly fibrous variety of *muscovite*.

SERPENTINE

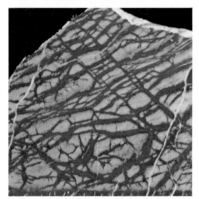

$Mg_3Si_2O_5(OH)_4$ Hydrous magnesium silicate. Monoclinic. Normally massive; also fibrous. White, yellow, brown or shades of green. Subresinous to greasy lustre. Conchoidal or splintery fracture. Hardness 2·5–4·0. SG 2·5–2·65.

Occurrence: Very widespread, particularly in altered *gabbros* and *ultrabasic* igneous rocks.

Important varieties: crystalline—*bastite*. Massive—*bowenite* and *ophiotite*. Lamellar—*antigorite*. Fibrous—*chrysotile* and *ligardite*. Amorphous—*deweylite*. Translucent—*precious stone*.

SERPENTINITE

Altered igneous rock of the *ultrabasic* clan. Normally a green or red compact rock with a banded or streaked texture. Original minerals replaced by varieties of *serpentine*.

Occurrence: England—Cornwall (Lizard Peninsula). Wales—Anglesey (Mona Complex). Canada—Stillwater Complex. Australia—New South Wales (Great Serpentine Belt).

SEYBERTITE

Synonym of *clintonite*.

SHACKANITE

Igneous rock of the *syenite* clan. A fine-grained porphyritic extrusive rock. Contains phenocrysts of *anorthoclase* and *augite* in a groundmass of *analcite*, *anorthoclase*, *orthoclase* and *apatite*.

Occurrence: Canada—British Columbia (Shackan Railway Station).

SHALE

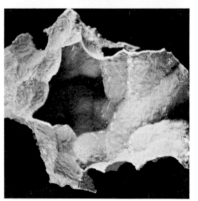

Fissile laminated type of *clay*.

SHEAR

Term used to describe sideways strain and resulting stress, as when one part of a mass tends to move relative to a neighbouring one.

SHEEN

Lustre.

SHELLY LIMESTONE

Descriptive term for a *limestone* rich in invertebrate macrofossils.

SHERIDANITE

$2[(Mg,Al)_6(Si,Al)_4O_{10}(OH)_8]$ Hydrous magnesium aluminium aluminosilicate. Monoclinic. Normally foliated masses. Pale green. Vitreous lustre. Good basal cleavage. Hardness 2·6–3·3. SG 2·7. Slowly decomposes in boiling sulphuric acid.

Occurrence: USA—Montana (Miles City in Custer County), Pennsylvania (Britons

Quarry in Chester County), Wyoming (Sheridan County).
A *chlorite*. Variety is *grochauite*.

SHINING

Quite bright.

SHONKINITE

Igneous rock of the *syenite* clan. A mottled, coarse-grained granular plutonic rock. Contains *augite, orthoclase, plagioclase, olivine* and *nepheline*.
Occurrence: USA—Montana (Square Butte in Highwood Mountains).

SHORTITE

$2[Na_2Ca_2(CO_3)_3]$ Sodium calcium carbonate. Orthorhombic. Tabular or prismatic crystals. Colourless or pale yellow. Vitreous lustre. Conchoidal fracture. Good cleavage in one direction. Hardness 3·0. SG 2·6. Decomposes in water with the separation of calcium carbonate.
Occurrence: USA—Wyoming (Green River in Sweetwater County).

SHOSHONITE

Igneous rock of the *gabbro* clan. A fine-grained porphyritic extrusive rock. Contains phenocrysts of *olivine* and *augite* in a groundmass of *labradorite, olivine, augite* and *leucite*.
Occurrence: USA—Wyoming (Shoshac River in Yellowstone National Park).

SHOT BORT

Variety of *diamond*. A milky-white to steel-grey spherical stone with radiating structure of great toughness.

SIBERITE

Reddish-violet variety of *tourmaline*.

SICKLERITE

$(Li,Mn,Fe)PO_4$ Lithium manganese iron phosphate. Orthorhombic. Normally mas-sive. Yellowish-brown to dark brown. Dull lustre. Light yellowish-brown to brown streak. Good cleavage in one direction. Hardness c4·0. SG 3·2–3·4. Soluble in hydrochloric acid.
Occurrence: USA—California (Pala in San Diego County). Western Australia—Wodgina. Finland—Tammela.
Variety is *ferri-sicklerite*.

SIDERITE

Synonym of *chalybite* and *hornblende*.

SIDERITE MUDSTONE

Type of *mudstone* rich in *chalybite*, normally found associated with *coal* deposits.

SIDERITIC LIMESTONE

Type of *limestone* enriched in *chalybite* by diagenetic processes.
Occurrence: England—the Midlands (Marlstone), Yorkshire (the Cornbrash).

SIDEROPHYLLITE

Black iron rich variety of *biotite*.

SIDEROPHYRE

Type of *meteorite*.

SIDEROPLESITE

Magnesium rich variety of *chalybite*.

SILCRETE

Synonym of *puddingstone*.

SILICATE

Extremely common radical, $-SiO_4^{4-}$ and variations.

SILICEOUS SINTER

White, grey or brown encrustations and concretions of *opal* formed around hot springs.
Variety is *geyserite*.

SILICIFIED

Having silica (SiO_2) introduced where it was not before as in the pores of a rock, or by displacement of existing material.

SILL

A sheet-like body of igneous rock which conforms to bedding or other planes.

SILLAR

Pyroclastic igneous rock of the *granite* clan. A white to salmon-red unconsolidated rock. Contains fragments of *rhyolite*,

Above: siliceous sinter

rhyolitic *pumice, biotite* and *glass*.
Occurrence: Peru—Arequipa region.

SILLIMANITE

$4[Al_2SiO_5]$ Aluminium silicate. Orthorhombic. Elongated crystals; also fibrous or columnar masses. White, brown, greyish-brown, greyish-green or olive-green. Vitreous lustre. Colourless streak. Uneven fracture. Perfect cleavage in one direction. Hardness 6·0–7·0. SG 3·23–3·24.
Occurrence: Scotland—the Highlands. USA—Massachusetts (Worcester), New York (York Town in Westchester County), North Carolina (Culsagee Mine in Macon County). Czechoslovakia—Bohemia (Moldau). Germany—Saxony (Freiberg). In regional metamorphic rocks.
Varieties are *fibrolite* (fibrous) and *westanite*.

SILT

Type of *aenaceous rock* with a grain size of between 0·005mm and 0·05mm.

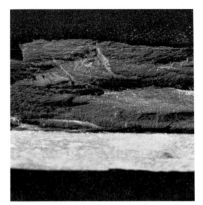

SILVER
4[Ag] Cubic. Cubic. Cubic or octahedral

Above and right: silt
Below: native silver

SILURIAN
See *geological time.*

crystals; also arborescent or wiry masses. Silver-white. Metallic lustre. Silver-white streak. Hackly fracture. No cleavage. Hardness 2·5–3·0. SG 10·1–11·1. Soluble in nitric acid.
Occurrence: USA—Arizona (Silver King Mine in Pinal County), Idaho (Silver City in Owyhee County), Montana (Butte). Canada—Ontario (Cobalt). Australia—New South Wales (Broken Hill). Germany—Saxony (Freiberg).

SILVER GLANCE

Synonym of *argentite*.

SILVITE

Synonym of *sylvine*.

SIMETITE

Deep red fluorescent variety of *amber*.

SINHALITE

4[MgAlBO$_4$] Magnesium aluminium borate. Orthorhombic. Normally rolled pebbles. Pale yellow to deep brown. Vitreous to adamantine lustre. No cleavage. Hardness 6·5. SG 3·47–3·6.
Occurrence: USA—New York (Warren County). Burma—gem gravels. Sri Lanka gem gravels.

SINTER

Mineral deposit formed around a hot spring by minerals being precipitated out of solution.
cf *siliceous sinter*.

SITAPARITE

Massive variety of *bixbyite* from India.

SKARN

A contact metamorphosed *limestone* or *dolomite*. Principal ore constituents, depending on locality, are *garnet epidote, hornblende, augite, serpentine* and *olivine*.
Occurrence: Worldwide.

SKEMMATITE or FERRIAN WAD

Variety of *wad*. Amorphous. Black. Metallic lustre. Chocolate-brown streak. Hardness 5·5–6·0. Easily soluble in hydrochloric acid with the evolution of chlorine.
Occurrence: USA—Idaho (Ada County), South Carolina (Iva in Anderson County). An alteration product of *pyroxmangite*.

SKLEROPELITE

General term used to describe massive dense uncleavable rocks produced by the low grade regional or contact metamorphism of *argillaceous rocks*.

SKOMERITE

Igneous rock of the *diorite* clan. A fine-grained porphyritic extrusive rock. Contains phenocrysts of *albite* in a groundmass of *augite* and *olivine*.
Occurrence: Wales—Pembrokeshire (Island of Skomer).

SKUTTERUDITE

8[CoAs$_3$] Cobalt arsenide. Cubic. Cubic crystals; also granular masses. Tin-white to silver-grey. Metallic lustre. Conchoidal to uneven fracture. Good cleavage in two directions. Hardness 5·5. SG 6·5. Soluble in nitric acid.
Occurrence: USA—Colorado (Horace Porter Mine in Gunnison County). France—Alsace (Markirch). Germany—Saxony (Schneeberg). Norway—Skutterud.

SLATE

Low grade regional metamorphic rock. A fine grained rock with perfect planar cleavage and enriched in *mica*.
Occurrence: Worldwide.
Metamorphism of *argillaceous rocks*.

SMALTITE

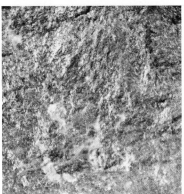

8[CoAs$_2$] Cobalt arsenide. Cubic. Cubic crystals; also massive. Tin-white to silver-grey. Metallic lustre. Conchoidal to uneven fracture. Good cleavage in two directions. Hardness 5·5. SG 6·5. Soluble in nitric acid.
Occurrence: England—Lancashire (Coniston). Scotland—Kirkcudbrightshire (Talnotry). Germany—Hesse (Bieber), Saxony (Schneeberg).
Variety is *cheleutite*.

SMECTITE or MONTMORILLONITE

Calcium magnesium aluminosilicate. A mixed *clay*. Massive. White, grey or shades of green. Dull lustre. Colourless streak. Hardness 1·0–2·0. SG 1·9–2·1.
Occurrence: Widespread as a constituent of *clays*.
Group member is *nontronite*.

SMITHSONITE or CALAMINE

2[ZnCO$_3$] Zinc carbonate. Hexagonal. Crystals rare. Normally botryoidal, reni-

form or granular masses. Greyish-white to dark grey, often tinted blue or green. Vitreous lustre. White streak. Uneven to subconchoidal fracture. Good cleavage in one direction. Hardness 4·0–4·5. SG 4·43. Soluble with effervescence in hydrochloric acid.

Occurrence: USA—California (Cerro Gardo district in Inyo County), New Mexico (Kelly in Magdalena County), Pennsylvania (Bamford in Lancaster County). Germany—Rhineland (Aachen). Greece—Laurium.

SMOKY QUARTZ or MORION

A fine smoky-yellow or smoky-brown variety of *quartz*.

SMYTHITE

Fe_3S_4 Iron sulphide. Hexagonal. Platy crystals. Black. Metallic lustre. Dark grey streak. Subconchoidal fracture. Perfect basal cleavage. Hardness 2·0—3·0. SG 4·06. Easily soluble in hydrochloric acid.
Occurrence: USA—Indiana (Bloomington in Monroe County).

SOAPSTONE

Synonym of *saponite* and *steatite*.

SOBRALITE

Synonym of *pyroxmanganite*.

SODALITE

$2[Na_4Al_3Si_3O_{12}Cl]$ Sodium aluminium silicate and chloride. Cubic. Dodecahedral crystals; also massive. Grey or white, often tinted green, blue or red. Vitreous lustre. Colourless streak. Conchoidal to uneven fracture. Good cleavage in two directions. Hardness 5·5–6·0. SG 2·14–2·3. Decomposes in hydrochloric acid with the separation of gelatinous silica.
Occurrence: USA—Massachusetts (Salem), Montana (Crazy Mountains). Italy—Monte Somma, Mount Vesuvius. Norway—Langesund Fiord.
Variety is *hackmanite*.

SODA NITRE

Synonym of *nitratine*.

SOGGENDALITE

Igneous rock of the *gabbro* clan. A fine-grained porphyritic hypabyssal rock. Contains phenocrysts of *pyroxene* in a groundmass of *labradorite* and *ilmenite*.
Occurrence: Norway—Soggendal.

SOLFATARIC

The action of sulphur when dissolved in a silicate melt, prolonging the period of crystallisation.

SOLID

Least energetic state of matter in which the constituent atoms have negligible translatory motion but vibrate about fixed positions. Usually these positions form a regular crystal lattice; occasionally there is no regularity, such substances being amorphous, and often viewed as supercooled liquids.

SOLID SOLUTION

A solid uniform mixture of two similar materials, there normally being possible the complete range from 100 per cent of the one to 100 per cent of the other. Alloys are everyday examples, quite common in mineralogy.

SOLIFLUCTION

Slow soil or scree flow downwards caused by action of contained water as it repeatedly freezes and melts.

SOLUBILITY

Measure of degree to which a (normally solid) solute will form a solution in a (normally liquid) solvent. Dependent on nature of solute and solvent, other substances present, and temperature.

SÖLVSBERGITE

Igneous rock of the *syenite* clan. A medium to fine-grained granular hypabyssal rock. Contains *sodaclase, aegirine* and *microcline*.
Occurrence: Norway—Lougenthal, Sölvsberget.

SOMMAITE

Igneous rock of the *syenite* clan. A medium to coarse-grained porphyritic plutonic rock. Contains phenocrysts of *augite* and *olivine* in a groundmass of *leucite, bytownite* and *orthoclase*.
Occurrence: Italy—Naples (Monte Somma). In ejected blocks.

SOUTH AFRICAN JADE

Massive gem variety of *grossular*.

SPANDITE

Name given to *garnet* intermediate in composition between *spessartite* and *andradite*.

SPANGITE

Synonym of *phillipsite*.

SPARAGMITE

Group name for the late pre-Cambrian *arkoses* of Scandinavia.

SPECIFIC GRAVITY

Measure of a substance's density in comparison with that of water. Values less than 1·0 indicate that samples float; greater values correspond to sinking.
Important measure in geology.
Also termed relative density.

SPECULAR HEMATITE

Black variety of *hematite* with fine rhom-

bohedral crystals and a splendid metallic lustre.

SPERRYLITE

4[PtAs$_2$] Platinum diarsenide. Cubic. Cubic crystals. Tin-white. Metallic lustre. Black streak. Conchoidal fracture. Poor cleavage. Hardness 6·0–7·0. SG 10·58.
Occurrence: USA—North Carolina (Franklin in Macon County), Wyoming (Rambler Mine in the Medicine Bow Mountains). Canada—Ontario (Vermilion Mine at Sudbury). South Africa—Transvaal (Potgietersrust).

SPESSARTITE

8[Mn$_3$Al$_2$Si$_3$O$_{12}$] Manganese aluminium silicate. Cubic. Dodecahedral crystals. Hyacinth-red to brownish-red. Vitreous lustre. Subconchoidal fracture. Hardness 7·0–7·5. SG 4·15–4·27.
Occurrence: USA—Connecticut (Branch-

ville), North Carolina (Salem), Virginia (Amelia County).
A *garnet*. Variety is *camptospessartite*.

SPHAEROSIDERITE

Globular concretions of fibrous *chalybite*.

SPHALERITE

Synonym of *blende*.

SPHENE or TITANITE

4[CaTiSiO$_5$] Calcium titanate and silicate. Monoclinic. Wedge shaped or prismatic crystals; also massive. Brown, grey, green, yellow or black. Adamantine to resinous lustre. White streak. Subconchoidal fracture. Good cleavage in two directions. Hardness 5·0–5·5. SG 3·4–3·56. Completely decomposes in sulphuric acid.
Occurrence: England—Devon (Tavistock). Scotland—Perthshire (Craig Cailleach). Northern Ireland—Newry (Crow Hill). USA—Maine (Sandford), New York (Pitcairn in St Lawrence County). Norway—Arendal. In *granites*, *gneiss* and *schists*.
Variety is *greenovite*.

SPHENOID

Wedge-shaped.

SPHERULITIC

Having structure in which globular masses (spherulites) of (usually) needle-like crystals are of importance. Spherulites are often

formed by devitrification.

SPICULE

A minute slender pointed hard body; one of the minute calcareous or siliceous bodies supporting the tissues of a number of invertebrates.

SPILITE

Igneous rock of the *gabbro* clan. A fine-grained vesicular submarine extrusive rock displaying *pillow lava* structure. Contains *albite*, *chlorite*, *epidote* and *calcite*.
Occurrence: England—Cornwall (Pentire Point), Devon (Chipley). Scotland—Ayrshire (Girvan). Wales—Anglesey (Mona Complex), Merionethshire (Cader Idris).

SPILOSITE

Synonym of *spotted slate*.

SPINEL

8[MgAl$_2$O$_4$] Magnesium aluminium oxide. Cubic. Octahedral crystals; also granular masses. Colour variable. Vitreous lustre. White streak. Conchoidal fracture. No cleavage. Hardness 7·5–8·0. SG c3·55.
Occurrence: Worldwide. In basic igneous rocks, metamorphosed aluminous rocks, metamorphosed *limestones*, granite *pegmatites*, high temperature ore veins and *placer deposits*. Varieties: gem *almandine*

3·13–3·2.
Occurrence: Scotland—Aberdeenshire (Peterhead). Republic of Ireland—Dublin (Killiney Bay). USA—Maine (Windham), Massachusetts (Chester), North Carolina (Stone Point in Alexander County).
A *pyroxene.* Varieties are *bikitaite, hiddenite* and *kunzite.*

SPOTTED SLATE or SPILOSITE

Contact metamorphic rock. A fine-grained rock with an impressed cleavage and clotted concentrations of cryptocrystalline iron oxides in a streaked matrix of *sericite, chlorite* and *quartz.*
Occurrence: Worldwide.

SPURRITE

$Ca_5Si_2O_8CO_3$ Calcium silicate and carbonate. Normally granular masses. White. Vitreous lustre. White streak. Uneven to splintery fracture. Good cleavage in one direction. Hardness $c5·0$. SG $c3·01$. Decomposes with effervescence in hydrochloric acid leaving gelatinous silica residue.
Occurrence: Northern Ireland—County Antrim (Scawt Hill). USA—California (Riverside).

STABLE

Not readily decomposed, with or without outside influence.

STAFFELITE

Green fibrous variety of *apatite.*
Also synonym of *francolite.*

STALACTITE

Massive concretionary mineral deposit of

Above: spinel. *Below:* spodumene

spinel, balas, rubicelle and *ruby spinel.* Iron rich—*ceylonite* and *chlorospinel.* Zinc rich—*gahnospinel.* Chromium rich—*picotite.*

SPLENDENT

Mirror-like.

SPODUMENE

$4[LiAlSi_2O_6]$ Lithium aluminium silicate. Monoclinic. Prismatic crystals; also cleavable masses. Greenish-white, greyish-white, yellowish-green, green or purple. Vitreous lustre. White streak. Uneven to subconchoidal fracture. Perfect cleavage in one direction. Hardness 6 5 7 0. SG

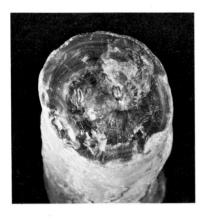

calcium carbonate formed by percolating solutions on the roofs of *limestone* caves.
Occurrence: England—Derbyshire (Castleton), Somerset (Cheddar Gorge), Yorkshire (Ingleborough). USA—Kentucky (Mammoth Caves). Belgium—the caves of Han. Yugoslavia—Carniola.

STALACTITIC

Associated with downward growth of calcium carbonate in certain caves in limestone. Similarly shaped.

STALAGMITE

Massive concretionary mineral deposit of calcium carbonate formed on the floors of limestone caves by dripping solutions.
Occurrence: See *stalactite*.

STANNITE or STANNINE

$2[Cu_2FeSnS_4]$ Copper iron tin sulphide. Tetragonal. Crystals rare, normally granular masses. Steel-grey to iron-black, tarnishing blue. Metallic lustre. Black streak. Uneven fracture. Indistinct cleavage. Hardness 4·0. SG 4·3–4·5. Decomposes in nitric acid with the separation of sulphur and tin oxide.
Occurrence: England—Cornwall (tin mines at Camborne, Redruth, St Agnes and St Austell). USA—South Dakota (the Black Hills). Australia—Tasmania (Zeehan).

STANNITE (of Breithaupt)

Synonym of *cassiterite*.

STAUROLITE

$(Fe,Mg)_4Al_{18}Si_8O_{46}(OH)_2$ Iron aluminium silicate with magnesium. Monoclinic. Prismatic crystals. Dark reddish-brown to brownish-black or yellowish-brown. Subvitreous lustre. Colourless or grey streak. Subconchoidal fracture. Good cleavage in one direction. Hardness 7·0–7·5. SG 3·65–3·75.
Occurrence: Scotland—Inverness-shire (Milltown). Republic of Ireland—Dublin (Killiney). USA—Connecticut (Litchfield), New York (Dover in Dutchess County), North Carolina (Franklin in Macon County). Germany—Bavaria (Aschaffenburg). In crystalline *schists*. Variety is *lusakite*.

STEATITE, FRENCH CHALK or SOAPSTONE

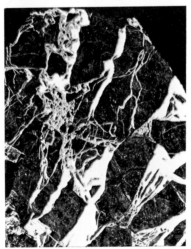

Coarse granular massive variety of *talc*.

STELLATE, -D

Star-like.

STEPHANITE

$4[Ag_5SbS_4]$ Silver antimony sulphide. Orthorhombic. Prismatic or tabular crystals; also massive. Iron-black. Metallic lustre. Iron-black streak. Subconchoidal to uneven fracture. Imperfect cleavage. Hardness 2·0–2·5. SG $c6·25$. Oxidises in nitric acid with the separation of sulphur and antimony oxide.
Occurrence: England—Cornwall (Wheal Boys Mine). USA—California (Grass Valley in Nevada County), Nevada (the Comstock Lode at Virginia City). Canada—Ontario (Cobalt). Czechoslovakia—Bohemia (Schemnitz). Germany—Saxony (Freiberg).

STEWARTITE

Variety of *diamond*. A magnetic *bort* with a high percentage of ash due to iron oxide content.

STIBICONITE

Synonym of *cervantite*.

STIBIOPALLADINITE

Pd_3Sb Palladium antimonide. Cubic. Cubic crystals. Silver-white to steel-grey. Metallic lustre. Uneven fracture. No cleavage. Hardness 4·0–5·0. SG 9·5. Soluble in hot sulphuric acid.
Occurrence: South Africa—Transvaal (Potgietersrust).

STIBNITE or ANTIMONITE

$4[Sb_2S_3]$ Antimony sulphide. Orthorhombic. Prismatic or acicular crystals; also columnar masses. Lead-grey, tarnishing black. Metallic lustre. Lead-grey streak. Subconchoidal fracture. Perfect cleavage in one direction. Hardness 2·0. SG 4·63. Soluble in hydrochloric acid.

found in the Meteor Crater in Arizona, USA.

Occurrence: USA—California (Hollister in San Benito County), Idaho (Coeur d'Alene in Shoshone County). Canada—Quebec (South Ham in Wolfe County). France—Haute-Loire (Lubilhac). Germany—Saxony (Freiberg).

STOLZITE

STICHTITE

$3[Mg_6Cr_2CO_3(OH)_{16} . 4H_2O]$ Hydrated magnesium chromium carbonate-hydroxide. Hexagonal. Massive. Lilac to rose-pink. Waxy to greasy lustre. Pale Lilac or white streak. Perfect cleavage in one direction. Hardness $1 \cdot 5–2 \cdot 0$. SG $2 \cdot 16$.
Occurrence: Canada—Quebec (Megantic Mine at Black Lake). Australia—Tasmania (Dundas).

STILBITE

$2[NaCa_2Al_5Si_{13}O_{36} . 14H_2O]$ Hydrated sodium calcium aluminium silicate. Monoclinic. Tabular crystals; also sheaf like aggregates. White. Vitreous lustre. Colourless streak. Uneven fracture. Perfect cleavage in one direction. Hardness $3 \cdot 5–4 \cdot 0$. SG $2 \cdot 09$. Decomposes in hydrochloric acid.

Occurrence: Scotland—Dunbartonshire and Stirlingshire (Kilpatrick Hills). Northern Ireland—County Antrim (Giant's Causeway). USA—New Jersey (Bergen Hill), New York (Phillipstown). Norway—Arendal. In *basalts*.
Varieties are *foresite* and *hypostilbite*. Also synonym of *heulandite*.

STILPNOMELANE

$K(Fe^{2+},Fe^{3+},Al)_{10}Si_{12}O_{30}(O,OH)_{12}$ Basic potassium iron aluminium silicate. Monoclinic. Foliated plates or fibrous masses. Black, greenish-black or yellowish-bronze. Pearly to vitreous lustre. Perfect cleavage in one direction. Hardness $3 \cdot 0–4 \cdot 0$. SG $c2 \cdot 8$.
Occurrence: New Zealand—Otago. Greenland—Kangerdlugssuaq. Sweden—Nordmark (Pen Mine).

STINKSTEIN

Synonym of *anthraconite*.

STISHOVITE

High density tetragonal form of silica

$4[PbWO_4]$ Lead tungstate. Tetragonal. Tabular crystals. Reddish-brown to brown, yellow, green or red. Resinous to subadamantine lustre. Colourless streak. Conchoidal to uneven fracture. Imperfect cleavage. Hardness $2 \cdot 5–3 \cdot 0$. SG $7 \cdot 9–8 \cdot 3$. Soluble in hydrochloric acid with the separation of yellow tungstic acid.
Occurrence: England (Force Craig Mine near Keswick). USA—Arizona (Primos Mine near Dragoon), Pennsylvania (Wheatley Lead Mines in Chester County). Canada—British Columbia (Cariboo district). Australia—New South Wales (Proprietary Mine at Broken Hill). Germany—Saxony (Zinnwald).

STONES

Silicate based *meteorites*.

STONY-IRONS

Silicate and metal based *meteorites*.

STRATIFIED

Bedded. See bed.

STREAK

Colour of substance when finely powdered, as when filed or used to scratch a hard unglazed plate.

STRENGITE

8[$FePO_4 \cdot 2H_2O$] Hydrated iron phosphate. Orthorhombic. Octahedral or tabular crystals; also crusts or aggregates. Colourless, red or violet. Vitreous lustre. White streak. Conchoidal fracture. Good cleavage in one direction. Hardness $c3 \cdot 5$. SG $c2 \cdot 87$. Soluble in hydrochloric acid.
Occurrence: USA—New Hampshire (Palermo near North Groton), Pennsylvania (Moore's Mill in Cumberland County), Virginia (Midvale in Rockbridge County). Germany—Bavaria (Pleystein). Portugal—Beira Province (Mangualde). Alteration product of *triphylite*.

STRESS

Force set up inside a mass as result of externally applied strains. Tend to compress, to stretch, or to shear.

STRIATED

Grooved.

STRONTIANITE

4[$SrCO_3$] Strontium carbonate. Orthorhombic. Prismatic or acicular crystals; also granular, fibrous or columnar masses. Colourless to grey. Vitreous lustre. Uneven to subconchoidal fracture. Good cleavage in one direction. Hardness $3 \cdot 5$. SG $3 \cdot 75$. Soluble in hydrochloric acid.
Occurrence: Scotland—Argyllshire (Strontian). USA—California (Strontium

Hills in San Bernardino County), New York (Schoharie in Schoharie County), Texas (Mount Bonnell in Travis County). Canada—British Columbia (Cariboo district). Austria—the Tyrol (Bixlegg). Germany—Saxony (Freiberg).

STRÜVERITE

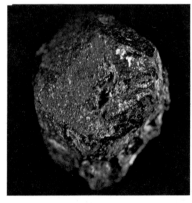

Black tantalum and niobium (columbium) enriched variety of *rutile*.
Also synonym of *chloritoid*.

STYLOLITES

Massive fibrous form of *calcite* developed in *limestones* and *marls* due to pressures on grain contacts during lithification.

SUBLIMATION

Process of condensation directly from gas to solid, i.e. without intermediate liquid state.

SUBSTITUTION

Replacement of one atom or group in a compound by another.

SUCCINITE

Amber coloured variety of *grossular*.

SUDBURITE

Igneous rock of the *gabbro* clan. A fine-grained granular extrusive rock. Contains *bytownite*, *hypersthene*, *augite* and *magnetite*.
Occurrence: Canada—Ontario (Sudbury).

SULPHATE

Radical derived from sulphuric acid; — SO_4^2.

SULPHUR

16[S_8] Orthorhombic. Tabular crystals; also massive. Shades of yellow. Resinous to greasy lustre. White streak. Conchoidal to uneven fracture. Poor cleavage. Hardness $1 \cdot 5$–$2 \cdot 5$. SG $2 \cdot 07$.
Occurrence: USA—Louisiana (Lake Charles), Texas (Freeport in Braznia County), Wyoming (Yellowstone National Park). Italy—Mount Etna, Naples (Solfatara). Spain—Cadiz (Conil).

SULPHURIC ACID

H_2SO_4; very strong, corrosive acid. Useful testing agent even when dilute.

SUNSTONE

Reddish-grey gem variety of *oligoclase*.

SUPERCOOLED

A liquid, such as *glass*, with a negligible rate of flow.

SUSSEXITE

Igneous rock of the *syenite* clan. A medium-grained porphyritic hypabyssal rock. Contains phenocrysts of *nepheline* in a groundmass of *nepheline, orthoclase, aegirine* and *perovskite*.
Occurrence: USA—New Jersey (Beemerville in Sussex County).

SYENITE

Igneous rock of the *syenite* clan. A medium to coarse-grained granular plutonic or hypabyssal rock. Contains *orthoclase, microcline, sodaclase, biotite* and *diopside*.
Occurrence: Worldwide. Associated with *granites*.

SYENITE CLAN

Igneous rocks containing between 55 per cent and 66 per cent silica. Free *quartz* rare. Alkali *feldspars* dominant over *plagioclase*. *Mica* or *pyroxene* is the commonest ferromagnesian constituent. Silica undersaturation leads to the formation of *feldspathoids*, normally *nepheline* or *leucite*.

SYENODIORITE

Igneous rock intermediate between the *syenite* and *diorite* clans. Alkali *feldspars* and *plagioclase* occur in roughly equal amounts.

SYENOGABBRO

Type of *gabbro* in which *orthoclase* occurs.

SYLVANITE

$2[AgAuTe_4]$ Silver gold telluride. Monoclinic. Prismatic or tabular crystals; also columnar or granular masses. Steel-grey to silver-white. Metallic lustre. Steel-grey to silver-white streak. Uneven fracture. Perfect cleavage in one direction. Hardness $1\cdot5-2\cdot0$. SG $8\cdot16$. Decomposes in nitric acid with the separation of gold.
Occurrence: USA—California (Melanes Mine in Calaveras County), Colorado (Cripple Creek in Telbi County). Canada—Ontario (Dame Mine at Porcupine). Western Australia—Kalgoorie. Rumania—Transylvania (Nagyág).

SYLVINE, SYLVITE or SILVITE

$4[KCl]$ Potassium chloride. Cubic. Cubic crystals; also granular or columnar masses. Colourless or white. Vitreous lustre. White streak. Uneven fracture. Perfect basal cleavage. Hardness $2\cdot0$. SG $c2\cdot0$. Bitter taste. Soluble in water.
Occurrence: USA—New Mexico and Texas (salt basins of the Permian era). Germany—Harz Mountains (Stassfurt). Chile—Tarapacá. Bedded *evaporite* deposits, associated with *halite* and *carnallite*.

SYLVITE

Synonym of *sylvine*.

SYNGENITE

$2[K_2Ca(SO_4)_2.H_2O]$ Hydrated potassium calcium sulphate. Monoclinic. Tabular or prismatic crystals; also crystalline crusts. Colourless, faintly yellowish or milky-white. Vitreous lustre. Conchoidal fracture. Perfect cleavage in two directions. Hardness $2\cdot5$. SG $2\cdot6$. Dissolves in water with the separation of *gypsum*.
Occurrence: USA—Hawaii (Haleakala on Mauri Island). Poland—Galicia (Kalisz).

SZECHENYIITE

Variety of *richterite* containing iron and aluminium.

TABULAR

See *crystal*.

TACHYDRITE or TACHHYDRITE

$CaMg_2Cl_6.12H_2O$ Hydrated calcium magnesium chloride. Hexagonal. Normally massive. Colourless or wax-yellow to honey-yellow. Vitreous lustre. Perfect cleavage in one direction. Hardness $2\cdot0$. SG $1\cdot67$. Bitter taste. Easily soluble in water.
Occurrence: Germany—Harz Mountains (Stassfurt), Saxony (Mansfield). Rare *evaporite*.

TACHYLYTE

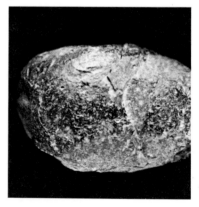

Name given to natural *glass* in extrusive igneous rocks.

TAENITE

Cubic nickel iron mineral (containing 12 per cent nickel) found in *meteorites*.

TAHITITE

Igneous rock of the *diorite* clan. A fine-grained porphyritic extrusive rock. Con-

tains phenocrysts of *haüyne* in a ground-mass of *analcite glass*, *augite*, *haüyne*, *orthoclase* and *leucite*.
Occurrence: Tahiti—Papenoo.

TAIMYRITE

Igneous rock of the *syenite* clan. A medium-grained granular extrusive rock. Contains *nosean*, *anorthoclase*, *sanidine* and *biotite*.
Occurrence: USSR—Siberia (Taimyr River).

TALASSKITE

Brown variety of *fayalite* containing 12 per cent ferric oxide.

TALC or AGALMATOLITE

$4[Mg_3Si_4O_{10}(OH)_2]$ Hydrous magnesium silicate. Monoclinic. Crystals rare, normally granular or foliated masses. White with greenish or greyish tints. Pearly lustre. White streak. Perfect basal cleavage. Hardness 1·0. SG 2·7–2·8. Feels greasy.
Occurrence: Very widespread, particularly in low grade metamorphic rocks.
Massive varieties are *potstone* and *steatite*. *Iron* rich variety is *minnesotaite*.

TALC-SCHIST

Type of *schist* containing *talc* as the

dominant schistose mineral.
Variety is *dolerine*.

TAMARAÏTE

Igneous rock of the *gabbro* clan. A fine-grained granular or porphyritic hypabyssal rock. Contains *augite* (phenocrysts when porphyritic), *hornblende*, *biotite*, *nepheline*, *plagioclase* and *orthoclase*.
Occurrence: Guinea—Island of Tamara (Topsaid Point).

TANTALITE

$4[(Fe,Mn)(Ta,Nb)_2O_6]$ Iron manganese tantalate and niobate (columbate). Orthorhombic. Prismatic or tabular crystals; also massive. Brownish-black to black. Submetallic to subresinous lustre. Dark red to black streak. Subconchoidal to uneven fracture. Good cleavage in one direction. Hardness 6·0–6·5. SG 7·95. Partially decomposes when evaporated with concentrated sulphuric acid.
Occurrence: Very widespread as accessory mineral in *granite pegmatites*. Forms a solid solution series with *columbite*.

TANTALUM

$2[Ta]$ Cubic. Minute cubic crystals; also irregular grains. Greyish-yellow. Metallic lustre. Hardness 6·0–7·0. SG 11·2.
Occurrence: USSR—Ural Mountains. Mongolia—Altai Mountains. Associated with gold deposits.

TAPIOLITE

$2[FeTa_2O_6]$ Iron tantalate. Tetragonal. Prismatic crystals. Black. Subadamantine to submetallic lustre. Cinnamon-brown to brownish-black streak. Uneven to subconchoidal fracture. No cleavage. Hardness 6·0–6·5. SG 7·9.
Occurrence: USA—Maine (Topsham in Sayadahoc County), South Dakota (Minnehaha Gulch in Custer County). Western Australia—Tabba Tabba. France—Haute-Vienne (Chanteloube). In *granite pegmatites*.
Variety is *mossite*.

TARAMITE

$(Ca,Na,K)_3(Fe^{2+},Fe^{3+})_5(Si,Al)_8O_{22}(OH)_2$ Hydrous calcium sodium potassium and iron aluminosilicate. Monoclinic. Prismatic crystals. Bluish-black. Vitreous lustre. Perfect cleavage in one direction. Hardness c5·0. SG 3·45. Decomposes in hot hydrochloric acid.
Occurrence: USSR—Ukraine (Wali-Tarama near Mariupol). In *nepheline syenite*.
An *amphibole*.

TARASPITE

Synonym of *dolomite*.

TARNISH

Change in surface colour and lustre caused by atmospheric corrosion.

TARNOWITZITE

Lead rich variety of *aragonite*.

TASMANITE

Sulphur rich resin. Amorphous. Scales. Reddish brown. Resinous lustre. Hardness 2·0. SG 1·18.
Occurrence: Australia—Tasmania (River Mersey).

TAVOLATITE

Igneous rock of the *syenite* clan. A fine-grained porphyritic extrusive rock. Contains phenocrysts of *leucite*, *haüyne* and *augite* in a groundmass of *leucite*, *haüyne*, *orthoclase*, *labradorite*, *biotite*, *aegirine-augite* and *garnet*.
Occurrence: Italy—Rome (Osteria di Tavolato).

TAWITE

Feldspathoid igneous rock. A coarse-grained granular plutonic rock. Contains *sodalite* and *aegirite*.
Occurrence: USSR—Kola Peninsula (Tawajok Valley).

TAWMAWITE

Chromium rich variety of *epidote*.

TAYLORITE

$4[(K,NH_4)_2SO_4]$ Potassium ammonium sulphate. Orthorhombic. Crystalline concretions. Yellowish-white. Dull lustre. Hardness 2·0. SG $c2·0$. Soluble in water. Tastes pungent and bitter.
Occurrence: Peru—Chinca Islands. In *guano* deposits.

TEALLITE

$2[PbSnS_2]$ Lead tin sulphide. Orthorhombic. Thin tabular crystals; also massive foliated aggregates. Greyish-black. Metallic lustre. Black streak. Perfect basal cleavage. Hardness 1·5. SG 6·36. Easily decomposes in hot concentrated hydrochloric acid.
Occurrence: Germany—Saxony (Himmelfürst Mine at Freiberg). Bolivia—Monserrat (El Salvador Mines).

TECTONIC

Associated with large-scale geological structures and phenomena, such as orogenesis.

TECTONITE

Name given to a rock deformed by dynamic metamorphism. The deformation is the response of the individual minerals in the rock to the stresses and strains acting upon them.

TEKTITES

Black or green natural *glass* containing 73 per cent silica. Button shaped, discoid or teat shaped. Fused appearance may be result of entering the Earth's atmosphere; possibly a result of *meteorite* impact on desert sands.
Occurrence: Australia—Tasmania (*australites*). Czechoslovakia—Bohemia (*moldavites*). Indochina (*indochinites*). Java (*javaites*).

TELLURITE

$8[TeO_2]$ Tellurium dioxide. Orthorhombic. Acicular crystals; also spherical masses. White. Subadamantine lustre. Perfect cleavage in one direction. Hardness 2·0. SG 5·9. Easily soluble in hydrochloric acid.

Occurrence: USA—Colorado (Keystone Mine in Boulder County), Nevada (Jefferson Canyon in Nye County). Rumania—Transylvania (Zalathna).

TELLURIUM

$3[Te]$ Hexagonal. Prismatic or acicular crystals; also columnar or granular masses. Tin-white. Metallic lustre. Grey streak. Perfect cleavage in one direction. Hardness 2·0–2·5. SG 6·1–6·3. Soluble in hot concentrated sulphuric acid.
Occurrence: USA—Colorado (Cripple Creek in Teller County), Nevada (Delamar in Linnah County). Western Australia—Kalgoorlie. Rumania—Transylvania (Faczebaja).

TENNANTITE

$8[Cu_3AsS_3]$ Copper arsenic sulphide. Cubic. Tetragonal crystals; also granular masses. Flint-grey to iron-black. Metallic lustre. Black or brown streak. Subconchoidal to uneven fracture. No cleavage. Hardness 3·7–4·5. SG 4·6–5·0. Decomposes in nitric acid with the separation of sulphur.
Occurrence: Widespread. In hydrothermal vein deposits.

TENORITE

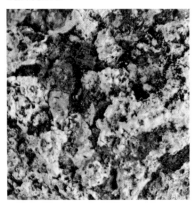

CuO Copper oxide. Monoclinic. Elongated lath-like crystals; also aggregates. Steel-grey to black. Metallic lustre. Conchoidal to uneven fracture. Poor cleavage. Hardness 3·5. SG 5·8–6·4. Easily soluble in hydrochloric acid.
Occurrence: England—Cornwall (Lostwi-

thiel). Scotland—Lanarkshire (Leadhills). USA—Montana (Butte), Nevada (Eureka), Oregon (Waldo in Josephine County). Germany—Westphalia (Daaden). Spain—Rio Tinto. Massive earthy variety is *melaconite*.

TEPHRITE

Igneous rock of the *gabbro* clan. A fine-grained porphyritic extrusive rock. Contains phenocrysts of *augite* and *labradorite* in a groundmass of *augite, labradorite, leucite* and *nepheline*.
Occurrence: Worldwide.

TEPHROITE

$4[Mn_2SiO_4]$ Manganese silicate. Orthorhombic. Crystals rare, normally crystalline masses. Shades of red and grey. Vitreous to greasy lustre. Pale grey streak. Subconchoidal fracture. Good cleavage in two directions. Hardness 5·5–6·0. SG 4·0–4·12. Gelatinises in hydrochloric acid.
Occurrence: USA—New Jersey (Sterling Hill). Sweden—Wermland (Pajsberg). An *olivine*. Variety is *picrotephroite*.

TERRA ROSSA

Type of *clay*. Contains *clay minerals*, ferric oxide, humas and calcium carbonate. Fine earthy texture and red colour with nodules of *limestone*.
Occurrence: Hungary. Yugoslavia—Dalmatia. Associated with *bauxite*.
Residual product of the weathering and disintegration of *limestones*.

TERTIARY

See *geological time*.

TESCHENITE

Igneous rock of the *gabbro* clan. A fine to coarse-grained granular hypabyssal rock. Contains *hornblende, augite, labradorite* and *analcite*.
Occurrence: Czechoslovakia—Bohemia (Paskan near Teschen).

TETRADYMITE

Bi_2Te_2S Bismuth tellurium sulphide. Hexagonal. Crystals rare, normally foliated or granular masses. Pale steel-grey. Metallic lustre. Pale steel-grey streak. Perfect cleavage in one direction. Hardness 1·5–2·0. SG c7·4.

Occurrence: USA—Arizona (Bradshaw City in Yavapai County), California (Cerro Gardo in Inyo County), Colorado (Red Cloud Mine in Boulder County), New Mexico (Organ district in Dona Ana County). Canada—British Columbia (West Kootenai). Western Australia—Dunallen Mine. Rumania—Csiklova. Sweden—Boliden. Associated with gold veins.

TETRAGONAL

See *crystal*.

TETRAHEDRITE

$8[Cu_3SbS_3]$ Copper antimony sulphide. Cubic. Tetragonal crystals; also massive. Flint-grey, iron-black or dull black. Metallic lustre. Black, brown or cherry-red streak. Subconchoidal to uneven fracture. No cleavage. Hardness 3·0–3·7. SG 4·8–5·1. Decomposes in nitric acid with the separation of sulphur and antimony oxide.
Occurrence: Very widespread in hydrothermal veins.
Varieties are *clinohedrite* and *freibergite*.

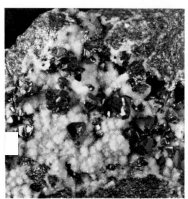

Above and previous column: tetrahedrite

TETRAHEDRON

Crystal form with four faces, in the cubic system.

TETRAKALSILITE

Synonym of *meionite*.

THENARDITE

$8[Na_2SO_4]$ Sodium sulphate. Orthorhombic. Tabular crystals; also crusts and efflorescences. Colourless, often tinted milky, yellowish, brownish or reddish. Vitreous lustre. Uneven to jagged fracture. Perfect cleavage in one direction. Hardness 2·5–3·0. SG c2·6. Faintly bitter taste.
Occurrence: Widespread in Western, North and South America, Southern Europe and North Africa. In *playa* lake deposits.

THERALITE

Igneous rock of the *gabbro* clan. A fine to coarse-grained granular plutonic rock.

Contains *augite, labradorite, nepheline, sodalite, olivine* and *biotite*.
Occurrence: Scotland—Ayrshire (Lugar Sill). USA—Montana (Gordon's Butte in the Crazy Mountains).

THERMAL METAMORPHISM

Same as contact metamorphism.

THERMOLUMINESCENT

A term applied to a substance which emits characteristic light when warmed. Not true luminescence (direct and immediate energy conversion of energy) but a form of phosphorescence (delayed emission, here triggered rather than caused by warmth).

THERMONATRITE

$4[Na_2CO_3.H_2O]$ Hydrated sodium carbonate. Othorhombic. Normally crusts or efflorescence. Colourless or white. Vitreous lustre. Poor cleavage. Hardness 1·0–1·5. SG c2·2. Bitter taste. Soluble in water.
Occurrence: USA—California (the Borax Lake area and Death Valley). Hungary—the Soda Lakes of Debrecz. Egypt. Sudan. Surface deposit in arid areas.

THINOLITE

Yellow or brown variety of *calcite*.
Occurrence: USA—Nevada. Australia. In *tufa* deposits.

THOLEIITE

Igneous rock of the *gabbro* clan. A medium-grained granular hypabyssal rock. Contains *labradorite, augite, ilmenite* and *glass*.
Occurrence: Germany—Rhineland (Tholei near Saar).

THOMSONITE

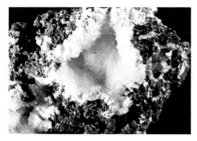

$4[NaCa_2Al_5Si_5O_{20}.6H_2O]$ Hydrated sodium calcium aluminosilicate. Orthorhombic. Crystals rare, normally columnar masses. White. Vitreous or pearly lustre. Colourless streak. Uneven to subconchoidal fracture. Perfect cleavage in one direction. Hardness 5·0–5·5. SG 2·3–2·4. Gelatinises in hydrochloric acid.
Occurrence: Scotland—Renfrewshire (Port Glasgow). USA—Arkansas (Magnet Cove in the Ozark Mountains), Colorado (Table Mountain near Golden). Italy—Monte Somma.
A *zeolite*. Varieties are *comptonite, faröelite, lintonite* and *ozarkite*.

THORIANITE

$4[ThO_2]$ Thorium dioxide. Cubic. Cubic crystals. Dark grey, brownish-black or black. Submetallic lustre. Grey to greenish-grey streak. Uneven to subconchoidal fracture. Poor cleavage. Hardness 6·5. SG 9·7. Soluble in sulphuric acid with the evolution of hydrogen.
Occurrence: USA—Pennsylvania (Easton). Sri Lanka—Sabaragamuwa Province (Balagoda). Madagascar—Betroka. In *pegmatites* and *placer deposits*.
Uranium rich variety is *uranothorianite*.

THORITE

$ThSiO_4$ Thorium silicate. Tetragonal. Prismatic or pyramidal crystals; also massive. Brownish-yellow to black. Vitreous to resinous lustre. Light orange to dark brown streak. Conchoidal fracture. Good cleavage in one direction. Hardness 4·5–5·0. SG 5·19–5·4. Gelatinises with hydrochloric acid.
Occurrence: USA—New York (Champlain). Norway—Brevik (Esmark). In syenitic *pegmatites*.
Varieties are orangite (orange coloured) and *uranothorite* (uranium rich).

THOROGUMMITE

$4[(Th,U)(SiO_4),(OH)_4]$ Thorium uranium silicate and hydroxide. Crystal system uncertain. Massive. Yellowish-brown. Dull lustre. Hardness 4·0–4·5. SG 4·43–4·54. Easily soluble in nitric acid.
Occurrence: USA—Texas (Llano County).
Alteration product of thorium and uranium minerals.

THORTVEITITE

$2[(Sc,Yt)_2Si_2O_7]$ Scandium yttrium silicate. Orthorhombic. Prismatic crystals. Greyish-green.
Occurrence: Norway—Saetersdalen (Iveland). In *pegmatite*

THUCHOLITE or TUCHOLITE

Mixture of hydrocarbons, ash and *uraninite*. Irregular nodules. Black Vitreous lustre. Conchoidal fracture. No cleavage. Hardness 3·5–4·0. SG *c*1·7.

Above: thortveitite

Occurrence: Widespread. Associated with *uraninite* deposits.

THULITE

Red variety of *zoisite*.

THURINGITE

$2[(Fe^{2+}, Fe^{3+}, Mg, Al)_6(Si, Al)_4O_{10}(O, OH)_8]$ Hydrous iron magnesium aluminium aluminosilicate. Monoclinic. Normally massive or scaly aggregates. Shades of green. Pearly to dull lustre. Pale green streak. Subconchoidal fracture. Good cleavage in one direction. Hardness 2·5. SG 3·15–3·19. Gelatinises in hydrochloric acid.
Occurrence: USA—Arkansas (Hot Springs), Pennsylvania (French Creek Mines in Chester County). Germany—Thuringia (Schmiedeberg).
A *chlorite*. Varieties are *klementite* and *mackensite*.

TIEMANNITE

4[HgSe] Mercury selenide. Cubic. Tetragonal crystals; also granular masses. Steel-grey to blackish-grey. Metallic lustre. Black streak. Uneven to conchoidal fracture. No cleavage. Hardness 2·5. SG 8·19.
Occurrence: USA—Utah (Marysvale in Piute County). Germany—Harz Mountains (Zorge).

TIGERS EYE

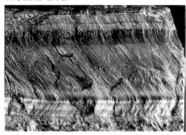

Above: tile ore with malachite
Below: tiger's eye, unpolished and polished

Silicified fibrous variety of *riebeckite*.

TILAÏTE

Igneous rock of the *gabbro* clan. A coarse-grained granular plutonic rock. Contains *diopside, olivine* and *bytownite*.
Occurrence: USSR—Ural Mountains (Tilaï-Kamen).

TILE ORE

A red or reddish-brown variety of *cuprite*. Earthy. Often contains iron oxide.

TILLEYITE

$Ca_5Si_2O_7(CO_3)_2$ Calcium silicate and carbonate. Monoclinic. Irregular grains. White. Perfect cleavage in one direction. SG c2·8.

Occurrence: Scotland—Inverness-shire (Muck). Northern Ireland—Carlingford. USA—California (Crestmore). Contact metamorphic mineral.

TILLITE

Type of *clay*. A fine stiff *clay* of variable composition which often contains rock fragments. Result of glacial erosion and redeposition. Limited to pre-Pleistocene age (*boulder clay*—Pleistocene glacial deposit).
Occurrence: Worldwide.

TILLOID

Name given to *clay* horizons which resemble *tillite* but are of non-glacial origin.
Occurrence: Congo—PreCambrian *turbidites*.

TIMAZITE

Altered igneous rock of the *diorite* clan. A fine-grained porphyritic extrusive rock. Contains phenocrysts of *hornblende* and *biotite* largely altered to *augite* and *magnetite* in a groundmass of *plagioclase* and *quartz*.
Occurrence: Yugoslavia—Serbia (Timok Valley).

TINCALCONITE

3[$Na_2B_4O_7 . 5H_2O$] Hydrated sodium

tetraborate. Hexagonal. Fine grained powder. White. Dull lustre.
Occurrence: USA—California (the Kramer district of Kern County and the Searles Lake area of San Bernardino County).

TINGUAITE

Igneous rock of the *syenite* clan. A fine-grained porphyritic hypabyssal rock. Contains phenocrysts of *orthoclase, nepheline* and *aegirine* in a groundmass of *orthoclase, nepheline, aegirine, leucite* and *katophorite*.
Occurrence: USA—Massachusetts (Manchester-by-the-Sea in Essex County), Montana (Bean Creek in Bearpaw Mountains). Norway—Hedrum. Brazil—Rio de Janeiro (Tinguá Mountains).

TIN PYRITES

Synonym of *stannite*.

TINZENITE

Synonym of *manganaxinite*.

TITANAUGITE

Titanium rich variety of *augite*.

TITANITE

Synonym of *rutile* and *sphene*.

TITANOHEMATITE

Titanium rich variety of *hematite*.

TOADS EYE TIN

Brown botryoidal or reniform variety of *cassiterite* with a radiating fibrous habit.

TODOROKITE

(Mn, Ba, Ca, Mg)Mn$_3$O$_7$.H$_2$O Hydrous manganese alkaline earths oxide. Monoclinic. Lath like crystals; also spongy reniform aggregates. Black. Metallic lustre. Perfect cleavage in two directions. Hardness 1·0–2·0. SG 3·67. Soluble in hydrochloric acid with the evolution of chlorine.
Occurrence: Japan—Hokkaido (Todoroki Mine).

TÖLLITE

Igneous rock of the *granite* clan. A fine-grained porphyritic hypabyssal rock. Contains phenocrysts of *hornblende, andesine, biotite, muscovite, quartz, orthoclase* and *garnet* in a groundmass of *quartz* and *feldspar*.
Occurrence: Austria—Tyrol (Töll near Meran).

TONALITE or QUARTZ DIORITE

Igneous rock of the *diorite* clan. A medium to fine-grained granular plutonic rock. Contains *quartz, oligoclase, biotite* and *hornblende*.
Occurrence: USA—California (Glennville), Connecticut (Barnes Hill near North Stonington). Canada—British Columbia (Dome Mountain near Vancouver). Australia—Granite Island in Encounter Bay. Austria—Tyrol (Monte Tonale).

TÖNSBERGITE

Igneous rock of the *syenite* clan. A red coarse-grained granular plutonic rock. Contains *orthoclase, andesine, pyroxene* and *biotite*.
Occurrence: Norway—Tönsberg.

TONSTEIN

Type of *clay*. A compact homogeneous rock. Contains *kaolinite* and *illite*.
Occurrence: Widespread. Associated with *coal* deposits.

TOPAZ

4[Al$_2$SiO$_4$(OH,F)$_2$] Hydrous aluminium fluosilcate. Orthorhombic. Prismatic crystals; also columnar or granular masses. Yellow or white, tinted green, blue or red.

Vitreous lustre. Colourless streak. Subconchoidal to uneven fracture. Perfect cleavage in one direction. Hardness 8·0. SG 3·4–3·65. Partially attacked by sulphuric acid.
Occurrence: England—Cornwall (St Michael's Mount). Scotland—Aberdeenshire (Cairngorm in the Cairngorm Mountains). USA—Colorado (Pikes Peak), Connecticut (Trumbull), Maine (Stoneham). Australia—Tasmania (Mount Bischoff). Norway—Fossum. Sweden—Finbo. In *granite* or *gneiss*.
Varieties are *pyrophysalite* (yellowish-white opaque crystals), *Brazilian ruby* (gemstone), and *pycnite* (columnar masses).

TOPAZITE

Altered igneous rock of the *granite* clan. A medium-grained granular hypabyssal rock. Contains *quartz* and *topaz* with minor *muscovite*.
Occurrence: Australia—Tasmania (Mount Bischoff).

TOPAZOLITE

Gem variety of *andradite* resembling *topaz*.

TORBANITE

Synonym of *boghead coal*.

TORBERNITE

$2|Cu(UO_2)_2(PO_4)_2.12H_2O|$ Hydrated copper uranium phosphate. Tetragonal. Tabular crystals; also foliated, micaceous or scaly aggregates. Shades of green. Vitreous to subadamantine lustre. Pale green streak. Perfect basal cleavage. Hardness 2·0–2·5. SG 3·22. Soluble in hydrochloric acid.
Occurrence: England—Cornwall (Redruth, St Agnes, St Austell, St Just). USA—New Mexico (White Signal district of Grant County), South Dakota (Hannibal Mine in Lawrence County), Utah (La Sal Mountains in San Juan County). South Australia—Mount Painter in the Flinders Range. Germany—Saxony (Schneeberg). Oxidation product of *uraninite*.

TORDRILLITE

Igneous rock of the *granite* clan. A fine-grained granular extrusive rock. Contains *quartz, sodaclase, albite* and minor ferromagnesian minerals.
Occurrence: USA—Alaska (Tordrillo Mountain), Colorado (Solomon Mine at Creed). Australia—New South Wales (Canbelego).

TOSCANITE

Igneous rock of the *granite* clan. A fine-grained porphyritic extrusive rock. Contains phenocrysts of *quartz* and *labradorite* in a groundmass of *sanidine, hypersthene, biotite, labradorite* and *quartz*.

Occurrence: Italy—Tuscany. Sweden—Helsingland (Dellen).

TOUCHSTONE

Synonym of *basanite*.

TOURMALINE or SCHORL

$(Na, Ca)(Li, Mg, Fe^{2+}, Al)_3(Al, Fe^{3+})_6B_3Si_6O_{27}(O,OH,F)_4$ Complex aluminium alkali metals iron and magnesium borosilicate. Hexagonal. Prismatic or acicular crystals; also massive. Normally black, brownish-black, bluish-black or reddish-black. Vitreous to resinous lustre. Colourless

Below: tourmaline

streak. Subconchoidal to uneven fracture. Poor cleavage. Hardness 7·0–7·5. SG 2·98–3·2. Unaffected by acids.
Occurrence: Worldwide. In *granite*, *gneiss*, *syenite*, *schists* and *crystalline limestones*.
Varieties are *achroïte*, *Brazilian emerald*, *Brazilian sapphire*, *dravite*, *indicolite*, *peridot*, *rubellite* and *verdelite*.

TRACHYANDESITE

Extrusive equivalent of *syenodiorite*.

TRACHYTE

Igneous rock of the *syenite* clan. A fine-

grained porphyritic extrusive rock. Contains phenocrysts of *sanidine* and *oligoclase* in a groundmass of *sanidine*, *albite*, *biotite* and *hornblende*.
Occurrence: Worldwide.
The extrusive equivalent of *syenite*.

TRACHYTOID

Physically or chemically similar to *trachyte* group of igneous rocks—for example, fine grained with a structure of parallel rods of *feldspars*.

TRAPEZOHEDRAL

See *crystal*.

TRAP ROCK

A type of rock in which either gas or oil may collect.

TRAVERTINE

Synonym of *calcareous sinter*.

TREMOLITE

$2[Ca_2Mg_5Si_8O_{22}(OH)_2]$ Hydrous calcium magnesium silicate. Monoclinic. Bladed crystals; also massive. White to dark grey. Vitreous lustre. Subconchoidal to uneven fracture. Perfect cleavage in one direction. Hardness 5·0–6·0. SG 2·9–3·2.
Occurrence: USA—New Jersey (London Grove in Chester County), New York (Gouverneur in St Lawrence County). Canada—Quebec (Litchfield in Pontiac County). Sweden—Wermland (Gulsjo). Switzerland—Canton Tessin (Campo-

longo).
In *crystalline limestones*, *gneiss*, *schists* and *granite*.
Varieties are *chrome-tremolite*, *hexagonite* and *winchite*.

TREMOLITE ASBESTOS

Massive fibrous variety of *tremolite*.

TREVORITE

$8[NiFe_2O_4]$ Nickel iron oxide. Cubic. Crystals rare, normally granular masses. Black to brownish-black. Metallic lustre. Brown streak. Poor cleavage. Hardness 5·0. SG c5·2. Strongly magnetic. Soluble with difficulty in hydrochloric acid.
Occurrence: South Africa—Transvaal (Bon Accord).

TRIASSIC

See *geological time*.

TRICLINIC

See *crystal*.

TRIDYMITE

High temperature hexagonal or orthorhombic polymorph of silica (*quartz* and *cristobalite*). Occurs in siliceous volcanic rocks, contact metamorphosed *sandstones* and stony *meteorites*.

TRIGONAL

See *crystal*.

TRIPESTONE

A concretionary variety of *anhydrite*.

TRIPHYLITE

$4[Li(Fe,Mn)PO_4]$ Lithium iron manganese phosphate. Orthorhombic. Crystals rare, normally massive. Brownish-grey or greenish-grey. Vitreous to subresinous lustre. Colourless to greyish-white streak. Uneven to subconchoidal fracture. Good cleavage in one direction. Hardness 4·0–5·0. SG c3·55. Soluble in hydrochloric acid.

Occurrence: USA—California (Pala in San Diego County), New Hampshire (Ruggles Mine at Grafton Centre). Canada—Manitoba (Pointe du Bois). Finland—Sukula. Germany—Bavaria (Hühnerkobel). In *granite pegmatites*.

TROCTOLITE or FORELLENSTEIN

Igneous rock of the *gabbro* clan. A coarse-grained granular plutonic rock. Contains *labradorite* and *olivine*.

Occurrence: England—Cornwall (Coverack Cove). Czechoslovakia—Silesia (Volpersdorf).

TROILITE

Variety of *pyrrhotite* commonly found in *iron meteorites*.

TRONA

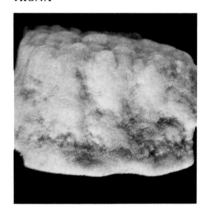

4[Na₃H(CO₃)₂.2H₂O] Hydrated sodium carbonate acid. Monoclinic. Crystals rare, normally fibrous or columnar masses. Colourless, grey or yellowish-white. Vitreous lustre. Uneven to subconchoidal fracture. Perfect cleavage in one direction. Hardness 2·5–3·0. SG 2·14. Bitter taste. Soluble in water.

Occurrence: USA—California (Searles Lake in San Bernardino County), Nevada (Fallen), Wyoming (Laramie). Hungary—Szegedin. Egypt—Nile Valley (Memphis). In saline lakes and an an efflorescence on deserts soils.

TROOSTITE

A manganese-bearing variety of *willemite*.

TROWLESWORTHITE

Igneous rock of the *syenite* clan. A red, coarse-grained granular plutonic rock. Contains *orthoclase, tourmaline* and *fluorite*.

Occurrence: England—Devon (Trowlesworthy).

Above: troostite

TRUNCATION

Replacement of an edge or corner of a crystal by a plane.

TSINGTAUITE

Igneous rock of the *granite* clan. A very fine-grained porphyritic hypabyssal rock. Contains phenocrysts of *microperthite* and *plagioclase* in a groundmass of granitic composition.

Occurrence: China—Kiautschou (Tsingtau).

TUCHOLITE

Synonym of *thucholite*.

TUFA

Carbonate rock. A porous concretionary deposit of calcium carbonate. Precipitated from spring or stream waters in *limestone* areas.

Occurrence: Worldwide. In *limestone* districts.

TUFF

Pyroclastic extrusive igneous rock composed of fragments with a diameter less than 2mm. Three varieties: *vitric tuff* (containing *glass* fragments), *lithic tuff* (con-

taining rock fragments) and *crystal tuff* (containing mineral fragments).

Occurrence: Worldwide.

TUNGSTATE

A salt of tungstic acid.

TUNGSTITE

H₂WO₄ Tungstic acid. Orthorhombic. Aggregates of acicular crystals; also earthy masses. Yellow or yellowish-green. Resinous to earthy lustre. Perfect basal cleavage. Hardness 2·5. SG c5·5.

Occurrence: England—Cornwall (Drakewall Mine near Callington), Cumberland (Carrock Fell). USA—Connecticut (Lanes Mine near Monroe), Idaho (Blue Wing district of Lemhi County). Canada—British Columbia (Kootenay Mine at Salmo). Australia—Tasmania (Ben Lomand).

Oxidation product of *wolframite*.

TURBIDITE

Arenaceous or *argillaceous rock* deposited by turbidite flow and displaying graded bedding, slump structures and bottom structures.

Occurrence: Worldwide.

TURGITE

Synonym of *hydrohematite*.

TURITE

Synonym of *hydrohematite*.

TURQUOISE or HYDRARGILLITE

CuAl₆(PO₄)₄(OH)₈.5H₂O Basic hydrated copper aluminium phosphate. Triclinic. Crystals rare, normally massive. Pale blue, greyish-green or bluish-green to apple-green. Vitreous to waxy lustre. White to pale green streak. Conchoidal to smooth fracture. Perfect basal cleavage. Hardness 5·0–6·0. SG 2·6–2·85. Soluble with difficulty in hydrochloric acid.

Occurrence: England—Cornwall (St. Austell). USA—California (Grove Mine in San Bernardino County), Colorado (Leadville in Lake County), Virginia (Lynch Station in Campbell County). Germany—Saxony (Messbach). Persia—

Kharasan (Madán).
Secondary weathering product of aluminium rich igneous or sedimentary rocks. Varieties are *chalcocite*, *chalcosiderite*, *chalchuite*, *faustite* and *rashleighite*.

TUXTLITE or DIOPSIDE-JADEITE

Pyroxene intermediate in composition and properties between *diopside* and *jadeite*.

TWINNED CRYSTAL

A composite crystal consisting of a number of usually equal and similar crystals united in reversed positions with respect to each other, either with symmetrical parts in contact (contact twin) or by the parts growing out of one another (penetration twin). Repeated twinning in one plane leads to the formation of polysynthetic twins.

TYUYAMUNITE

$4[Ca(UO_2(_2(VO_4)_2 . 10H_2O]$ Hydrated calcium uranium vanadate. Orthorhombic. Scales or laths; also massive. Yellow to greenish-yellow. Adamantine to waxy lustre. Perfect basal cleavage. Hardness $c2·0$. SG $3·67-4·35$. Soluble in hydrochloric acid.
Occurrence: USA—Colorado (Paradox Valley in Montrose County), Utah (Henry Mountains in Garfield County). USSR—Turkestan (Tyuya Muyun).

UGRANDITE GARNETS

Group name for *uvarovite*, *grossular* and *andradite*.

UINTAHITE

Complex hydrocarbons. Amorphous. Massive. Black. Vitreous lustre. Dark brown streak. Conchoidal fracture. Hardness $2·0-2·5$. SG $c1·07$.
Occurrence: USA—Utah (Uintah Valley near Fort Duchesne).

ULEXITE

$2[NaCaB_5O_8 . 8H_2O]$ Hydrated sodium calcium borate. Triclinic. Capillary or acicular crystals; also massive aggregates. Colourless to white. Vitreous to silky lustre. Uneven fracture. Perfect cleavage in one direction. Hardness $2·5$. SG $c2·0$. Decomposes in hot water.
Occurrence: USA—California (Death Valley in Inyo County), Nevada (Esmeralda County). Canada—Nova Scotia (Windsor). Chile—Tarapacá. In *playa* lake deposits.

ULLMANNITE

$4[NiSbS]$ Nickel sulphide and antimonide. Cubic. Cubic crystals. Steel-grey to silver-white. Metallic lustre. Greyish-black streak. Uneven fracture. Perfect basal cleavage. Hardness $5·0-5·5$. SG $6·65$. Decomposes in nitric acid.
Occurrence: England—Durham (Brancepeth Colliery), Northumberland (Settlingstones Mine near Fourstones). Australia—New South Wales (Broken Hill). Germany—Westphalia (Siegen).
Cobalt rich variety is *willyamite*.

ULRICHITE

Igneous rock of the *syenite* clan. A fine-grained porphyritic hypabyssal rock. Contains phenocrysts of *nepheline*, *sanidine*, *barkevikite* and *olivine* in a groundmass of *sanidine*, *hornblende* and *aegirine*.
Occurrence: New Zealand—Dunedin (Portobello).
Also synonym of *uraninite*.

ULTRABASIC

Igneous rocks containing less than 45 per cent silica. No *quartz* or *feldspar* developed; only ferromagnesian minerals. Main rock types: *peridotite*, *pyroxenite* and *picrite*.

ULTRAMARINE

Synonym of *lazurite*.

UMPTEKITE

Igneous rock of the *syenite* clan. A fine to coarse-grained granular plutonic rock. Contains *orthoclase*, *microcline-microperthite*, *sodalite*, *arfvedsonite*, *aegirine*, *biotite* and *nepheline*.
Occurrence: USSR—Kola Peninsula (Umptek).

UNAKITE

Altered igneous rock of the *granite* clan. A coarse-grained granular plutonic rock affected by dynamic metamorphism. Contains *orthoclase*, *quartz* and *epidote*.
Occurrence: USA—North Carolina (Great Smoky Mountains of the Unaka Range), Virginia (Milam's Gap at Blue Ridge near Luray).

UNCOMPAHGRITE

Feldspathoid igneous rock. A bluish-grey coarse-grained granular plutonic rock. Contains *melilite, augite, andradite* and *magnetite*.
Occurrence: USA—Colorado (Iron Hill near Powderhorn in Gunnison County).

UNCTUOUS

Fatty; oily; smooth and greasy in texture or appearance.

UNDERSATURATED

See saturated.

URALIAN EMERALD

Gem variety of *andradite*.

URALITE

Actinolite pseudomorphed after *pyroxene*.

URALITISATION

The alteration of an original *pyroxene* in an igneous rock to a mass of fibrous *amphibole*.

URANINITE, ULRICHITE or URANPECHERZ

UO_2 Uranium dioxide. Cubic. Octahedral crystals; also massive. Black, brownish-black, greyish-black or greenish-black. Submetallic to greasy lustre. Brownish-black to olive-green streak. Uneven to conchoidal fracture. Hardness 5·0–6·0. SG

c10·8. Slowly attacked by hydrochloric acid.
Occurrence: Worldwide. Varieties are *broeggerite, cleveite, pitchblende* and *nasturan.*

URANOPHANE or URANOTILE

$Ca(UO_2)_2Si_2O_7 \cdot 6H_2O$ Hydrous calcium uranium silicate. Orthorhombic. Acicular crystals; also massive. Yellow. Vitreous lustre. Hardness 2·0–3·0. SG 3·81–3·9. Soluble in warm hydrochloric acid with the separation of silica.
Occurrence: USA—North Carolina (the Mica Mines of Mitchell County). Czechoslovakia—Silesia (Kupferberg). Germany—Saxony (Schneeberg).

URANOTHORIANITE

Uranium rich variety of *thorianite.*

URANOTHORITE

Uranium rich variety of *thorite.*

URANOTILE

Synonym of *uranophane.*

URANPECHERZ

Synonym of *uraninite.*

URTITE

Feldspathoid igneous rock. A medium-grained granular plutonic rock. Contains *nepheline, aegirine* and *apatite.*
Occurrence: USSR—Kola Peninsula (Lujavr-Urt).

USSINGITE

$Na_2AlSi_3O_8OH$ Basic sodium aluminium silicate. Triclinic. Granular masses. Reddish-violet. Vitreous lustre. Perfect cleavage in one direction. Hardness 6·0–7·0. SG c2·5. Soluble in hydrochloric acid.
Occurrence: Greenland—Kangerdluarsuk. USSR—Kola Peninsula (Alluaiva). In *pegmatite.*

UTAHLITE

Compact nodular variety of *variscite.*

UVAROVITE

$8|Ca_3Cr_2Si_3O_{12}|$ Calcium chromium silicate. Cubic. Rhombdodecahedral crystals. Emerald-green. Vitreous lustre. Greenish-white streak. Hardness 7·5. SG 3·42.

Occurrence: Scotland—Shetlands (Unst). Spain—Pyrenees (Venasque). USSR—Ural Mountains (Kyshtymsk).
A *garnet*.

VAESITE

NiS_2 Nickel sulphide. Cubic. Octahedral crystals. Grey.
Occurrence: Congo—Kasompi Mine.

VALBELLITE

Igneous rock of the *ultrabasic* clan. A fine-grained granular hypabyssal rock. Contains *bronzite, olivine, hornblende* and *magnetite*.
Occurrence: Italy—Piedmont (Val Bello).

VALENCIANITE

Synonym of *adularia*.

VALENTINITE

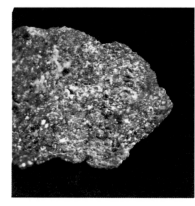

$4[Sb_2O_3]$ Antimony trioxide. Orthorhombic. Prismatic or tabular crystals; also massive. Colourless to white. Adamantine lustre. White streak. Perfect cleavage in one direction. Hardness 2·5–3·0. SG 5·76. Soluble in hydrochloric acid.
Occurrence: USA—California (Picahotes Mine in San Benito County), Oregon (Ochoco district in Crook County). Canada—Quebec (South Ham in Wolfe County). Germany—Saxony (Freiberg). Oxidation product of *stibnite*.

VANADATE

A salt of vanadic acid.

VANADINITE

$2[Pb_5(VO_4)_3Cl]$ Lead chloride and vanadate. Hexagonal. Prismatic crystals. Shades of red or yellow. Subresinous to subadamantine lustre. White or yellow streak. Uneven to conchoidal fracture. Hardness 2·75–3·0. SG 6·88. Soluble in hydrochloric acid with the precipitation of lead chloride.
Occurrence: Scotland—Dumfriesshire

(Wanlockhead). USA—Arizona (Old Yuma Mine in Pima County), New Mexico (Hillsboro in Sierra County). Austria—Carinthia. USSR—Ural Mountains (Berescovsk).
Oxidation product of lead deposits.
Variety is *endlicheite*.

VARIOLITE

Igneous rock of the *diorite* clan. A greyish-green rock containing rounded bodies (varioles) in a very fine-grained dense groundmass. Contains varioles of *plagioclase, augite* and *ilmenite* in a groundmass of *augite* and devitrified *glass*.
Occurrence: USSR—Lake Onega (Jalguba).

VARISCITE

$8[AlPO_4.2H_2O]$ Hydrated aluminium phosphate. Orthorhombic. Crystals rare, normally massive. Green. Vitreous to waxy lustre. White streak. Uneven to splintery

fracture. Good cleavage in one direction. Hardness 3·5–4·5. SG c2·57.
Occurrence: USA—Arkansas (Montgomery in Garland County), Utah (Fairfield in Utah County). Western Australia—Ninghanboun Hills. Germany—Saxony (Mersbach).
Varieties are *amatrice* and *utahlite*.

VARNSINGITE

Igneous rock of the *gabbro* clan. A coarse-grained granular *pegmatite*. Contains *albite, augite, apatite* and *sphene*.
Occurrence: Sweden—Nordinga district.

VARVE CLAY

Graded laminated *clay* produced on the floors of glacial lakes. 'Coarse' layer during the summer with an inflow of detritus and 'fine' layer during the winter with the settling of finer particles in the frozen over lake.
Occurrence: Widespread in glaciated areas.

VARVITE

Name given to a graded laminated *clay* which resembles *varve clay* but was produced by non-glacial processes.

VATERITE

Synonym of *calcite*.

VAUGNERITE

Igneous rock of the *diorite* clan. A coarse-grained granular hypabyssal rock. Contains *andesine, biotite, hornblende, apatite* and *quartz*.
Occurrence: France—Lyons (Vaugneray).

VEIN

Sheet of mineral or minerals intruded into small faults, fissures and joints, often restricted to those of economic significance. Many varieties, depending on situation, attitude and extent.

VELARDEÑITE

Synonym of *gehlenite*.

VENANZITE

Synonym of *euktolite*.

VERD ANTIQUE

Green impure variety of *calcite* with *serpentine* used as a gemstone.

VERDELITE

Green variety of *tourmaline*.

VERITE

Igneous rock of the *ultrabasic* clan. A black pitch like, fine-grained porphyritic extrusive rock. Contains phenocrysts of *olivine* and *phlogopite* in a groundmass of *phlogopite*, *diopside* and a little *glass* of *sanidine* composition.
Occurrence: Spain—Cabo de Gata (Vera). Variety is *fortunite*.

VERMICULAR

Worm-like.

VERMICULITE

$2[(Mg,Fe^{2+},Fe^{3+},Al)_{6-7}(Si,Al)_8O_{20}(OH)_4 \cdot 8H_2O]$ Basic hydrated magnesium iron aluminium aluminosilicate. Monoclinic. Foliated scales. Colourless, yellow, green or brown. Pearly lustre. Perfect basal cleavage. Hardness 1·5. SG 2·3. Decomposes in hydrochloric acid.
Occurrence: England—Lancashire (Walney Island). USA—Massachusetts (Milbury near Worcester), Pennsylvania (West Chester).

VESICLES

See vesicular.

VESICULAR

Containing small roughly spherical cavities originally formed by bubbles of trapped gas, sometimes later filled by another mineral. *cf* amygdaloidal.

VESUVIANITE

Synonym of *idocrase*.

VIBERTITE

Synonym of *bassanite*.

VINTLITE

Igneous rock of the *diorite* clan. A greenish-black, fine-grained porphyritic hypabyssal rock. Contains phenocrysts of *hornblende, epidote, oligoclase* and *quartz* in a groundmass of *plagioclase, hornblende, diopside, bronzite* and *biotite*.
Occurrence: Austria—Tyrol (Vintl near Klausen).

VIOLANE

Dark violet-blue fibrous variety of *diopside*.

VIOLARITE

$8[FeNi_2S_4]$ Iron nickel sulphide. Cubic. Octahedral crystals; also massive. Greyish-violet. Metallic lustre. Uneven to subconchoidal fracture. Poor cleavage. Hardness 4·5–5·6. SG *c*4·8. Decomposes in nitric acid with the separation of sulphur.
Occurrence: USA—California (Julian).

VIRIDINE

Synonym of *manganandalusite*.

VISCOSITY

Measure of resistance to flow in a liquid or gas, so with significance when applied to magmas and lavas in motion.

VISCOUS

High measure of resistance to flow in a liquid or gas.

VISÉITE

$2[NaCa_5Al_{10}Si_3P_5O_{30}(OH,F)_{18} \cdot 16H_2O]$ Hydrous aluminium calcium sodium fluosilicophosphate. Cubic. Massive. White. Hardness 3·0–4·0. SG 2·2.
Occurrence: Belgium—Visé.
A *zeolite*.

VITRAIN

Constituent of *coal*. Lenticular. Brownish-black. Vitreous lustre. Conchoidal fracture. Brittle.

VITREOUS

Glass-like.

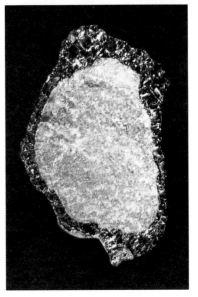

Above: vitrain

VITRIC TUFF

Type of *tuff* composed mainly of *glass* fragments.

VIVIANITE

$2[Fe_3(PO_4)_2 \cdot 8H_2O]$ Hydrated iron phosphate. Monoclinic. Prismatic crystals; also earthy masses. Colourless; oxidises to pale blue or greenish-blue. Vitreous or dull lustre. Colourless to bluish-white streak. Fibrous fracture. Perfect cleavage in one direction. Hardness 1·5–2·0. SG *c*2·68. Easily soluble in hydrochloric acid.

VOGESITE

Igneous rock of the *syenite* clan. A fine-grained porphyritic hypabyssal rock. Contains phenocrysts of *hornblende* and *diopside* in a groundmass of *orthoclase*, *hornblende* and *diopside*.
Occurrence: France—Vosge Mountains (Ardlautal).

VOLCANIC

Associated with activity of volcanos.

VOLCANIC BRECCIA

See *breccia* (*volcanic*).

VOLCANIC CONGLOMERATE

See *conglomerate* (*volcanic*).

VOLCANIC DUST

Pyroclastic igneous rock with a grain size of less than 0·25mm.

VOLCANIC GLASS

Constituent of extrusive igneous rocks. Rapid chilling solidifies magma before minerals can crystallise, producing natural *glass*.
Occurrence: Worldwide. In groundmass of extrusive rocks. Occasionally forms whole rock (*obsidian* and *pitchstone*).

VOLCANIC ROCKS

Synonym of extrusive igneous rocks.

VOLCANITE

Igneous rock of the *diorite* clan. A fine-grained porphyritic extrusive rock. Contains phenocrysts of *anorthoclase*, *andesine* and *augite* in a groundmass of *feldspar*, *augite* and *glass*.
Occurrence: Italy—Lipari Islands (volcano).

VOROBYEVITE, MORGANITE or ROSTERITE

Sodium and caesium rich variety of *beryl*.

VREDENBURGITE

Orientated intergrowth of *jacobsite* and *hausmannite*.

VULPINITE

A granular variety of *anhydrite* found at Vulpino in Lombardy, Italy.

VULSINITE

Igneous rock of the *syenite* clan. A mottled grey, fine-grained porphyritic extrusive rock. Contains phenocrysts of *sodaclase*, *anorthite* and *augite* in a groundmass of *sodaclase*, *labradorite* and *augite*.
Occurrence: Italy—Vulsinian district (Bolsena).

WACKESTONE

Textural name for type of *limestone*.

WAD

Variable group of hydrated oxides of manganese, cobalt, copper, iron, aluminium or barium. Amorphous. Earthy or

compact masses. Black, bluish-black or brownish-black. Dull lustre. Variable streak. Hardness 1·0–6·5. SG 2·8–4·4.
Occurrence: Worldwide. Alteration product of primary ores in oxidising conditions in bog, lake or shallow marine environment.
Varieties are *lithiophorite* and *skemmatite*.

WARDITE

$Na_4CaAl_{12}(PO_4)_8(OH)_{18}.6H_2O$ Basic hydrated sodium calcium aluminium phosphate. Tetragonal. Pyramidal crystals; also massive. Colourless, pale green or bluish-green. Vitreous lustre. Perfect basal cleavage. Hardness 5·0. SG 2·87. Soluble with difficulty in hydrochloric acid.
Occurrence: USA—New Hampshire (Beryl Mountain near West Andover), Utah (Fairfield in Utah County). France—Département Creuse (Montebras).

WAVELLITE or HYDRARGILLITE

$2[Al_6(PO_4)_4(OH)_6.9H_2O]$ Basic hydrated aluminium phosphate. Orthorhombic. Crystals rare, normally globular aggre-

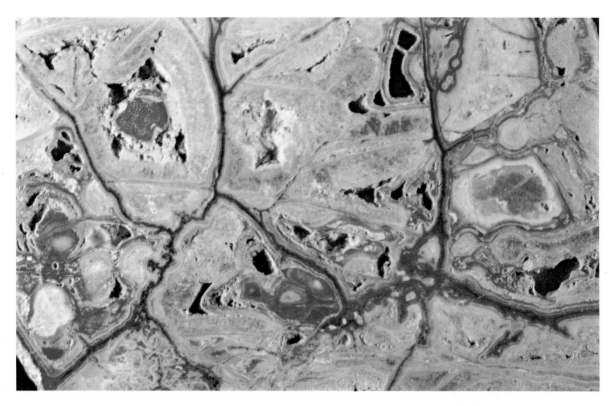

Above: wardite. *Below:* wavellite

gates, crusts or stalactite-like. Greenish-white, green or yellow. Vitreous lustre. White streak. Uneven to subconchoidal fracture. Perfect cleavage in one direction. Hardness 3·25–4·0. SG 2·36. Easily soluble in hydrochloric acid.
Occurrence: England—Devon (Barnstaple). Republic of Ireland—County Cork (Kinsale), County Tipperary (Clonmel). USA—Arkansas (Hot Springs in Garland County), Florida (Dunellen in Marion County), Pennsylvania (General Trimble's Mine in Chester County). Germany—Saxony (Frankenberg).

WAVELLITE (of Dewey)

Synonym of *gibbsite* (*of Torrey*).

WAX OPAL

Yellow variety of *opal* with a waxy lustre.

WEATHERING

Initial stage of denudation, wearing down of rocks on the surface without their being moved (see erosion) by temperature effects, wind, precipitation, living agents, and chemical action.

WEBSTERITE

Igneous rock of the *ultrabasic* clan. A coarse-grained granular plutonic rock. Contains *hypersthene* and *diopside*.
Occurrence: USA—Maryland (Johnny Lake Road in Baltimore County), North Carolina (Webster).

WEHRLITE

Igneous rock of the *ultrabasic* clan. A course-grained granular plutonic rock. Contains *olivine*, *dialage* and *hornblende*.
Occurrence: New Zealand—South Westland (Lower Cascade Valley). Hungary—Szarvaskö.

WEINSCHENKITE

Variety of *hornblende* enriched in ferrous iron and aluminium.

WEISELBERGITE

Igneous rock of the *diorite* clan. A fine-grained porphyritic extrusive rock. Contains phenocrysts of *plagioclase* and *augite* in a groundmass of *augite*, *plagioclase* and *glass*.
Occurrence: Germany—Saar/Nahe region (Weiselberg near St Weidel).

WELDED TUFF

Synonym of *ignimbrite*.

WELLSITE

$(Ba,Ca,K_2)Al_2Si_3O_{10} \cdot 3H_2O$ Hydrated barium calcium potassium aluminosilicate. Monoclinic. Prismatic crystals. Colourless to white. Vitreous lustre. No cleavage. Hardness 4·0–4·5. SG 2·28–3·37.
Occurrence: USA—North Carolina (Buck Creek Corundum Mine in Clay County). A *zeolite*.

WENNEBERGITE

Igneous rock of the *syenite* clan. A fine-grained porphyritic hypabyssal rock. Contains phenocrysts of *orthoclase*, *biotite* and *quartz* in a groundmass of *apatite*, *sphene* and *glass*.
Occurrence: Germany—Ries (Wenneberg).

WERNERITE

Calcium sodium aluminosilicate sulphate carbonate and chloride. Formula uncertain. Tetragonal. Pyramidal crystals; also massive. White, often tinted red, green or blue. Vitreous to pearly lustre. Colourless streak. Subconchoidal fracture. Distinct cleavage in two directions. Hardness 5·0–6·0. SG 2·66–2·73. Imperfectly decomposes in hydrochloric acid.
Occurrence: USA—New York (Two Ponds in Orange County), Pennsylvania (Elizabeth Mine in Chester County). Norway—Arendal.
A *scapolite*.

WESTANITE

Brick-red radiating crystalline variety of hydrous *sillimanite*.

WHITE PYRITES

Synonym of *marcasite*.

WHITNEYITE

Variety of native *copper* containing arsenic.

WICHTISITE

Igneous rock of the *gabbro* clan. A very fine-grained hypabyssal rock composed entirely of *glass* of gabbroic composition.

Occurrence: Finland—Wichtis Parish (Kukkaran).

WILKEITE

$2[Ca_5(Si,P,S)_3O_{12}(O,OH,F)]$ Complex basic calcium silicate. Hexagonal. Prismatic crystals. Pink or yellow. Subresinous lustre. Imperfect basal cleavage. Hardness 5·0. SG c3·2. Easily soluble in hydrochloric acid with a residue of flocculent silica.
Occurrence: USA—California (Crestmore in Riverside County). In *crystalline limestone*.

WILLEMITE

$6[Zn_2SiO_4]$ Zinc silicate. Hexagonal. Prismatic crystals; also massive. White or greenish yellow. Vitreous or resinous lustre. Colourless streak. Conchoidal to uneven fracture. Cleavage visible in one direction. Hardness 5·5. SG 3·89–4·18. Gelatinises in hydrochloric acid.
Occurrence: USA—New Jersey (Mine Hill at Franklin Furnace), New Mexico (Merritt Mine in Socorro County). Belgium—Liege (Altenberg).
Variety is *troostite*.

WILLIAMSITE

Impure apple-green lamellar variety of *antigorite*.

WILLYAMITE

Cobalt rich variety of *ullmannite*.

WILSONITE

Magnesium potassium aluminosilicate. Formula uncertain. Crystal system uncertain. Massive. Reddish-white to red. Vitreous lustre. Hardness 3·5. SG c2·8.
Occurrence: USA—New York (Antwerp in St Lawrence County). Canada—Bathurst.
An altered *scapolite*.

WILUITE

Synonym of *grossular* and *idocrase*.

WINCHITE

Blue variety of *tremolite*.

WINDSORITE

Igneous rock of the *granite* clan. A fine-grained granular *aplite*. Contains *microperthite, orthoclase, oligoclase, quartz* and *biotite*.
Occurrence: USA—Vermont (Little Asatrey Mountain near Windsor).

WISERINE

Synonym of *anatase*.

WITHAMITE

Red or yellow variety of *epidote* from Glencoe in Scotland.

WITHERITE

4[BaCO$_3$] Barium carbonate. Orthorhombic. Pyramidal crystals; also massive. Colourless to milky white. Vitreous lustre. White streak. Uneven fracture. Good cleavage in one direction. Hardness 3·0–3·5. SG c4·3. Soluble in hydrochloric acid.
Occurrence: England—Cumberland (Alston Moor), Durham (Morrison Mine at Annfield Plain), Northumberland (Fallowfield near Hexham). USA—Arizona (Castle Dome in Yuma County), California (Platina in Shasta County). Canada—Ontario (Thunder Bay district).

WOLFRAM

Synonym of *wolframite*.

WOLFRAMITE or WOLFRAM

2[(Fe,Mn)WO$_4$] Iron manganese tungstate. Monoclinic. Prismatic crystals; also massive. Dark grey, brownish-black or black. Submetallic to metallic lustre. Reddish-brown, brownish-black streak. Uneven fracture. Perfect cleavage in one direction. Hardness 4·0–4·5. SG c7·4. Decomposes in hot concentrated hydrochloric acid.
Occurrence: Worldwide. In *granite pegmatites*, contact metamorphic rocks and hydrothermal veins. Main ore of tungsten.

WOLLASTONITE

6[CaSiO$_3$] Calcium silicate. Monoclinic. Tabular crystals; also massive. White. Vitreous lustre. White streak. Uneven fracture. Perfect cleavage in one direction. Hardness 4·5–5·0. SG 2·8–2·9. Decomposes in hydrochloric acid with the separation of silica.
Occurrence: Northern Ireland—Mountains of Mourne (Dunmore Head). USA—New York (Diana in Lewis County), Pennsylvania (Attleborough in Buck County). Canada—Quebec (Morin). Finland—Pargas.

WOODENDITE

Igneous rock of the *syenite* clan. A dull black, fine-grained granular extrusive rock. Contains *augite, enstatite, olivine* and *magnetite* in a *glass* of *feldspar* composition.
Occurrence: Australia—Victoria (Woodend).

WOOD OPAL

Yellow or brown variety of *opal* petrifying wood.

WOODRUFFITE

$(Zn,Mn^{2+})_2Mn_5^{4+}O_{12}\cdot 4H_2O$ Hydrated manganese zinc oxide. Crystal system uncertain. Botryoidal masses. Black. Dull lustre. Brownish-black streak. Conchoidal fracture. Hardness 4·5. SG 3·71.
Occurrence: USA—New Jersey (Sterling Hill).

WOOD TIN

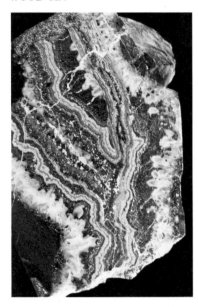

Botryoidal or reniform variety of *cassiterite* with a radiating fibrous habit.

WULFENITE

4|PbMoO₄| Lead molybdate. Tetragonal. Tabular crystals; also granular masses. Orangish-yellow to yellow. Resinous to adamantine lustre. White streak. Subconchoidal to uneven fracture. Good cleavage in one direction. Hardness 2·75–3·0. SG 6·5–7·0. Soluble in concentrated sulphuric acid.

Occurrence: USA—Arizona (Red Cloud Mine in Yuma County), Nevada (Eureka in Eureka County), New Mexico (Organ Mountains in Dona Ana County), Pennsylvania (Wheatley Mine in Chester County). Australia—New South Wales (Broken Hill). Germany—Saxony (Zinnwald).

WURTZITE

2|ZnS| Zinc sulphide. Hexagonal. Pyramidal or prismatic crystals; also fibrous or columnar masses. Brownish-black. Resinous lustre. Brown streak. Cleavage visible in one direction. Hardness 3·5–4·0. SG 3·98. Soluble in hydrochloric acid with the evolution of hydrogen sulphide.
Occurrence: England—Cornwall (Liskeard). USA—Montana (Butte), Utah (Frisco in Beaver County). Czechoslovakia—Bohemia (Mies).

WYOMINGITE

Feldspathoid igneous rock. A dull pinkish-red, fine-grained porphyritic extrusive rock. Contains phenocrysts of *phlogopite* in a groundmass of *leucite* and *diopside*.
Occurrence: USA—Wyoming (Fifteen Mile Spring in the Leucite Hills).

XANTHITANE

Synonym of *anatase*.

XANTHITE

A yellowish-brown variety of *idocrase*.

XANTHOPHYLLITE

Aluminium rich variety of *clintonite*.

XANTHOSIDERITE (of Schmid)

Synonym of *goethite*.

XENOLITH

Inclusion of earlier material such as country rock in an igneous mass.

XENOMORPHIC

Without external crystal appearance yet crystalline internally.
See metamict.

XENOTIME

4|YtPO₄| Yttrium phosphate. Tetragonal. Prismatic crystals. Yellowish-brown to reddish-brown, red or yellow. Vitreous lustre. Pale brown, yellowish or reddish streak. Uneven to splintery fracture. Good cleavage in one direction. Hardness 4·0–5·0. SG 4·4–5·1.
Occurrence: USA—Colorado (Cheyenne Canyon in El Paso County). Norway—Arendal. Brazil—Minas Gerais (Dattas).

XONOTLITE

$CaSiO_3\cdot 2H_2O$ Hydrated calcium silicate. Monoclinic. Massive. White, bluish-grey or pink. Splintery fracture. Hardness 6·5. SG c2·71. Decomposes in hydrochloric acid with the separation of silica.
Occurrence: Scotland—Isle of Mull (Kilfinnichan). Mexico—Tetela de Xonotla.

YAMASKITE

Igneous rock of the *gabbro* clan. A coarse-

grained granular plutonic rock. Contains *hornblende, titanaugite* and *anorthite*.
Occurrence: Canada—Quebec (Mount Yamaska in the Monteregian Hills).

YATALITE

Igneous rock of the *diorite* clan. A coarse-grained granular *pegmatite*. Contains *uralite, albite, titanite* and *quartz*.
Occurrence: South Australia—The Hundreds of Yatala.

YELLOW GROUND

A superficial oxidised deposit covering *blue ground*.
Occurrence: South Africa—Kimberley.

YENITE

Synonym of *ilvaite*.

YODERITE

$4[(Al,Mg,Fe)_2Si_2(O,OH)_5]$ Basic aluminium magnesium iron silicate. Monoclinic. Lath shaped crystals. Purple. Indistinct cleavage. Hardness 6·0. SG 3·39.
Occurrence: Tanzania—Kangwa (Mantia Hill).

YTTROCERITE

Variety of *fluorite* with cerium replacing calcium and minor amounts of yttrium present.

YTTROFLUORITE

Granular massive variety of *fluorite* with yttrium replacing calcium.

YTTROTITANITE

Synonym of *keilhauite*.

YUGAWARALITE

$Ca_4Al_7Si_{20}O_{54} \cdot 14H_2O$ Hydrated calcium aluminium silicate. Monoclinic. Tabular crystals. Colourless or white. Vitreous lustre. Imperfect cleavage. Hardness 4·5. SG $c2\cdot2$.
Occurrence: Japan—Kanagawa Prefecture (Yugawara Hotspring).
A *zeolite*.

YUKSPORITE

Calcium strontium barium and alkalis silicate and fluoride. Formula uncertain. Crystal system uncertain. Fibrous or lamellar masses. Red.
Occurrence: USSR—Kola Peninsula (Yukspor).

ZARATITE or EMERALD NICKEL

$Ni_3CO_3(OH)_4 \cdot 4H_2O$ Basic hydrated nickel carbonate. Cubic. Massive. Emerald-green. Vitreous to greasy lustre. Green streak. Conchoidal fracture. Hardness 3·5. SG 2·57–2·69. Easily soluble with effervescence in heated hydrochloric acid.
Occurrence: Scotland—Shetlands (Swinaness near Unst). USA—Pennsylvania (Lows Mine in Lancaster County). Australia—Tasmania (Heazlewood).

ZEOLITES

Mineral group. Hydrated aluminosilicates

of alkali metals and alkaline earths. Secondary minerals formed by the alteration of *feldspars*. Occur in cavities in igneous rocks and in ore deposits. Main members: *analcite, chabazite, edingtonite harmotome, heulandite, laumontite, mesolite, natrolite, phillipsite, scolecite, stilbite* and *thomsonite*.

ZINC BLENDE

Synonym of *blende*.

ZINCITE

$2[ZnO]$ Zinc oxide. Hexagonal. Crystals rare, normally massive. Orangish-yellow to red. Subadamantine lustre. Orangish-yellow streak. Conchoidal fracture. Perfect cleavage in one direction. Hardness 4·0. SG $c5\cdot7$. Soluble in hydrochloric acid.
Occurrence: USA—New Jersey (Franklin). Australia—Tasmania (Heazlewood Mine). Italy—Tuscany (Bottino).

ZINCKENITE

$12[Pb_6Sb_{14}S_{27}]$ Lead antimony sulphide. Hexagonal. Prismatic crystals; also massive. Steel-grey. Metallic lustre. Steel-grey streak. Uneven fracture. Indistinct cleavage. Hardness 3·0–3·5. SG 5·3. Soluble in hot hydrochloric acid with the evolution of hydrogen sulphide and the separation of lead chloride.
Occurrence: USA—Colorado (Brobdig-

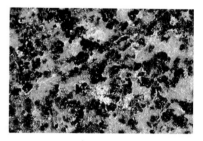

ZOBTENITE

Variety of *flaser gabbro* from Germany.

ZOISITE

nag Mine in San Juan County), Nevada (Morey in Nye County). Australia—Tasmania (Magnet Mine at Dundas). Germany—Harz Mountains (Wolfsberg).

ZINNWALDITE

$2[K_2(Li, Fe, Al)_6(Si, Al)_8O_{20}(F, OH)_4]$ Basic potassium lithium iron and aluminium fluo-aluminosilicate. Monoclinic. Tabular crystals. Pale violet, yellow or brown. Pearly lustre. Perfect basal cleavage. Hardness $2 \cdot 5$–$3 \cdot 0$. SG $2 \cdot 62$–$3 \cdot 2$.
Occurrence: England—Cornwall (St Just). Germany—Saxony (Zinnwald). A *mica*.

ZIPPEITE

$2[(UO_2)_3(SO_4)_2(OH)_2.8H_2O]$ Basic hydrated uranium sulphate. Orthorhombic. Crystals rare, normally crusty or earthy masses. Orangish-yellow. Dull to silky lustre. Perfect cleavage in one direction.
Occurrence: USA—Colorado (Gilpin County), Utah (Fruita in Wayne County). Czechoslovakia—Bohemia (Joachimsthal).

ZIRCON

$4[ZrSiO_4]$ Zirconium silicate. Tetragonal. Prismatic crystals; also irregular masses. Colourless, often tinted yellow, green, red or brown. Adamantine lustre. Colourless streak. Conchoidal fracture. Cleavage visible in one direction. Hardness $7 \cdot 5$. SG $4 \cdot 68$–$4 \cdot 7$.
Occurrence: Widespread. In *granites, syenites, schists* and *crystalline limestones*. Varieties are *cyrtolite, hyacinth, jargon* and *zirconite*.

ZIRCONITE

A grey or brown variety of *zircon*.

$4[Ca_2Al_3Si_3O_{12}OH]$ Hydrous calcium aluminium silicate. Orthorhombic. Prismatic crystals; also massive. Greyish-white, grey, greenish-grey, green, yellowish-brown or red. Vitreous lustre. Colourless streak. Uneven to subconchoidal fracture. Perfect cleavage in one direction. Hardness $6 \cdot 0$–$6 \cdot 5$. SG $3 \cdot 25$–$3 \cdot 37$.
Occurrence: Scotland—Inverness-shire (Glen Urquhart), Ross-shire (Loch Garve). USA—North Carolina (Cullakenee Mine in Clay County), Pennsylvania (West Bradford in Chester County). Norway—Arendal.
Varieties are *aryolite* and *thulite*.

ZUNYITE

$4[Al_{13}Si_5O_{20}(OH,F)_{18}Cl]$ Basic aluminium fluosilicate and chloride. Cubic. Tetrahedral crystals. Colourless. Vitreous lustre. Hardness $7 \cdot 0$. SG $2 \cdot 9$.
Occurrence: USA—Colorado (Zuni Mine in San Juan County).

ZUSSMANITE

$K(Fe^{2+},Mg,Mn,Al)_{13}(Si,Al)_{18}O_{42}(OH)_{14}$ Basic aluminosilicate of potassium and iron. Hexagonal. Tabular crystals. Pale green. Perfect cleavage in one direction.
Occurrence: USA—California (Laytonville district of Mendocino County). In metamorphosed *shales, ironstones* and *limestones*.

LIST OF ILLUSTRATIONS

The colour illustrations are listed below in the order in which they appear in the book. Each picture is identified by page number and position (indicated by a letter). The letters give the order in which the pictures appear in the columns of text; for two or three column pictures, the position is taken as being that of the left hand edge.

Other information provided is the linear magnification, the source of the specimen and its origin (where known, this is given as accurately as possible). The sources of specimens are indicated by letters after the magnification:

a National Museum of Wales, Cardiff

b County Museum, Truro, Cornwall

c Glenjoy Lapidary Supplies, Wakefield, Yorkshire

d Rocks & Minerals, Cheltenham, Gloucestershire

e Other sources

48b	**Amethyst (double terminated crystals)** Rhodesia	x1.30 d
48c	**Sceptre Amethyst** Rhodesia	x1.30 d
48d	**Amygdaloidal Basalt** Rhodesia	x1.08 d
48e	**Amygdaloidal Basalt** Northern Ireland	x2.02 c
49a	**Analcite** Poona, India	x4.82 d
49b	**Analcite** Portrush, Northern Ireland	x0.99 c
50a	**Anapaite** Spain	x2.02 d
50b	**Anatase (brown crystals) with albite** St. Gothard, Switzerland	x1.73 a
50c	**Ancylite** Ravalli County, Montana	x1.15 a
50d	**Andalusite** Dover Mine, Mineral County, Nevada	x1.26 d
50e	**Andesite** Somerset	x2.02 c
51a	**Andradite** Durango, Mexico	x1.21 d
51b	**Andradite (massive)** Unrecorded	x1.69 c
52a	**Andradite** Pyrenees	x2.02 a
52b	**Anglesite (bluish black)** Yuma County, Arizona	x2.02 c
52c	**Anglesite** Sonora, Mexico	x0.84 d
52d	**Anhydrite** Unrecorded	x1.69 c
52e	**Anhydrite** Pyramid Lake, Washoe County, Nevada	x1.24 d
53a	**Ankerite** Unrecorded	x5.17 d
53b	**Annabergite** Cottonhood Canyon, Stillhater Ranges, Nevada	x1.28 a
54a	**Anorthite (white) with pyroxene** Grass Valley, Nevada County, California	x0.95 d
54b	**Anorthoclase** Unrecorded	x1.80 c
54c	**Anthophyllite** Balmat, New York State	x1.48 d
54d	**Anthracite** South Wales	x2.02 c
55a	**Antigorite (pseudomorphing forsterite)** Vikersund, Norway	x0.95 a
55b	**Native Antimony** Allemont	x1.54 a
55c	**Antimony in pegmatite** Lovelock, Pershing County, Nevada	x1.89 d
55d	**Antozonite** Wölsendorf, Bavaria, Germany	x1.80 a
55e	**Apache Tears (black) in perlite** Superior, Arizona	xc1.50 c
56a	**Apatite** Unrecorded	x1.91 c
56b	**Aplite** Devonshire	x2.02 c
56c	**Apophyllite** Poona, India	x1.04 c
57a	**Aquamarine** Brazil	x1.86 c
57b	**Green Aragonite** Mexico	x0.48 c
57c	**Aragonite** Spain	x1.32 d
57d	**Aragonite** Aragon, Spain	x2.02 d
58a	**Arduinite** Val di Fossa, Trento, Italy	x1.43 a
58b	**Arfvedsonite (black) with eudialyte** West Greenland	x1.69 a
58c	**Argentite** Guanajuato, Mexico	x2.02 d
59a	**Arsenopyrite** St. Agnes, Cornwall	x2.28 b
59b	**Native Arsenic** Harz Mountains, Germany	x1.58 d
59c	**Arsenolite** Bissoe, Cornwall	x1.46 a
60a	**Arsenopyrite** Devonshire	x1.37 c
60b	**Arsenopyrite with calcite** Parc Mine, Llanwrst, Denbighshire	x1.62 a
60c	**Artinite** San Benito County, California	x1.75 c
60d	**Asbestos** Unrecorded	x1.37 c
60e	**Asbestos** Globe, Arizona	x2.02 c
60f	**Blue Asbestos** Windhoek, South-west Africa	x1.04 d
61a	**Asphalt** Val-de-Travers, France	x1.56 a
61b	**Asphaltite (anthraxolite)** Vermillion Lake, Sudbury, Ontario, Canada	x1.52 a
61c	**Asphaltum** Limmer, Hanover, Germany	x1.71 a
61d	**Astrophyllite** USSR	x1.15 c
61e	**Atacamite** Mexico	x2.02 d
61f	**Augite** Norway	x1.35 d
62a	**Aurichalcite** Mapimi, Mexico	x1.56 c
62b	**Austinite** Toele County, Utah	x1.67 d
62c	**Austinite (crystal detail)** Toele County, Utah	x3.58 d
62d	**Autunite** Devonshire	x1.48 d
63a	**Axinite** Baja, California	x1.69 c
63b	**Azurite** Santa Rita, Mexico	x1.98 c
63c	**Azurite** Unrecorded	x0.86 c
63d	**Babingtonite** New Britain, Connecticut	x1.08 a
63e	**Baddeleyite (massive)** Minas Gerais, Brazil	x1.24 a
63f	**Bakerite** Boron, Kern County, California	x0.93 a
64a	**Ballstone** Durham	x0.90 c
64b	**Barkevikite** San Benito County, California	x1.75 d
65a	**Banded Agate** Botswana	x4.62 c
65b	**Barytes (variety oakstone)** Derbyshire	x0.73 c
65c	**Barytes Rose with quartz** Leadhills, Scotland	x0.62 c
66a	**Barytes with chalcopyrite** Germany	x0.48 c
66b	**Barytes Rose** Norman, Oklahoma	x1.69 c
66c	**Red Barytes** Washington County, Missouri	x0.57 d
66d	**Cawk (variety of barytes)** Unrecorded	x1.73 c
66e	**Barytocalcite** Unrecorded	x1.58 c
66f	**Basalt** Shropshire	x2.02 c
66g	**Bastite** Lizard, Cornwall	x0.63 a
67a	**Bauxite** Arkansas	x0.88 c
67b	**Bayldonite** Penberthy Croft Mines St. Hillary, Cornwall	x1.86 b

68a	**Bementite (massive-brown)** Olympic Mountains, Washington	x1.58 a
68b	**Benitoite (white) with neptunite** Headwaters, San Benito River, California	x1.71 a
69a	**Berthierine** Charles, Alsace, France	x1.50 a
69b	**Bertrandite on mica with hyolite** Unrecorded	x2.02 d
69c	**Beryl** Mozambique	x0.96 d
69d	**Yellow Beryl** Norway	x0.45 d
69e	**Berzelianite (black) in calcite** Skrikerum, Sweden	x1.80 a
69f	**Berzeliite** Sweden	x1.80 a
70a	**Bikitaite** Rhodesia	x1.65 a
70b	**Bindheimite** Trevinnick Mine, St. Kew, Cornwall	x0.95 b
70c	**Biotite** Prieska, South Africa	x0.97 d
71a	**Metallic Bismuth** Upper Lias, Whitby, Yorkshire	x3.06 a
71b	**Bismuth (silver white)** Cornwall	x1.67 a
72a	**Bismuthinite** Carrock Fell, Cumberland	x1.37 a
72b	**Bismuthinite with chalcopyrite** East Pool Mine, Illogan, Cornwall	x1.46 b
72c	**Bismutite (massive)** Brejauba, Minas Gerais, Brazil	x1.91 a
72d	**Bituminous Coal** South Wales	x2.02 c
72e	**Bixbyite on tetrahedrite** Postmansburg, South Africa	x1.84 d
73a	**Blende** Shropshire	x0.69 c
73b	**Blende** Cardiganshire	x1.35 a
73c	**Blende** Galena, Illinois	x1.44 d
73d	**Blende with quartz** Roughton Gill, Cumberland	x1.65 d
73e	**Blue Ground** Kimberley, South Africa	x1.44 a
74a	**Blue John** Castleton, Derbyshire	x4.94 c
75a	**Bog Iron Ore** Unrecorded	x1.30 a
75b	**Bone Bed** Gloucestershire	x2.02 c
75c	**Boracite** Turkey	x1.60 a
75d	**Chambersite (manganese rich boracite)** Venice Salt Dome, Louisiana	x2.02 d
75e	**Borax** Esmeralda County, Nevada	x1.58 a
75f	**Bornite** Ajo, Arizona	x2.02 c
76a	**Bornite (peacock ore)** Pima County, Arizona	x5.64 d
77a	**Plumosite (variety of boulangerite)** Wheal Boys, St. Endellion, Cornwall	x2.02 b
77b	**Boulangerite (massive)** Gard, France	x2.02 a
77c	**Bournonite (greyish blue)** Cornwall	x0.68 a
78a	**Braunite** Tirodi Mines, Central Provinces, India	x1.84 a
78b	**Bravoite on pyrite** Rico, Colorado	x1.37 a
78c	**Brazilianite** Arrasuahy, Minas Gerais, Brazil	x1.58 a
79a	**Breunnerite in schist** Bolzano, Italy	x0.99 a
79b	**Breunnerite (crystal detail)** Bolzano, Italy	x2.09 a
79c	**Brewsterite** Scotland	x1.91 a
79d	**Brewsterite (crystal detail)** Scotland	x4.08 a
79e	**Brochantite** Pima County, Arizona	x1.69 d
80a	**Bromyrite (white)** Unrecorded	x1.98 a
80b	**Brookite** Arkansas	x1.73 d
80c	**Brucite** Cedar Hill, Pennsylvania	x1.35 d
81a	**Bytownite (green) in basalt** Eycott Hill, Cumberland	x1.02 a
81b	**Cacoxenite (yellowish white acicular crystals) on limonite** Rotlaüfchen, Germany	x1.91 a
81c	**Calaverite** Cripple Creek, Colorado	x1.65 a
82a	**Dogtooth Calcite** Cumberland	x2.46 c
82b	**Dogtooth Calcite (side view)** Cumberland	x0.80 c
83a	**Calcite flowers** Oskaloosa, Iowa	x0.73 d
83b	**Blue Calcite** Switzerland	x0.97 c
83c	**Phantom Calcite** Chihuahua, Mexico	x1.21 c
83d	**Banana Calcite** Charcas, Mexico	x0.89 d
83e	**Cona Calcite** Charcas, Mexico	x1.28 d
83f	**Cave Crystal (variety of calcite)** Yuma County, Arizona	x1.44 c
83g	**Iceland Spar (variety of calcite)** USSR	x1.78 c
84a	**Caledonite** Leadhills, Scotland	x1.78 a
84b	**Californite** Big Bar Station, Butte County, California	x1.76 a
84c	**Campylite** Dry Ghyll Mine, Coldbeck Fells, Cumberland	x1.11 b
85a	**Campylite** Cumberland	x1.93 a
85b	**Cancrinite in syenite** Litchfield, Connecticut	x1.46 a
86a	**Carnallite** Alberta, Canada	x1.11 d
86b	**Carnelian** South-west Africa	x1.33 c
86c	**Carnotite in sandstone** San Rafael Swells, Utah	x1.48 a
86d	**Carpholite (slice)** Meuville, Ardennes, Belgium	x0.78 a
87a	**Cassiterite** Trevaunance Mine, St. Agnes, Cornwall	x3.54 b
87b	**Cassiterite** Cornwall	0.80 c
87c	**Cassiterite** Kinta Mine, Penang, Malaysia	x4.32 d
88a	**Cassiterite** Kinta Mine, Penang, Malaysia	x2.02 d
88b	**Catlinite** near Eatontown, New Jersey	x1.15 a
88c	**Catlinite** near Eatontown, New Jersey	x1.15 a
88d	**Cats Eye (unpolished)** India	x1.71 c
88e	**Cats Eye (polished)** India	x1.80 c
89a	**Celestine** Madagascar	x1.21 c
89b	**Celestine** Yate, Gloucestershire	x0.95 d
89c	**Barytocelestine** Cardiff, Glamorganshire	x1.56 d
90a	**Celestine** Yate, Gloucestershire	x2.12 c

90b	**Celestine (bluish) with aragonite**	x1.28 d	98b	**Chlorite**	x1.52 d
	Charcas, Mexico			Skye, Scotland	
90c	**Celsian**	x1.30 a	99a	**Chloritoid in schist**	x1.26 a
	Benallt Mine, Rhiw, Aberdaron, Carnarvonshire			Leeds, Quebec, Canada	
91a	**Cerargyrite**	x1.28 a	99b	**Chondrodite and spinel in calcite**	x1.37 a
	New South Wales, Australia			Franklin, New Jersey	
91b	**Cerussite**	x1.30 b	99c	**Chromite**	x0.82 d
	St. Minver, Cornwall			Rhodesia	
91c	**Cerite (massive)**	x1.62 a	99d	**Chromite**	x1.58 c
	Bastnäs, Sweden			Philippine Islands	
91d	**Cerussite**	x1.41 c	100a	**Chrysoberyl**	x2.02 a
	Tsumeb, West Africa			Ceylon	
92a	**Cervantite**	x2.02 d	100b	**Chrysocolla**	x1.26 c
	Sonora, Mexico			Arizona	
92b	**Ceylonite in ejected block**	x2.02 a	100c	**Kupfer Pecherz (chrysocolla)**	x1.19 a
	Monte Somma, Italy			Mexico	
92c	**Chabazite**	x1.82 d	100d	**Chrysolite**	x1.75 c
	Kibblehouse, Pennsylvania			Unrecorded	
92d	**Chalcanthite**	x1.28 d	100e	**Chrysolite (massive)**	x1.26 a
	Mexico			USA	
92e	**Chalcedony Rose**	x0.78 d	101a	**Chrysoprase (massive)**	x0.91 c
	Mozambique			Australia	
93a	**Rainbow Chalcedony**	x3.40 d	101b	**Chrysoprase**	x0.93 c
	Prieska, South Africa			Australia	
93b	**Chalcedony**	x1.39 c	101c	**Chrysotile**	x1.73 a
	Cornwall			Unrecorded	
93c	**Torolite (chrome chalcedony)**	x1.32 d	101d	**Chrysotile**	x1.91 d
	Rhodesia			Thurman, New York State	
93d	**Cherry Chalcedony**	x0.77 c	102a	**Cinnabar**	x1.57 d
	USSR			Humbolt County, Nevada	
94a	**Chalcocite**	x0.91 c	102b	**Cinnabar**	x2.02 d
	Shropshire			Spain	
94b	**Chalcocite (redruthite)**	x1.15 a	102c	**Citrine**	x0.86 c
	Bulno Esperanza, Antafagusta			Unrecorded	
94c	**Chalcocite**	x0.78 b	103a	**Citrine**	x2.24 c
	Tin Croft Mine, Illogan, Cornwall			Brazil	
94d	**Chalcophanite**	x1.26 a	104a	**Clay**	x2.02 c
	Unrecorded			Devonshire	
94e	**Chalcophyllite**	x1.58 a	104b	**Clay ironstone concretion**	1.44 a
	Cornwall			Unrecorded	
94f	**Chalcophyllite with cuprite**	x0.78 b	104c	**Clay ironstone concretion (weathered honeycomb)**	x0.86 a
	Wheal Gorland, Gwennap, Cornwall			Unrecorded	
94g	**Chalcopyrite**	x0.96 c	104d	**Clay ironstone**	x1.43 a
	Devonshire			Coal Measures, Hanley, Staffordshire	
95a	**Chalcopyrite on dolomite**	x2.32 c	105a	**Clay Slate**	x2.02 c
	USA			Cornwall	
95b	**Chalcopyrite on barytes**	x1.53 c	105b	**Cleavelandite with yellow beryl**	x0.84 c
	Germany			Norway	
95c	**Chalcosiderite**	x1.58 b	105c	**Clinochlore**	x0.78 a
	West Phoenix Mine, Linkinhorne			West Chester, Pennsylvania	
95d	**Chalcosiderite (crystal detail)**	x3.39 b	105d	**Clinochlore**	x0.99 a
	West Phoenix Mine, Linkinhorne			Vikersund, Norway	
95e	**Chalcotrichite**	x1.60 a	105e	**Clinoclase with quartz**	x1.78 a
	San Miguel Mine, Huelvei, Spain			Wheal Unity, Cornwall	
95f	**Chalk**	x1.69 d	106a	**Clinohedrite in schefferite**	x1.53 a
	Amesbury, Wiltshire			Franklin, New Jersey	
95g	**Chalybite with cryolite**	x1.84 a	106b	**Clinohumite in calcite**	x1.52 a
	Arksuk Fjord, Greenland			Franklin, New Jersey	
95h	**Chalybite on quartz**	x0.86 b	106c	**Clinozoisite**	x1.32 a
	Carn Brea Mine, Cornwall			Sonora, Mexico	
95i	**Chalybite with quartz**	x0.70 c	106d	**Clinozoisite**	x1.63 d
	Unrecorded			Mozambique	
96a	**Chalybite**	x2.04 b	107a	**Cobaltite (massive)**	x1.37 a
	Carn Brea, Cornwall			Sweden	
96b	**Chamosite**	x1.26 a	107b	**Cobaltite**	x1.75 a
	Nucic, Czechoslovakia			Espanola, Ontario, Canada	
97a	**Banded Chert**	x0.97 c	107c	**Colemanite**	x2.88 d
	Australia			Kern County, California	
97b	**Chert**	x1.02 c	108a	**Collophane**	x4.82 c
	Monkton, Devon			Leadhills, Lanarkshire, Scotland	
97c	**Chiastolite**	x1.06 c	109a	**Collinsite in quercyite**	x1.93 d
	Australia			Francois Lake, Colorado	
97d	**Chiastolite Slate**	x2.02 c	109b	**Colophonite**	x2.02 a
	Devonshire			Unrecorded	
97e	**Chibinite**	x0.61 c	109c	**Coloradoite**	x1.46 a
	USSR			Colorado	
98a	**Childrenite**	x1.67 a	109d	**Columbite**	x1.80 a
	Cornwall			Ulefoss Norway	

109e	**Colusite**	x1.56 a
	Colusa Mines, Butte, Montana	
110a	**Conglomerate**	x2.02 c
	Hertfordshire	
110b	**Connellite**	x4.32 b
	Unrecorded	
110c	**Cookeite (massive) with quartz crystals**	x1.95 d
	Mount Apatite, Auburn, Maine	
111a	**Cookeite with quartz crystals**	x1.69 d
	Mount Apatite, Auburn, Maine	
111b	**Copiapite**	x2.02 a
	Chile	
111c	**Copper**	x3.66 a
	Houghton County, Michigan	
111d	**Copper**	x1.95 d
	Santa Rita, New Mexico	
111e	**Copper in selenite**	x2.02 c
	Pima County, Arizona	
112a	**Cordierite**	x1.91 c
	Unrecorded	
112b	**Cordierite (massive)**	x1.37 a
	Iilijärvi, Kisko, Finland	
112c	**Cordierite (brown)**	x0.97 a
	Bjordammen, Norway	
112d	**Cornetite**	x1.71 a
	Lubumbashi, Congo—Kinshasa	
113a	**Cornwallite**	x1.10 a
	Potts Gill Mine, Caldbeck Fells, Cumberland	
113b	**Coronadite**	x1.35 a
	Roughton Gill, Caldbeck Fells, Cumberland	
113c	**Corundum (greyish black)**	x1.37 d
	Gallatin County, Montana	
113d	**Corundum**	x1.91 c
	Unrecorded	
114a	**Covelline**	x1.80 a
	Butte, Montana	
114b	**Covelline**	x1.58 a
	Summitville, Rio Grande County, Colorado	
114c	**Crandallite with wardite**	x0.67 a
	Fairfield, Utah	
115a	**Cristobalite**	x2.02 d
	San Bernardino County, California	
115b	**Crocoite**	x1.06 a
	Molotov, USSR	
115c	**Crocoite**	x1.91 d
	Tasmania, Australia	
115d	**Crocoite (crystal detail)**	x2.30 a
	Molotov, USSR	
115e	**Crossite**	x2.02 d
	Centra Costa County, California	
117a	**Cryolite with chalybite**	x1.84 a
	Arksuk Fjord, Greenland	
118a	**Cuprite**	x4.96 c
	Unrecorded	
119a	**Cyanotrichite**	x1.93 d
	Arizona	
119b	**Damourite**	x0.88 a
	Williamstown, South Australia	
120a	**Danalite**	x1.10 b
	Botallack, St. Just in Penwith, Cornwall	
120b	**Danburite**	x0.78 d
	Charcas, Mexico	
120c	**Daphnite in quartz**	x1.35 a
	Caradon Mines, St. Cleer, Cornwall	
120d	**Datolite**	x1.28 b
	Parc Bean Cove, Mullion, Cornwall	
121a	**Davidite**	x1.32 a
	Radium Hill, Olary, South Australia	
121b	**Dawsonite**	x2.02 a
	Miniera del Siele, Tuscany, Italy	
121c	**Dawsonite (crystal detail)**	x4.82 a
	Miniera del Siele, Tuscany, Italy	
121d	**Delafossite (dark reddish brown)**	x1.50 a
	Bisbee, Arizona	
121e	**Delafossite (crystal detail)**	x3.20 a
	Bisbee, Arizona	
122a	**Demantoid on asbestos**	x1.62 a
	Emarese, Aosta, Italy	

122b	**Descloizite**	x0.84 c
	Tsumeb, South-west Africa	
123a	**Desert Rose**	x3.12 c
	Arizona	
124a	**Deweylite**	x1.56 a
	Chester County, Pennsylvania	
124b	**Diabantite in diabase**	x1.22 a
	Loudoun County, Virginia	
124c	**Diabase**	x1.71 a
	Loudoun County, Virginia	
124d	**Diaboleite**	x1.97 a
	Mammoth Mine, Arizona	
124e	**Diallage**	x2.02 a
	Unrecorded	
125a	**Diamond**	x1.91 c
	Unrecorded	
125b	**Diaspore**	x1.26 d
	Dover Mine, Mineral County, Nevada	
125c	**Diatomite**	x1.71 a
	Toome, County Londonderry, Northern Ireland	
125d	**Dickite**	x1.24 d
	Hamburg, Pennsylvania	
125e	**Digenite**	x2.02 d
	Tsumeb, South-west Africa	
125f	**Digenite with pyrite**	x1.21 c
	Butte, Montana	
126a	**Black Star Diopside**	x1.46 c
	India	
126b	**Diopside**	x0.48 c
	Brazil	
126c	**Diopside**	x1.17 c
	Congo	
126d	**Diopside (crystal detail)**	x3.75 c
	Congo	
126e	**Dioptase**	x1.52 d
	Guachib, South Africa	
126f	**Diorite**	x1.69 c
	Cumberland	
127a	**Dolerite**	x1.58 c
	Durham	
127b	**Dolomite**	x2.02 c
	Leicestershire	
127c	**Striped Dolomite**	x0.99 c
	Brazil	
127d	**Dolomite with chalcopyrite**	x.1.08 c
	USA	
127e	**Domeykite**	x1.76 a
	Mohawk Mine, Michigan	
128a	**Dravite**	x0.82 c
	Australia	
128b	**Dufrenite**	x2.02 a
	Unrecorded	
128c	**Duftite**	x0.70 d
	Tsumeb, South-west Africa	
129a	**Dumortierite**	x0.78 c
	Brazil	
129b	**Dumortierite (massive)**	x0.61 d
	Mozambique	
129c	**Serpentinised Dunite**	x1.22 a
	The Lizard, Cornwall	
129d	**Dysanalyte (black)**	x1.78 a
	Arkansas	
129e	**Dysanalyte (crystal detail)**	x3.80 a
	Arkansas	
130a	**Dyscrasite**	x1.65 a
	Unrecorded	
130b	**Dysluite**	x2.02 a
	Ogdensburgh, Sussex County, New Jersey	
130c	**Edenite (grey)**	x1.28 a
	Franklin, New Jersey	
131a	**Eclogite**	x3.61 d
	Rhodesia	
131b	**Elaterite**	x1.56 a
	Derbyshire	
132a	**Electrum**	x2.02 d
	Kittitas County, Washington	
132b	**Emerald**	x3.86 c
	Zillertal, Austria	

132c	**Emery**	x2.02	a
	Naxos, Greece		
132d	**Enargite**	x1.35	a
	Montana		
133a	**Enargite on pyrite**	x1.75	d
	Quiruvilca, Peru		
134a	**Epidote on quartz**	x1.04	c
	Unrecorded		
134b	**Epidote (crystal detail)**	x2.22	c
	Unrecorded		
134c	**Epsomite**	x1.54	a
	Treharris, Glamorgan		
134d	**Erinite**	x2.02	a
	Cornwall		
135a	**Erythrite**	x1.06	c
	Morocco		
135b	**Erythrite (crystal detail)**	x3.38	c
	Morocco		
135c	**Euclase**	x2.02	a
	San Diego County, California		
136a	**Eucryptite**	x2.02	a
	Fort Victoria, Rhodesia		
136b	**Eudialyte with arfvedsonite**	x1.65	a
	West Greenland		
136c	**Eudialyte**	x1.54	a
	Brevig, Norway		
136d	**Eudialyte**	x0.73	c
	USSR		
136e	**Eulytine**	x1.52	b
	Restormel Royal Iron Mine, Cornwall		
137a	**Euxenite**	x2.02	d
	Unrecorded		
137b	**Evansite (greyish)**	x1.48	a
	Creu Olorde, Barcelona, Spain		
137c	**Faröelite with apophyllite**	x1.54	a
	Cave of Nalsö, Isle of Farö, Sweden		
138a	**Faröelite**	x1.95	a
	Cave of Nalsö, Isle of Farö, Sweden		
138b	**Fassaite**	x1.71	a
	Helena, Montana		
138c	**Faustite with chrysocolla**	x1.91	d
	Copper King Mine, Eureka County, Nevada		
139a	**Feldspar crystals**	x2.38	d
	Mozambique		
139b	**Fayalite**	x1.35	a
	Rockport, Massachusetts		
140a	**Feldspar (salmon brown) with acmite**	x0.99	a
	Wassau, Wisconsin		
140b	**Ferberite**	x1.71	a
	Spain		
140c	**Fergusonite**	x1.21	a
	Ytterby, Sweden		
141a	**Ferrierite**	x2.02	a
	Unrecorded		
141b	**Ferrimolybdite**	x1.75	a
	Lehmi County, Idaho		
143a	**Flint**	x0.86	a
	Gravesend, Kent		
143b	**Flint nodule**	x1.00	a
	Charlton, Gloucestershire		
143c	**Flos-ferri**	x1.43	c
	Austria		
144a	**Fluocerite (purple)**	x1.73	a
	Dalarme, Sweden		
144b	**Fluorite**	x1.58	c
	Yuma County, Arizona		
144c	**Fluorite**	x2.02	d
	Arizona		
144d	**Fluorite**	x0.65	c
	Unrecorded		
144e	**Fluorite (octahedral)**	x0.80	d
	Illinois		
144f	**Fluorite**	x0.70	b
	Hights Mine, Weardale, Durham		
145a	**Fluorite**	x2.02	d
	Illinois		
145b	**Fluorite**	x1.39	b
	Cornwall		
145c	**Fluorite**	x1.04	e
	Durango, Mexico		
146a	**Forsterite with spinel**	x2.02	a
	Crestmore, California		
146b	**Foshagite**	x1.10	a
	Crestmore, California		
146c	**Fouqueite**	x1.17	a
	Timmins, Ontario, Canada		
146d	**Granular Fowlerite**	x1.46	a
	Franklin, New Jersey		
147a	**Franckeite**	x2.02	a
	San Jose Mine, Oruro, Bolivia		
147b	**Francolite**	x1.28	a
	Fowey Consols, Cornwall		
147c	**Franklinite with willemite and calcite**	x1.37	d
	Franklin, New Jersey		
147d	**Freibergite (massive)**	x1.62	a
	Animas, Chocaya, Bolivia		
148a	**Fuchsite in schist**	x1.75	a
	Outokumpu, Finland		
148b	**Fuchsite**	x1.77	d
	Virginia		
148c	**Verdite (variety of fuchsite)**	x0.93	c
	South Africa		
148d	**Fuller's Earth (attapulgite)**	x0.72	a
	Attapulgus, Georgia		
148e	**Fuller's Earth (attapulgite)**	x0.72	a
	Attapulgus, Georgia		
148f	**Gabbro**	x1.54	c
	Cornwall		
149a	**Gadolinite in feldspar**	x1.63	a
	Sweden		
149b	**Galena**	x2.00	b
	Wheal Hope, St. Agnes, Cornwall		
149c	**Gahnite**	x2.02	d
	Rose, Massachusetts		
150a	**Galena**	x1.32	d
	Tri-state, Kansas		
150b	**Galena**	x0.99	c
	Mendip Hills, Somerset		
150c	**Galena**	x1.18	c
	North Yorkshire		
151a	**Garnet**	x2.73	c
	Zillertal, Austria		
151b	**Garnierite**	x1.35	a
	New Caledonia		
152a	**Gehlerite**	x1.06	a
	Luna County, New Mexico		
153a	**Genthite**	x4.42	a
	Unrecorded		
153b	**Germanite with tennantite**	x1.75	a
	Tsumeb, South-west Africa		
154a	**Gersdorffite**	x1.56	a
	Goslar, Germany		
154b	**Geyserite**	x1.84	a
	Sulphur Point, Rotorua, New Zealand		
154c	**Gibbsite (massive)**	x2.02	a
	Unrecorded		
154d	**Gilbertite in china clay**	x1.80	a
	St. Austell, Cornwall		
154e	**Gillespite**	x2.02	a
	Rush Creek Deposit, Esquire Mine, Fresno County, California		
155a	**Gismondite**	x1.46	a
	Capo di Bove, Rome, Italy		
155b	**Glauberite**	x2.02	a
	Camp Verde, Arizona		
155c	**Glauberite**	x1.35	c
	Camp Verde, Arizona		
155d	**Glauconite**	x1.30	a
	Thorshaun, Stromso, Sweden		
156a	**Glaucophane**	x1.26	d
	San Benito County, Mexico		
156b	**Gmelinite**	x0.84	c
	Portrush, County Antrim, Northern Ireland		
156c	**Gneiss**	x2.02	c
	Argyllshire		

156d	**Gneissoid Granite**	x1.48 d		166b	**Hedenbergite**	x1.02 a	
	Rhodesia				Unrecorded		
157a	**Goethite**	x4.22 c		167a	**Heliotrope**	x0.99 c	
	Chihuahua, Mexico				India		
157b	**Goethite**	x0.84 d		167b	**Hematite**	x0.89 c	
	Macon, Georgia				Cumberland		
157c	**Goethite**	x0.88 b		167c	**Botryoidal Hematite**	x0.86 d	
	Royal Iron Mine, Restormel, Cornwall				Mozambique		
158a	**Goethite**	x1.32 d		167d	**Hematite**	x0.84 e	
	Cumberland				Cumberland		
158b	**Gold nugget**	2.60 b		167e	**Hematite (brush iron ore)**	x0.97 d	
	Carnon Valley Tin Stream, Cornwall				Unrecorded		
158c	**Gold**	x1.58 b		167f	**Hemimorphite**	x0.86 c	
	Transylvania, Rumania				Derbyshire		
158d	**Gold**	x1.20 c		167g	**Herderite**	x1.04 a	
	Cornwall				Concord, New Hampshire		
159a	**Gold**	x1.54 b		168a	**Herkimer Diamonds**	x1.73 d	
	Transylvania, Rumania				Middleville, New York State		
159b	**Gossan (duricrust)**	x0.78 d		168b	**Hessonite**	x1.39 c	
	Windhoek, South-west Africa				Canada		
160a	**Granite**	x2.02 c		168c	**Hessonite (crystal detail)**	x2.96 c	
	Devon				Canada		
160b	**Granite**	x1.43 e		168d	**Hetaerolite on chalcophanite**	x1.75 a	
	Cornwall				Franklin, New Jersey		
160c	**Granite with autunite**	x1.48 d		168e	**Heterogenite**	x1.71 a	
	Devon				Kambore Mine, Congo		
160d	**Granite**	x1.43 e		169a	**Heulandite**	x3.06 a	
	Cornwall				Faeroe Islands		
160e	**Granite Aplite**	x1.06 c		169b	**Hexagonite**	x1.56 d	
	Devon				New York State		
160f	**Granodiorite**	x1.67 c		169c	**Hexagonite (crystal detail)**	x3.34 d	
	Leicestershire				New York State		
161a	**Graphite**	x1.75 c		170a	**Hollandite**	x0.80 a	
	Unrecorded				Washoe County, Nevada		
161b	**Graphite**	x1.39 a		170b	**Holmquistite**	x1.56 a	
	Unrecorded				Lac Malartic, Quebec, Canada		
161c	**Graphite Schist**	x1.30 d		170c	**Hornblende in basalt**	x1.06 d	
	New Territories, Hong Kong				Devon		
161d	**Gratonite**	x2.02 a		170d	**Weinschenkite (variety of hornblende)**	x1.73 d	
	Cerro de Pasco, Peru				Rockbridge County, Virginia		
161e	**Greenockite**	x1.58 a		170e	**Hornblendite**	x2.02 c	
	Unrecorded				Worcestershire		
162a	**Greywacke**	x2.02 c		170f	**Hornfels**	x2.02 c	
	South Wales				Devon		
162b	**Grossular**	x1.99 c		170g	**Hortonolite**	x1.75 a	
	South West Africa				Monroe, Orange County, New York		
162c	**Griphite**	x2.02 a		171a	**Howeite**	x1.73 d	
	Pennington County, South Dakota				Laytownville, California		
163a	**Grossular**	x2.02 a		171b	**Howlite**	x1.04 d	
	Belstone Consols, Okehampton, Devon				Arizona		
163b	**Gummite**	x1.26 a		171c	**Howlite (massive)**	x1.43 a	
	Congo				Wentworth, Nova Scotia, Canada		
163c	**Gypsum**	x0.88 c		171d	**Huebnerite**	x1.19 a	
	Edlington Brick Works, Doncaster, Yorkshire				Henderson, North Carolina		
163d	**Halite**	x1.78 c		171e	**Humite**	x2.02 a	
	Salton Sea, California				Vesuvius, Italy		
164a	**Halite**	x2.78 d		172a	**Hyalite in feldspar**	x2.02 a	
	Imperial County, California				Unrecorded		
164b	**Halite**	x2.02 c		172b	**Hyalite**	x2.02 d	
	Overland, Australia				Unrecorded		
164c	**Halloysite**	x1.35 d		172c	**Hydromagnetic and artinite with serpentine**	x1.84 d	
	Kowloon, Hong Kong				Staten Island, New York City		
164d	**Hancockite**	x1.35 a		173a	**Hydrozincite**	x1.69 d	
	Franklin Furnace, New Jersey				Black Rock, Arizona		
165a	**Hardystonite (massive) with franklinite**	x1.69 a		173b	**Hydrozincite**	x2.02 d	
	Franklin Furnace, New Jersey				Durango, Mexico		
165b	**Harmotome**	x1.02 a		173c	**Hypersthene**	x1.30 a	
	Hartz, Germany				Labrador, Canada		
165c	**Hastingsite with magnetite**	x2.02 a		173d	**Iddingsite**	x1.50 a	
	Haytor Vale, Devon				California		
165d	**Hatchettite**	x1.35 a		174a	**Idocrase**	x1.62 a	
	Cymmer Colliery, Penarth, Glamorgan				Arviron, France		
165e	**Hatchettolite**	x1.95 a		174b	**Ilmenite**	x1.46 a	
	Norway				Norway		
165f	**Hausmannite**	x2.02 a		174c	**Ilvaite**	x1.71 a	
	Unrecorded				Unrecorded		
166a	**Hauyne**	x2.02 a		174d	**Indicolite**	x1.91 c	
	Monte Somma, Italy				Unrecorded		

175a	Indochinite	x2.23 a
	Thailand	
176a	**Shelly Ironstone**	x1.67 a
	Whitby, Yorkshire	
176b	**Itacolumite**	x1.28 a
	Brazil	
177a	**Jade**	x0.60 c
	USSR	
177b	**Jamesonite**	x0.90 c
	Unrecorded	
177c	**Jamesonite (crystal detail)**	2.88 c
	Unrecorded	
177d	**Jarosite (massive)**	x1.67 a
	Javilla Mountains, New Mexico	
177e	**Red Jasper**	x1.06 c
	South Africa	
177f	**Jasper**	x1.17 d
	Preiska, South Africa	
177g	**Jasper (diaspro)**	x1.06 c
	Morocco	
177h	**Jasper (jasplite)**	x0.97 d
	Windhoek, South West Africa	
177i	**Jasper (cricolite)**	x1.04 c
	Unrecorded	
178a	**Jasper (mookite)**	x0.46 c
	Australia	
178b	**Blue Jasper**	x0.60 c
	Kola Peninsular, USSR	
178c	**Jeferisite**	x1.06 a
	Libby, Montana	
178d	**Jet**	x0.93 c
	Whitby, Yorkshire	
178e	**Jordanite**	x1.67 a
	Switzerland	
178a	**Kalinite**	x0.95 d
	Silver Peak, Esmeralda County, Nevada	
179b	**Kalinite**	x1.69 d
	Silver Peak, Esmeralda County, Nevada	
179c	**Pigs Eggs (kaolinised feldspar)**	x0.65 b
	North Goonbarrow Clay Pit, Cornwall	
179d	**Kaolinised Granite**	x1.56 d
	Kowloon, Hong Kong	
180a	**Keilhauite**	x1.69 a
	Arendal, Norway	
180b	**Kermesite**	x1.30 b
	Bräunsdorf, Saxony, Germany	
181a	**Knebelite**	x1.46 a
	Stallberg Mine, Sweden	
181b	**Koppite in carbonatite**	x1.35 a
	Sove Mine, Illefoss, Norway	
182a	**Kunzite**	x1.15 c
	Brazil	
182b	**Kyanite**	x1.90 c
	USSR	
183a	**Labradorite displaying labradorescence**	x3.98 d
	Labrador, Canada	
183b	**Labradorite**	x0.93 c
	Unrecorded	
183c	**Langbeinite**	x1.48 a
	Carlsbad, New Mexico	
183d	**Langite**	x0.80 b
	Levant Mine, St. Just in Penwith, Cornwall	
184a	**Laumontite**	x1.37 c
	Poona, India	
184b	**Lava**	x1.04 d
	Caldera Bandama, Canary Islands	
184c	**Lawsonite**	x4.04 a
	Sonoma County, California	
185a	**Lazulite in sandstone**	x1.62 a
	Lincoln County, Georgia	
185b	**Lazurite**	x1.28 c
	USSR	
185c	**Lazurite**	x1.58 c
	Afghanistan	
185d	**Lepidocrocite**	x1.76 a
	Bieber, Hessen, Germany	
185e	**Lepidolite**	x1.35 c
	Norway	

186a	**Lepidomelane**	x1.54 a
	Faraday Township, Ontario, Canada	
186b	**Leucite in lava**	x0.91 a
	Rome, Italy	
186c	**Liebethenite**	x1.76 a
	Phoenix Mine, Cornwall	
186d	**Liebethenite (massive)**	x1.98 a
	Western Australia	
187a	**Lignite**	x1.02 d
	Collie, Western Australia	
187b	**Limonite**	x1.48b
	Botallack Mine, St. Just in Penwith, Cornwall	
187c	**Lignite**	x1.67 d
	Devon	
187d	**Dolomitised Limestone**	x2.02 c
	Somerset	
187e	**Limestone**	x2.02 c
	Somerset	
187f	**Limestone**	x1.52 d
	Khyber Pass, North-west Pakistan	
188a	**Crinoidal Limestone**	x2.02 c
	Somerset	
188b	**Limonite**	x1.18 c
	Somerset	
188c	**Linarite**	x1.54 b
	Leadhills, Lanarkshire, Scotland	
188d	**Linarite**	x2.02 a
	Chuquicamata, Chile	
188e	**Linarite (crystal detail)**	x4.81 a
	Chuquicamata, Chile	
188f	**Linnaeite**	x1.98 a
	Rhondda, Glamorgan	
188g	**Liroconite**	x1.28 b
	Wheal Gorland, Cornwall	
188h	**Listwanite**	x0.70 c
	USSR	
189a	**Litharge**	x1.87 d
	Patagonia, Arizona	
189b	**Lithiophilite**	x1.48 c
	Unrecorded	
189c	**Lithomarge**	x1.78 a
	Unrecorded	
189d	**Lodestone**	x1.67 c
	Iron Mountain, Utah	
189e	**Loellingite**	x2.02 a
	USA	
190a	**Lovozerite**	x0.60 c
	USSR	
191a	**Magnesian Limestone**	x2.02 c
	Yorkshire	
191b	**Magnetite with dolomite**	x1.80 c
	Fresno County, California	
191c	**Magnetite in barytes**	x1.93 c
	Iron County, Utah	
192a	**Magnetite (octahedral crystals)**	x4.82 a
	Unrecorded	
192b	**Magnetite**	x1.00 d
	Ma-on-shan Mine, Kowloon, Hong Kong	
192c	**Malachite**	x1.08 d
	Zacatecas, Mexico	
192d	**Malachite**	x1.32 c
	Zacatecas, Mexico	
192e	**Malachite (crystal detail)**	x2.83 c
	Zacatecas, Mexico	
192f	**Malachite**	x0.80 c
	Yavapai County, Arizona	
192g	**Malachite (polished)**	x1.15 d
	Congo	
193a	**Malachite (unpolished)**	x6.12 d
	Congo	
194a	**Manasseite and hydrotalc in serpentine**	x1.78 a
	Dypingdal, Vikersund, Norway	
194b	**Manganapatite**	x2.02 c
	Strickland Quarry, Portland, Connecticut	
194c	**Manganite**	x1.67 d
	Hotazel	
194d	**Manganophyllite**	x1.69 a
	Benallt Mine, Rhiw, Aberdaron, Carnarvonshire	

194e	**Iona Marble**	x0.94 c		204d	**Montebrasite**	x1.64 a
	Iona, Scotland				Creuze, France	
194f	**Carrara Marble**	x1.43 d		204e	**Monticellite with calcite**	x1.91 a
	Carrara, Italy				Fassathal, Tyrol, Austria	
195a	**Marcasite**	x1.37 d		204f	**Montmorillonite**	x1.60 d
	Mississipi				Greenwood, Maine	
195b	**Margarite**	x1.08 d		205a	**Montroseite**	x1.65 a
	Chester, Massachusetts				Monument Valley, Arizona	
195c	**Ephesite (variety of margarite)**	x1.35 d		205b	**Moonstone**	x1.69 c
	Postmansburg, South Africa				India	
195d	**Marialite**	x2.02 a		205c	**Moroxite**	x3.76 a
	Pianura, Naples, Italy				Arendal, Norway	
195e	**Mariposite**	x0.73 c		205d	**Moss Agate**	x1.06 c
	Brazil				India	
195f	**Markfieldite**	x1.56 d		205e	**Pink Moss Agate**	x0.48 c
	Leicestershire				India	
195g	**Marl**	x2.02 c		205f	**Green Moss Agate**	x1.04 c
	Gloucestershire				Unrecorded	
195h	**Dendrites on Marl**	x2.02 d		205g	**Mottramite**	x2.02 d
	California				Unrecorded	
196a	**Martite**	x1.97 d		206a	**Mudstone**	x2.02 c
	Rhodesia				Gloucestershire	
196b	**Maucherite**	x1.13 a		206b	**Murmanite**	x0.90 c
	Marbella, Spain				USSR	
196c	**Meionite**	x2.02 a		206c	**Muscovite**	x1.69 c
	Monte Somma, Italy				Maricopa County, Arizona	
197a	**Melaconite**	x1.73 a		206d	**Nagyagite**	x1.84 a
	Bolivia				Unrecorded	
197b	**Melanterite**	x3.10 b		207a	**Natrolite**	x0.55 c
	260 Fathom Level, Robinson Section,				USSR	
	South Crofty Mine, Camborne, Cornwall			207b	**Natrolite**	x0.63 c
198a	**Meliphanite**	x1.91 a			County Antrim, Northern Ireland	
	Norway			207c	**Spreustein (variety of natrolite)**	x0.60 c
198b	**Mercury with cinnabar and pyrite**	x3.76 d			USSR	
	Socrates Mine, Sonoma County, California			207d	**Natrolite (crystal detail)**	x2.02 c
198c	**Merwinite with spurrite**	x1.00 a			County Antrim, Northern Ireland	
	Marble Canyon, Texas			207e	**Nepheline**	x1.80 a
198d	**Mesolite**	x1.26 a			Bancroft, Ontario, Canada	
	Faeroe Islands			207f	**Nepheline with biotite**	x2.02 a
199a	**Mica Schist**	x1.19 e			Monte Somma, Italy	
	Brittany, France			208a	**Nephrite Jade**	x0.60 c
199b	**Mica garnet schist**	x1.08 e			British Columbia, Canada	
	Brittany, France			208b	**Nepunite**	x2.02 a
199c	**Mica**	x1.04 c			San Benito County, California	
	Norway			208c	**Niccolite**	x2.02 b
199d	**Rose Mica**	x0.86 d			Pengelly Mine, St. Ewe, Cornwall	
	Sandston, Virginia			208d	**Niccolite**	x0.61 c
200a	**Sphere Mica**	x2.92 c			USSR	
	Norway			208e	**Nigerite**	x2.02 a
200b	**Microcline**	x1.80 c			Egbe District, Kabba Province, Nigeria	
	Unrecorded			208f	**Nitratine**	x2.02 a
201a	**Microcline**	x0.89 a			Chile	
	Colorado			209a	**Thunder Egg (type of nodule)**	x0.80 d
201b	**Microganite**	x2.02 c			Madras, Oregon	
	Cornwall			209b	**Thunder Egg (section)**	x1.72 d
201c	**Microlite**	x2.02 a			Madras, Oregon	
	USA			209c	**Norbergite**	x1.71 a
201d	**Microperthite**	x1.60 a			Norberg, Sweden	
	Golden Point Quarry, Luxulyan, Cornwall			210a	**Novacekite**	x2.38 d
201e	**Millerite**	x1.84 d			Chihuahua, Mexico	
	Narvoo, Illinois			210b	**Nosean in phonolite**	x1.69 a
202a	**Millerite with linnaeite**	x4.19 a			Riedan, Laacher See, Germany	
	Powell Duffryn Colliery, Bargoed, Wales			210c	**Novaculite**	x1.22 a
202b	**Mimetite**	x2.36 b			Hot Springs, Arkansas	
	Wheal Alfred, Phillack, Hayle, Cornwall			210d	**Obsidian**	xc0.91 c
203a	**Mimetite**	x1.35 c			California	
	Chihuahua, Mexico			211a	**Mahogany Obsidian**	x0.76 d
203b	**Mimetite and wulfenite**	x1.54 d			Hurricane, Utah	
	Maricopa County, Arizona			211b	**Green Obsidian**	x0.76 d
203c	**Minium**	x1.58 a			Ascension Islands	
	Zmeinogorsk Mine, Siberia, USSR			211c	**Snowflake Obsidian**	x1.15 c
203d	**Moldavite**	x2.02 a			USA	
	Netolice, Czechoslovakia			211d	**Oligoclase**	x1.15 a
204a	**Molybdenite**	x0.74 c			Sweden	
	Wisconsin			211e	**Olivenite**	x0.84 b
204b	**Ferrimolybdenite**	x1.56 c			Gwennap, Cornwall	
	Lehmi County, Idaho			211f	**Olivenite**	x1.15 b
204c	**Monazite**	x0.97 a			Tin Croft Mine, Illogan, Redruth, Cornwall	
	Ivelard, Norway					

212a	**Olivenite (crystal detail)**	x2.46	b
	Tin Croft Mine, Illogan, Redruth, Cornwall		
212b	**Olivine**	x1.32	a
	Monte Somma, Italy		
212c	**Olivine**	x0.92	c
	Unrecorded		
212d	**Onyx**	x0.53	c
	Italy		
212e	**Oolitic Limestone**	x2.02	c
	Gloucestershire		
212f	**Opal**	x1.69	d
	Guanajuato, Mexico		
213a	**Opal**	x4.96	c
	Australia		
214a	**Green Opal**	x1.69	c
	Brazil		
214b	**Pink Opal with cinnabar**	x1.15	c
	USA		
214c	**Mexican Black Opal**	x1.75	c
	Mexico		
214d	**Ophicalcite**	x0.48	c
	USSR		
215a	**Orthoclase**	x1.35	d
	Harris, Scotland		
215b	**Orthoclase**	x2.00	c
	Unrecorded		
215c	**Ottrelite in schist**	x1.37	a
	Leeds, Quebec, Canada		
215d	**Ozokerite**	x0.93	a
	Unrecorded		
216a	**Paligorskite**	x1.30	a
	Metaline Falls, Washington		
216b	**Pandermite (massive)**	x1.52	a
	Panderma, Sea of Marmora, Turkey		
216c	**Paracelsian**	x1.38	a
	Benallt Mine, Rhiw, Aberdaron, Wales		
217a	**Paragonite (massive) with staurolite**	x1.73	a
	Switzerland		
217b	**Pargasite in Tiree marble**	x1.32	a
	Tiree, Scotland		
217c	**Peat**	x2.02	c
	Somerset		
217d	**Pectolite**	x1.56	b
	Dean Quarry, St. Keverne, Cornwall		
218a	**Pegmatite**	x2.02	c
	Cornwall		
218b	**Penninite**	x2.02	a
	Zermatt, Switzerland		
218c	**Pentlandite**	x0.54	c
	USSR		
218d	**Periclase**	x1.39	a
	Monte Somma, Italy		
218e	**Pericline**	x1.48	a
	Binnenthal, Switzerland		
218f	**Peridot**	x2.02	a
	Lashaine, Monduli, Tanzania		
218g	**Peridotite**	x4.31	c
	New Zealand		
219a	**Peridotite**	x2.02	c
	New Zealand		
219b	**Peristerite**	x1.32	a
	Hylba, Ontario, Canada		
219c	**Perlite**	x1.75	c
	Superior, Arizona		
219d	**Perovskite in carbonatite**	x1.78	a
	Magnet Cove, Arkansas		
219e	**Perthite**	x1.71	a
	Bathurst, Nova Scotia, Canada		
219f	**Petalite**	x1.52	a
	Sweden		
219g	**Petrified Wood (silicified)**	x2.02	a
	Texas		
219h	**Petrified Wood**	x1.10	c
	Australia		
220a	**Petrified Wood**	x0.34	c
	USA		
220b	**Petzite**	x1.86	a
	Colorado		
220c	**Phacolite in basalt**	x1.56	a
	Collingwood, Victoria, Australia		
220d	**Pharmacosiderite**	x1.00	b
	Gunnislake Old Mine, Cornwall		
220e	**Feldspar Phenocrysts**	x1.67	d
	Silvermine Bay, Lantau, Hong Kong		
220f	**Phillipsite**	x1.62	a
	Monte Somma, Italy		
220g	**Phlogopite**	x0.95	d
	Gould Lake Mine, Kingston, Ontario, Canada		
221a	**Phosgenite**	x1.50	a
	Monteponi, Sardinia, Italy		
221b	**Phosphuranylite**	x1.35	a
	Katherine, Northern Territory, Australia		
221c	**Phyllite**	x2.02	c
	Cornwall		
221d	**Picrolite**	x1.17	d
	Cedar Hill		
222a	**Piemontite**	x2.02	a
	Peavine Mountain, near Reno, Nevada		
222b	**Pigeonite**	x1.32	a
	Belmont Quarry, Loudoun County, Virginia		
222c	**Pinite in pegmatite**	x0.88	a
	Castle-an-dinas, Cornwall		
222d	**Pisolites**	x1.04	d
	Leckhampton, Gloucestershire		
222e	**Pisolitic Limestone**	x2.02	c
	Gloucestershire		
222f	**Pitchblende**	x1.80	d
	Cornwall		
223a	**Pitchstone**	x2.02	c
	Scotland		
223b	**Plasma**	x0.99	a
	Cambay, India		
223c	**Platinum**	x2.02	a
	Ural Mountains, USSR		
223d	**Platinum (detail)**	x4.32	a
	Ural Mountains, USSR		
223e	**Plattnerite**	x1.80	d
	Diamond Mine, Eureka County, Nevada		
224a	**Pollucite**	x1.73	a
	Rhodesia		
224b	**Polybasite**	x1.75	a
	Unrecorded		
224c	**Polyhalite**	x1.56	a
	Carlsbad, New Mexico		
225a	**Porcellanite**	x1.11	c
	Australia		
225b	**Flower Rock Porphyry**	x1.02	d
	Texada Island, British Columbia, Canada		
225c	**Powellite**	x2.02	a
	Black Mountain District, California		
225d	**Green Prase**	x0.93	c
	Austria		
225e	**Prehnite**	x1.17	c
	Dumbarton		
225f	**Priceite**	x1.30	a
	Death Valley, California		
226a	**Proustite**	x2.02	a
	Unrecorded		
226b	**Proustite in calcite**	x2.02	a
	Unrecorded		
226c	**Pseudomalachite**	x0.78	b
	Gunnislake Old Mine, Cornwall		
226d	**Pseudomalachite (crystal detail)**	x1.67	b
	Gunnislake Old Mine, Cornwall		
227a	**Psilomelane**	x1.02	e
	Unrecorded		
227b	**Stalactitic psilomelane**	x1.06	a
	Wassaw, Ghana		
227c	**Pumice**	x1.06	c
	Lipari Islands		
227d	**Pumice**	x1.15	e
	Italy		
227e	**Pumpellyite**	x1.41	a
	Woodbridge, Connecticut		
227f	**Purpurite**	x0.59	d
	Unrecorded		

228a	**Pyargyrite**	x2.02 a	
	Mexico		
228b	**Pyargyrite**	x1.65 a	
	Unrecorded		
228d	**Pyrite**	x0.57 b	
	Wheal Kitty, St. Agnes, Cornwall		
228e	**Pyrite**	x1.02 e	
	Italy		
228f	**Pyrite in quartz**	x0.89 c	
	Unrecorded		
228g	**Pyrochlore**	x1.39 a	
	Hybla, Ontario, Canada		
228h	**Pyrochlore in calcite**	x1.15 a	
	Otca, Quebec, Canada		
229a	**Pyrolusite**	x0.90 c	
	New Mexico		
229b	**Pyrolusite**	x0.88 c	
	Mendip Hills, Somerset		
229c	**Pyromorphite**	x0.70 c	
	Leadhills, Lanarkshire, Scotland		
229d	**Pyromorphite**	x2.02 d	
	Dumfries, Scotland		
229e	**Pyromorphite**	x1.37 d	
	Shoshone County, Idaho		
229f	**Pyromorphite (crystal detail)**	x2.92 d	
	Shoshone County, Idaho		
229g	**Pyrophyllite**	x2.02 d	
	Mariposa County, California		
230a	**Pyroxene**	x1.13 a	
	Unrecorded		
230b	**Pyrrhotite**	x2.02 a	
	Monte Somma, Italy		
230c	**Pyrrhotite (massive)**	x1.78 a	
	Panorama Mine, Barmouth, Merionethshire		
230d	**Quartz**	x0.63 d	
	Rhodesia		
230e	**White Quartz**	x0.75 e	
	Cumberland		
230f	**Quartz Stalactite**	x1.50 d	
	Rhodesia		
230g	**Green Quartz**	x0.61 d	
	Mozambique		
230h	**Blue Adventurine Quartz**	x0.56 c	
	India		
230i	**Quartz and agate**	x1.06 c	
	Germany		
230j	**Quartz with oil filled inclusions**	x0.97 a	
	Tylerybont Quarry, Pontsticill, Breconshire		
231a	**Rutilated Quartz**	x3.95 c	
	Brazil		
232a	**Quartzite**	x2.02 c	
	Shropshire		
232b	**Rammelsbergite**	x1.71 a	
	South Lorrain, Ontario, Canada		
233a	**Rashleighite**	x1.15 a	
	Stenales, near St. Austell, Cornwall		
233b	**Realgar**	x1.80 c	
	Getchel Mine, Humboldt County, Nevada		
233c	**Realgar**	x2.02 a	
	Hungary		
233d	**Realgar and orpiment**	x0.83 c	
	USA		
233e	**Red Ochre**	x1.69 c	
	Unrecorded		
233f	**Retinalite**	x1.58 a	
	Unrecorded		
234a	**Rhodochrosite**	x2.18 c	
	Argentine		
234b	**Rhodochrosite**	x0.70 c	
	Unrecorded		
234c	**Rhodolite**	x1.67 a	
	Jackson County, North Carolina		
234d	**Rhodonite**	x0.67 c	
	Australia		
235a	**Rhyolite**	x1.58 d	
	Hurricane, Utah		
235b	**Rhyolite**	x1.84 c	
	Shropshire		

235d	**Ripidolite on quartz**	x1.52 d	
	St. Clair, Pennsylvania		
235c	**Ripidolite**	x1.58 a	
	Chester, Vermont		
236a	**Rosasite**	xc1.48 e	
	Durango, Mexico		
236b	**Rosasite**	x3.15 e	
	Durango, Mexico		
236c	**Roscoelite**	x2.02 d	
	Unrecorded		
236d	**Rose Quartz**	x1.00 c	
	South Africa		
236e	**Rubellite**	x1.54 a	
	Unrecorded		
236f	**Ruby**	x3.38 c	
	USSR		
237a	**Ruby in zoisite**	x3.62 c	
	Zambia		
237b	**Ruby Spinel**	x2.02 a	
	Ceylon		
237c	**Red Rutile**	x1.35 c	
	Glen Lochay, Scotland		
237d	**Sahlite**	x1.37 a	
	Arendal, Norway		
238a	**Sandstone**	x1.26 d	
	Northern Ireland		
238b	**Sandstone**	x2.02 c	
	Yorkshire		
238c	**Bituminous Sandstone**	x1.62 d	
	Davenport, Santa Cruz, California		
238d	**Calcareous Sandstone**	x2.02 c	
	Dorset		
238e	**Ferruginous Sandstone**	x2.02 c	
	Wiltshire		
238f	**Zebra standstone**	x0.63 d	
	Michigan		
238g	**Sanidine in trachyte**	x1.41 a	
	Drachenfels, Rhineland, Germany		
239a	**Saponite**	x0.80 a	
	Boylestone Quarry, Renfrewshire, Scotland		
239b	**Sapphire**	x1.48 d	
	Rhodesia		
239c	**Sapphire**	x1.80 c	
	Unrecorded		
239d	**Star Sapphire**	x2.02 d	
	Ratnapurna, Ceylon		
239e	**Sardonyx**	x1.06 c	
	India		
239f	**Satin Spar**	x1.04 c	
	Nottingham		
240a	**Scapolite (massive)**	x1.32 d	
	Boardman's Bridge, Connecticut		
240b	**Manalite (variety of scapolite)**	x1.43 d	
	Meldon, Okehampton, Devon		
240c	**Scheelite**	x1.56 d	
	Korea		
240d	**Schefferite**	x1.56 a	
	Langban, Sweden		
240e	**Schefferite with clinohedrite**	x1.46 a	
	Franklin, New Jersey		
240f	**Schist**	x0.86 c	
	Zillertal, Austria		
240g	**Hornblende Schist**	x1.69 d	
	Malvern Hills, Worcestershire		
240h	**Hornblende Schist**	x2.02 c	
	Cornwall		
241a	**Schorlomite (massive)**	x1.80 a	
	Magnet Cave, Arkansas		
241b	**Scolecite**	x0.80 c	
	India		
241c	**Scoria**	x1.65 d	
	Mount Wellington, Auckland, New Zealand		
241d	**Scorodite on quartz**	x1.80 d	
	Gold Hill, Tooele County, Utah		
241e	**Scorodite (crystal detail)**	x3.85 d	
	Gold Hill, Tooele County, Utah		
241f	**Scorodite with carminite and quartz**	x1.67 d	
	Gold Hill, Tooele County, Utah		

242a	Selenite Forth Worth, Texas	x0.93 d	251a	Stalactite Ocean Island, Pacific	x0.88 d
242b	Selenite Rose Mexico	x0.48 d	251b	Stannite Carn Brea Mine, Cornwall	x1.17 b
242c	Peach Blossom Selenite Charcas, Mexico	x0.80 d	251c	Staurolite Scotland	x0.41 c
242d	Selenite Adelaide, Australia	x0.44 c	251d	Steatite Canada	x0.75 c
242e	Selenite France	x0.75 c	251e	Steatite in serpentine Ruan Minor, Lizard, Cornwall	x0.72 b
242f	Senarmontite Constantine, Algeria	x1.54 a	251f	Stephanite Mina Clina, Tresfunitas, Chile	x1.43 a
242g	Sepiolite with calcite Hyble, Ontario, Canada	x1.30 c	252a	Stibnite Felsobanya, Rumania	x0.41 c
243a	Septarian Nodule Hurricane, Utah	x1.56 d	252b	Stichtite in serpentine Swaziland	x1.67 d
244a	Serendibite Warren County, New York State	x1.56 a	252c	Stilbite Poona, India	x0.76 c
244b	Sericite Ogilby, California	x1.17 a	252d	Stilbite Poona, India	x0.50 c
244c	Serpentine Cornwall	x1.28 c	252e	Stilpnomelane with quartz and pyrite Longvale Quarry, Mendocino County, California	x1.56 d
244d	Serpentine Cornwall	x2.02 c	252f	Stolzite Broken Hill, New South Wales, Australia	x1.15 a
244e	Slickenside Serpentine Ma-on-shan, Hong Kong	x1.35 d	252g	Strengite Waldgirmes, Germany	x1.89 a
244f	Serpentine The Lizard, Cornwall	x0.50 b	253a	Strontianite Westphalia, Germany	x1.19 d
244g	Shale Dorset	x2.02 c	253b	Calcium rich Strontianite Winfield, Pennsylvania	x1.69 d
244h	Zinc Shale Tsumeb, South-west Africa	x0.54 d	253c	Strontianite Winfield, Pennsylvania	x1.37 d
245a	Sicklerite Bull Moose Mine, Custer County, South Dakota	x1.37 a	253d	Strontianite (crystal detail) Winfield, Pennsylvania	x2.92 d
245b	Siliceous Sinter Rotorna, New Zealand	x1.69 d	253e	Strüverite Madagascar	x2.02 a
245c	Sillimanite Chester, Connecticut	x1.19 a	253f	Sulphur Sicily	x0.53 c
246a	Siltstone Kaghan Valley, West Pakistan	x1.75 d	253g	Sunstone India	x1.80 c
246b	Silver Zacatecas, Mexico	x3.82 b	254a	Syenite Drachenfels, Germany	x0.77 c
246c	Siltstone Gloucestershire	x2.02 c	254b	Syenite Silver Bay, Hong Kong	x1.30 d
246d	Silver Durango, Mexico	x1.41 d	254c	Sylvanite Colorado	x2.02 a
247a	Silver Canada	x1.54 d	254d	Sylvine Carlsbad, New Mexico	x2.02 d
247b	Silver Unrecorded	x1.35 c	254e	Tachylyte Trap Rock, Cyprus	x1.78 a
247c	Skutterudite Bou Azzer, Morocco	x1.37 a	255a	Talc Unrecorded	x1.78 c
247d	Smaltite Unrecorded	x2.02 a	255b	Talc Schist Shetland Islands	x2.02 c
247e	Smaltite (massive) Coleman, Ontario, Canada	x1.30 a	255c	Tarnowitzite Unrecorded	x2.02 d
248a	Smithsonite Mendip Hills, Somerset	x1.26 c	255d	Tasmanite Mersey River, Tasmania	x1.80 a
249a	Specular Hematite Cumberland	x1.06 c	256a	Teallite (massive) Montserrat, Bolivia	x2.02 a
249b	Specular Hematite Rhodesia	x0.99 d	256b	Tennantite with pyrite Gunnislake, Cornwall	x1.54 d
249c	Spessartite Tsiliazina, Madagascar	x2.02 a	256c	Tendrite Unrecorded	x1.58 a
249d	Sphaerosiderite Clackamas County, Oregon	x1.78 d	256d	Tephroite with zincite Franklin, New Jersey	x2.02 a
249e	Sphene in Urtite USSR	x1.72 c	257a	Tetrahedrite Quiruvilca, Mexico	x1.69 d
249f	Spinel with chondrodite in calcite Franklin, New Jersey	x2.02 a	257b	Tetrahedrite in quartz Huanchaca, Bolivia	x1.13 a
249g	Red Spinel Unrecorded	x1.91 c	257c	Tetrahedrite (Falherz) Huanchaca, Bolivia	x1.26 a
250a	Spinel India	x4.31 a	257d	Thenardite Camp Verde, Arizona	x1.22 a
250b	Spodumene Brazil	x1.80 c	257e	Thinolite Washoe County, Nevada	x0.93 d
250c	Spodumene Brazil	x1.52 c	258a	Thomsonite Portrush, County Antrim	x0.99 c

258b	**Thomsonite (crystal detail)**	x3.15 c
	Portrush, County Antrim	
258c	**Thorianite**	x2.02 a
	Ceylon	
258d	**Thorite (massive)**	x1.24 a
	Brevig, Norway	
258e	**Thortvitite**	x2.02 a
	Southern Norway	
258f	**Thulite with clinozoisite**	x1.37 d
	Mozambique	
259a	**Tile Ore with malachite**	x0.80 b
	Gwennap, Cornwall	
259b	**Tiemannite with chalcopyrite**	x2.02 a
	Clausthal, Harz, Germany	
259c	**Tigers Eye (unpolished)**	x0.62 c
	South Africa	
259d	**Tigers Eye (polished)**	x0.62 c
	South Africa	
260a	**Tincalconite**	x2.02 d
	Kern County, California	
260b	**Topaz**	x0.70 b
	USSR	
260c	**Topaz**	x1.80 c
	Mexico	
260d	**White Topaz**	x1.50 c
	Unrecorded	
261a	**Torbernite**	x1.91 d
	Sonora, Mexico	
261b	**Torbernite**	x1.28 d
	Mexico	
261c	**Torbernite**	x0.99 b
	Gunnislake Old Mine, Cornwall	
261d	**Tourmaline in quartz**	x1.39 c
	Brazil	
262a	**Tourmaline in quartz**	x0.70 c
	Brazil	
262b	**Schorl (tourmaline)**	x1.30 d
	Cornwall	
262c	**Tourmaline**	x2.02 d
	Brazil	
262d	**Tourmaline**	x1.80 d
	Brazil	
262e	**Trachyte**	x1.91 c
	Somerset	
262f	**Tremolite**	x1.43 d
	New York State	
262g	**Triphylite**	x1.43 a
	Ross Mine, Custer, South Dakota	
263a	**Trona**	x1.50 a
	Inyo County, California	
263b	**Troosite**	x1.46 a
	Franklin, New Jersey	
263c	**Tuff**	x2.02 c
	Somerset	
263d	**Tuff**	x1.52 d
	Yuma, Arizona	
264a	**Turquoise**	x2.02 d
	Lynch Station, Virginia	
264b	**Turquoise**	x1.58 c
	Arizona	
264c	**Uintahite**	x1.10 a
	Uintah Mountains, Utah	
264d	**Ulexite**	x1.71 c
	Boron, California	
264e	**Unakite**	x1.06 c
	Virginia	
265a	**Uralite**	x0.95 a
	Salida, Colorado	
265b	**Uraninite**	x1.54 d
	Gold Butte, Nevada	
265c	**Uraninite**	x1.15 d
	Orphan Boy Mine, Gold Canyon, Arizona	
265d	**Uranophane**	x1.08 d
	Mozambique	
265e	**Uranophane**	x1.80 a
	Congo	
265f	**Urtite**	x0.54 c
	USSR	

265g	**Uvarocite**	x1.30 a
	Orford, Quebec, Canada	
266a	**Valentinite on stibnite**	x1.71 a
	Cornwall	
266b	**Vanadinite**	x3.71 c
	Old Yuma Mine, Pima County, Arizona	
266c	**Vanadinite**	x0.74 d
	Mozambique	
266d	**Variscite**	x1.78 d
	Utah	
266e	**Variscite**	x1.91 d
	Arkansas	
267a	**Vermiculite**	x1.76 a
	Connecticut	
267b	**Vitrain**	x1.32 a
	South Pit, Bedwas, Monmouthshire	
267c	**Vitrain**	x1.32 a
	South Pit, Bedwas, Monmouthshire	
267d	**Vivianite**	x1.06 c
	Australia	
268a	**Vogesite**	x0.59 d
	Unrecorded	
268b	**Vorobyevite**	x1.08 c
	Brazil	
268c	**Vulpinite**	x1.13 c
	Bergamo, Italy	
268d	**Wad**	x0.86 b
	Upton Pine, near Exeter, Devon	
268e	**Wavellite**	x1.52 c
	Pencil Bluff, Arkansas	
269a	**Wardite with Grandallite**	x1.86 a
	Fairfield, Utah	
269b	**Wavellite**	x1.15 c
	Garland County, Arkansas	
270a	**Wernerite**	x1.32 a
	Renfrew County, Ontario, Canada	
270b	**Wernerite (massive)**	x1.39 a
	Boston, Massachusetts	
270c	**Willemite (reddish brown)**	x1.37 a
	Franklin, New Jersey	
270d	**Willemite**	x1.41 a
	Muldiva, New Zealand	
270e	**Green Willemite**	x1.58 a
	Franklin, New Jersey	
270f	**Williamsite**	x1.48 a
	The Lizard, Cornwall	
270g	**Willyamite**	x0.88 c
	USSR	
270h	**Wilsonite**	x1.75 a
	Templeton, Canada	
271a	**Withamite in andesite**	x1.67 a
	Cutting on New Road, ½ mile west of 'Meeting of Three Waters', Glencoe, Scotland	
271b	**Witherite**	x1.58 c
	Unrecorded	
271c	**Wolframite**	x0.78 c
	Cornwall	
271d	**Wolframite**	x0.85 c
	South Crofty Mine, Cornwall	
271e	**Wolframite**	x0.31 b
	Pannasqueira Mine, Beira Baiza, Portugal	
271f	**Wollastonite**	x1.73 d
	Yuma County, Arizona	
271g	**Wood opal**	x0.58 c
	Australia	
272a	**Wood tin**	x1.02 b
	Unrecorded	
272b	**Wulfenite**	x1.67 d
	Chihuahua, Mexico	
272c	**Wulfenite**	x1.78 d
	Red Cloud Mine, Yuma County, Arizona	
272d	**Wulfenite (crystal detail)**	x3.76 d
	Red Cloud Mine, Yuma County, Arizona	
272e	**Wurtzite**	x1.65 a
	Pribram, Czechoslovakia	
272f	**Xenotime**	x2.02 à
	Tvedestrand, Norway	

287

273a	**Yoderite**	x1.82 a
	Mautia Hill, Tanzania	
273b	**Zaratite with kaolinite**	x1.48 a
	Ronna Mine, Kladno, Czechoslovakia	
273c	**Zaratite on pentlandite**	x1.73 a
	Unrecorded	
273d	**Zeolite**	x1.65 a
	India	
273e	**Zincite with tephroite**	x2.02 a
	Franklin, New Jersey	
273g	**Zincite**	x1.30 a
	Franklin, New Jersey	